M000250132

HVAC Systems
Design
Handbook

Other HVAC Titles of Interest

HVAC Systems Design Handbook

Roger W. Haines
Laguna Hills, California

C. Lewis Wilson
Heath Engineering Company
Salt Lake City, Utah

Third Edition

McGraw-Hill

New York   San Francisco   Washington, D.C.   Auckland   Bogotá
Caracas   Lisbon   London   Madrid   Mexico City   Milan
Montreal   Nw Delhi   San Juan   Singapore
Sydney   Tokyo   Toronto

Library of Congress Cataloging-in-Publication Data

Haines, Roger W.
    HVAC systems design handbook / Roger W. Haines, C. Lewis Wilson.
    —3rd ed.
        p.   cm.
    Includes index.
    ISBN 0-07-025963-1 (alk. paper)
    1. Heating—Equipment and supplies—Design and construction.
    2. Ventilation—Equipment and supplies—Design and construction.
    3. Air conditioning—Equipment and supplies—Design and
    construction.   I. Wilson, C. Lewis.   II. Title.
    TH7345.H35   1998
    697—dc21   97-39188
                                                                            CIP

McGraw-Hill

A Division of The *McGraw-Hill* Companies

Copyright © 1998 by The McGraw-Hill Companies, Inc. All rights re-
served. Printed in the United States of America. Except as permitted
under the United States Copyright Act of 1976, no part of this publication may be reproduced or distributed in any form or by any means,
or stored in a data base or retrieval system, without the prior written
permission of the publisher.

    3 4 5 6 7 8 9 0   DOC/DOC   9 0 3 2 1 0 9

ISBN 0-07-025963-1

*The sponsoring editor for this book was Harold B. Crawford, the
editing supervisor was Frank Kotowski, Jr., and the production
supervisor was Pamela A. Pelton. It was set in Century Schoolbook by
Pro-Image.*

Printed and bound by R. R. Donnelley & Sons Company.

McGraw-Hill books are available at special quantity discounts to use
as premiums and sales promotions, or for use in corporate training programs. For more information, please write to the Director of Special
Sales, McGraw-Hill, 11 West 19th Street, New York, NY 10011. Or contact your local bookstore.

 This book is printed on recycled, acid-free paper containing
a minimum of 50% recycled, de-inked fiber.

Information contained in this work has been obtained by The Mc-
Graw-Hill Companies, Inc. ("McGraw-Hill") from sources believed to
be reliable. However, neither McGraw-Hill nor its authors guarantee
the accuracy or completeness of any information published herein and
neither McGraw-Hill nor its authors shall be responsible for any errors, omissions, or damages arising out of use of this information. This
work is published with the understanding that McGraw-Hill and its
authors are supplying information but are not attempting to render
engineering or other professional services. If such services are required, the assistance of an appropriate professional should be sought.

Contents

Preface

This book was begun with the aim of distilling nearly fifty years of experience in HVAC into a short, succinct treatise. The book is written primarily for the HVAC designer in a counsulting or contracting office. It can be read with profit by anyone in the industry, including maintenance and operating people, building managers, and sales engineers. It is not written as the basis for a college or vocational school curriculum, although there may be some material here which would be useful in such a situation.

The second and third editions of the book are an effort to update and expand the work of the first edition, which found an apparent following in the HVAC industry. The purposes and expected users of this book remain the same.

The book is an overview of the HVAC design process with emphasis on the things we think are important, such as orderly procedures and good communication. While basic theory cannot be and has not been ignored, it is presented here as simply as possible, with emphasis on applications. We have tried always to answer the questions of how and why because it is only with that kind of understanding that we can extrapolate our experience into new areas.

Throughout the book there are many references for further study. Use these resources and others when you need them. True professionals understand not only what they did, but why they did it and why it is important. We hope this book will help you to a better understanding.

Roger W. Haines
C. Lewis Wilson

Acknowledgments

It is impossible to remember or acknowledge all the people who have contributed to my education over these sixty odd years. My father taught me the sheet-metal trade. The faculty at Iowa State University gave me good theoretical training. All the people I've dealt with in my work experience have helped — other engineers, contractors, manufacturers' representatives, clients. I still remember the owner who took pity on a poor apprentice and taught me how to properly file a screwdriver tip. My many friends in ASHRAE taught me much through formal and informal discussion.

A few names must be mentioned. Ted Neubauer was my first model of a truly professional engineer. John Blossom introduced me to the problem-solving process described in Chap. 1. Ralph Thompson and Doug Hittle taught me electronics and many other things. Frank Govan wrote the sections on boilers in Chap. 10. Don Bahnfleth taught me how to write reports. Frank Bridgers and Don Paxton gave me my first job as an engineer, along with basic training in design and professional attitudes.

Finally, as always, I could accomplish nothing without the support, encouragement, and especially, patience of Wilma, my wife of fifty-seven years.

Roger W. Haines

My association with Roger Haines goes back to 1964 when I, as a young engineering student, found summer employment with Bridgers and Paxton Consulting Engineers in Albuquerque, New Mexico. Roger was the Chief Design Engineer and was patient with and helpful to a rank beginner. With Roger, I too give thanks to Frank Bridgers and Don Paxton, especially to Frank, for mentorship, as engineer and businessman, and as a gentleman in daily living, and for his encouragement with others to go on to graduate school at Purdue University after graduation from Brigham Young University. I am indebted to those who have let me teach in an adjunct role over many years. The teacher learns twice, once in the getting and once in the giving. I am indebted to partners at Heath Engineering Company who have helped, learned, taught, and tolerated for more than twenty-five years.

I acknowledge my good wife Grace, and children Emily, Michael, and Amanda, the lights of my life. Grace served as clerk-typist for the second edition and has accepted the impact of a spare-time writing schedule on an already busy life.

Sincere thanks are given to Roger Haines and his wife Wilma for the opportunity to be a part of this work.

C. Lewis Wilson

Introduction

I.1 Definition and Purpose of HVAC Systems

Air conditioning is defined as the simultaneous control of temperature, humidity, radiant energy, air motion, and air quality within a space for the purpose of satisfying the requirements of comfort or a process. Not included in the definition, but often required, is control of the pressure in the conditioned space relative to adjacent areas. Another factor which becomes important in many applications is the noise level associated with the air conditioning equipment.

Most people equate air conditioning with cooling but, as the definition states, air conditioning is a great deal more than that. Comfort must also be defined—a difficult task because the sensation of comfort is subjective and varies with the individual and the level of activity. Cleanliness relates to the broad subject of indoor air quality, which includes not only dust and dirt but also gaseous contaminants, viruses, and bacteria.

It becomes quickly evident that to accomplish true air conditioning is not all that simple, and in some industrial or institutional applications, it may be very difficult.

The purpose of this book is to discuss various air conditioning design procedures and systems and to give the reader the tools necessary to understand and solve many air conditioning problems. For simplicity, the acronym HVAC (heating, ventilating, and air conditioning) is used unless only one of these factors is being discussed. Throughout the book, frequent reference is made to the American Society of Heating, Refrigerating, and Air-Conditioning Engineers (ASHRAE) Handbooks,[1] the primary and authoritative reference books for the HVAC and refrigeration industries. ASHRAE is the primary source of information and research in these fields. The reference book entitled *Industrial Ventilation: A Manual of Recommended Practice*[2] is also definitive in many applications.

I.2 Engineering as a Business

One of the fallacies of engineering education in an earlier day, perhaps even at present, is the failure to recognize and teach that engineers are business-people first and technicians second. There is a corollary to this concept in the adage "The person who knows *how* usually winds up working for the person who knows *why*."

A simplified version of the fundamentals of business identifies the three-legged milkstool of *marketing, management,* and *production.* Most engineers become proficient in production, i.e., in the mastery and use of technical knowledge, but only some engineers are effective managers, and few engineers are successful marketers.

This brief expression at the beginning of this book begs the reader to remember the importance of satisfying the customer (be it client, boss, or public served) and of conducting the work in a technically correct but also profitable manner so that the engineering wage is paid with something left over to encourage a repeat engagement.

In the context of this book, being an engineer in a businesslike way implies designing HVAC systems which achieve the desired level of comfort and performance with an economical combination of first costs and subsequent operating and maintenance costs, all brought to pass in a timely and efficient design process. The businesslike engineer is also communicative and easy to work with, traits which sell the product of the engineering effort.

I.3   HVAC System Design

"HVAC system design is a process (intellectual) commonly involving teamwork and iteration, which leads to a device, system, and/or process (physical) that satisfies a need".[3]

The design of HVAC systems is based on scientific principles of mathematics, physics, and chemistry as developed into discussions of thermodynamics, fluid mechanics, heat transfer, and psychrometrics. Brief discussions of these fundamental topics are given in the chapters, and additional reference material is listed at the end of each chapter. Beyond the raw science, however, are a great many empirical and experience factors which modify the calculated data. These turn the HVAC design process into *art* as well as *science.* This book offers both procedure and tricks of the trade and encourages the reader to develop insights which will lead to intuitive understanding of many engineering problems.

I.4   Computers in HVAC Design

The contribution of computers to the welfare and accomplishments of people over the past fifty years is phenomenal. Short of the ability to reason and to exhibit a sense of right and wrong, the desktop or laptop computer processing gigabytes of information in nanoseconds, challenges and often, though not always, surpasses the human mind in information management. Even the handheld calculator is a miracle in the context of a previous generation.

The contribution of computers to today's HVAC industry is remarkable. From the concept and product research laboratory, to the engineering office computer-aided drafting (CAD) function, to the massive building energy analysis program, to the automated ductwork fabrication machine, to the controller which runs the HVAC or process equipment, computers play a role in every step of the design and construction process. But amid this worship of the inanimate wonder, several restraining factors must be kept in mind.

1. The computer can handle more data than we might ever pursue were we limited to manual calculations. Therefore our solutions may seem more precise. But we must remember the GIGO (garbage in, garbage out) rule. We must not tout output accuracy to 10 significant places when the input only had 3 significant places, nor should we be too excited about volumes of information when all we want to do is differentiate between a small, medium, or large piece of system equipment.

2. Computers are managed by software programs which employ algorithms (logical sequences of calculation) to manipulate and present information. The programs are created by and are only as good as the programmer. Programs may reflect the biases of a sponsor, and programs must first be evaluated as if they appeared on an editorial page. Many programs are developed with a profit motive; others come from a not-for-profit environment. Program cost is no indication of value.

3. Since the strength of the computer in quickly manipulating input information to a useful output form is also a weakness if the input is erroneous, there is no substitute for the knowledgeable provider of input and the interpreter of output. Engineering experience is more valuable than ever, and the liabilities consequent to error are greater than ever, as projects get larger, more complicated, and more costly, and are more responsible for human life and well-being.

4. It is still possible and often practical to design HVAC and industrial process systems without extensive computer involvement. Many examples in this book are presented with manual calculations so that the reader will fully understand the sequences and processes involved.

I.5   Need for Orderly Procedures

Abraham Lincoln's alleged composition and editing of the Gettysburg Address on the back of an envelope was an oratorical success, but a bad example for the technical professions. In today's highly technical, regulated, and litigious society, there is no substitute for organized analysis and documented design effort. Detailed, orderly design records, with underlying assumptions clearly stated and with explanations of how and why decisions were made, are worth their weight in lawyers when inevitable questions arise. Good design notes and calculations make for easy checking. They also help when there is a change in assignment and someone else takes over or supplements the work. Experience has shown that most details of a design procedure cannot be recalled accurately after a lapse of 6 to 12 months. Yet it is typically at or beyond that time that questions arise in the field, in operation, or in review of a system for expansion or alteration, or when system performance is challenged as perceived system performance fails to meet expectation.

For purposes of consistency and effective use of time, many design procedures are standardized with the use of forms and formats, computer programs, spreadsheets and the like. Such standardization serves the design process by compacting the repetitive and mundane, thereby allowing more time for creativity in solving the unique issues of the assignment. This book describes some useful procedures, but leaves great latitude for individual designers to develop their own design methods.

References

1. ASHRAE Handbooks, four volumes, one volume republished each year. Available from American Society of Heating, Refrigeration, and Air-Conditioning Engineers, 1791 Tullie Circle NE, Atlanta, GA 30329. Other references in this text will specify year of publication, volume, title, and page or chapter.
2. *Industrial Ventilation: A Manual of Recommended Practice*, 21st edition, American Conference of Governmental Industrial Hygienists Inc. (ACGIH).
3. F. W. Incropera, Purdue University, annual newsletter, 1992.

HVAC Systems
Design
Handbook

HVAC Engineering Fundamentals: Part 1

1.1 Introduction

This chapter is devoted to "fundamental" fundamentals — certain principles which lay the foundation for what is to come. Starting with the original author's suggested thought process for analyzing typical problems, the reader is then exposed to a buzzword of our time: *value engineering*. Next follows a discussion of codes and regulations, political criteria which constrain potential design solutions to the bounds of public health and welfare (and sometimes to special interest group sponsored legislation). The final sections of the chapter offer a brief review of the basic physics of heating, ventilating, and air conditioning (HVAC) design in discussions of fluid mechanics, thermodynamics, heat transfer, and psychrometrics. Numerous classroom and design office experiences remind us of the value of continuous awareness of the physics of HVAC processes in the conduct of design work.

1.2 Problem Solving

Every HVAC design involves, as a first step, a problem-solving process, usually with the objective of determining the most appropriate type of HVAC system for a specific application. It is helpful to think of the problem-solving process as a series of logical steps, each of which must be performed in order to obtain the best results. Although there are various ways of defining the process, the following sequence has been found useful:

1. *Define the objective.* What is the end result desired? For HVAC the objective usually is to provide an HVAC system which will control

the environment within required parameters, at a life-cycle cost compatible with the need. Keep in mind that the cost will relate to the needs of the process. More precise control of the environment almost always means greater cost.

2. *Define the problem.*   The problem, in this illustration, is to select the proper HVAC systems and equipment to meet the objectives. The problem must be clearly and completely defined so that the proposed solutions can be shown to solve the problem.

3. *Define alternative solutions.*   Brainstorming is useful here. There are always several different ways to solve any problem. If remodeling or renovation is involved, one alternative is to do nothing.

4. *Evaluate the alternatives.*   Each alternative must be evaluated for effectiveness and cost. Note that "doing nothing" always has a cost equal to the opportunity, or energy, or efficiency "lost" by not doing something else.

5. *Select an alternative.*   Many factors enter into the selection process—effectiveness, cost, availability, practicality, and others. There are intangible factors, too, such as an owner's desire for a particular type of equipment.

6. *Check.*   Does the selected alternative really solve the problem?

7. *Implement the selected alternative.*   Design, construct, and operate the system.

8. *Evaluate.*   Have the problems been solved? The objectives met? What improvements might be made in the next design?

1.3   Value Engineering*

Value analysis or *value engineering* (VE) describes a now highly sophisticated analytical process which had its origins in the materiel shortages of World War II. In an effort to maintain and increase production of war-related products, engineers at General Electric developed an organized method of identifying the principal function or service to be rendered by a device or system. Then they looked at the current solution to see whether it truly met the objective in the simplest and most cost-effective way, or whether there might be an alternative approach that could do the job in a simpler, less costly, or more durable way. The results of the value engineering process now permeate our lives, and the techniques are pervasive in business. Consider our improved automobile construction methods, home appliances, and the like as examples. Even newer technologies such as those pertaining to television and computers have been improved by

---

*Compare this section with Sec. 1.2.

quantum leaps by individuals and organizations challenging the status quo as being inadequate or too costly.

Alphonso Dell'Isolo is generally credited as being the man who brought value engineering to the construction industry, which industry by definition includes HVAC systems. Dell'Isolo both "wrote the book"[1] and led the seminars which established the credibility of the practice of value engineering in engineering firms and client offices across the land.

There is a national professional society called SAVE (Society of American Value Engineers), headquartered in Smyrna, Georgia. The society certifies and supports those who have an interest in and commitment to the principles and practices of the VE process.

Value engineering in construction presumes an issue at hand. It can be a broad concern such as a system, or it can be a narrow concern such as a device or component. The *VE process* attacks the status quo in four phases.

1. *Gather information.* Clearly and succinctly identify the purpose(s) of the item of concern. Then gather information related to performance, composition, life expectancy, use of resources, cost to construct, the factors which comprise its duty, etc. Make graphs, charts, and tables to present the information. Identify areas of high cost in fabrication and in operation. Understand the item in general and in detail.

2. *Develop alternatives.* First ask the question, Do we even need this thing, this service at all? Or are we into it by habit or tradition? If the function is needed, then ask, How else could we accomplish the same objective? Could we reasonably reduce our expectation or acceptably reduce the magnitude of our effort? Could we eliminate excess material (make it lighter or smaller)? Could we substitute a less expensive assembly? Could we eliminate an element of assembly labor? Could we standardize a line of multisize units into just a few components?

In this phase, we learn not to criticize, not to evaluate, for the "crazies" spawn the "winners." "Don't be down on what you are not up on." Be creative and open-minded. Keep a written record of the ideas.

3. *Evaluate the alternatives.* Having developed ideas for different ways of doing the same thing, now evaluate the objective and subjective strengths and weaknesses of each alternative. Study performance versus cost — cost both to construct and to operate. Look for the alternative which will work as well or better for the least overall cost. This will often be a different solution from the original.

4. *Sell the best solution.* This ties back into a weakness of many engineers and designers: They have great ideas, but they have a hard

time getting these ideas implemented. By first understanding the purpose of a device or system, then producing good data to understand current performance, and finally developing an alternative with documented feasibility, the sales effort is greatly supported.

Gas forced-air furnaces are an example of an HVAC unit which has been improved over time by value engineering. The purpose of the furnace now, as before, is to use the chemical energy of a fuel to warm the environment, i.e., to heat the house. But there is a world of difference between the furnace of the 1930s, with its cast-iron or heavy-metal refractory-lined firebox and 4-ft-diameter bonnet, and the high-technology furnaces of today. Size is down, capacity is up, weight is down, relative cost is down, fuel combustion efficiency is up, and reliability is debatably up.

Variable-speed drives for pumps and fans are devices which have been improved to the point of common application. The operating-cost advantages of reduced speed to "match the load" have been known and used in industry for a long time, but technology has taken its time to develop reliable, low-cost, variable-speed controllers for commercial motors, such as variable-frequency drives now used in HVAC applications.

If value engineering seems to share some common analytical technique with Sec. 1.2 on problem solving, the dual presentation is intentional. Both discussions are approaches to solving problems, to improving service. The first is an interpretation of a mentor's example, the second is a publicly documented, formal procedure. The HVAC system designer will benefit greatly if she or he can commit to an analytical thought process which defines the problem, proposes solutions, identifies the optimum approach, and finally presents the solution in a credible and compelling way.

1.4   Codes and Regulations

No HVAC designer should undertake a design task without first having an awareness of and hopefully a working familiarity with the various codes and ordinances which govern and regulate building construction, product design and fabrication, qualification of engineers in practice, etc. Codes generally are given the force of law on the basis of protecting the public safety and welfare. Penalties may be applied to those who violate established codes, and the offending installation may be condemned and regarded as unsuitable for use by enforcement authorities. As young design practitioners, we were advised to "curl up with a good code book" until we became thoroughly familiar with its precepts.

Codes are particularly definitive regarding a building's structural integrity, electrical safety, plumbing sanitation, fuel-fired equipment and systems, fire prevention detection and protection, life safety and handicapped accessibility in buildings, energy conservation, indoor air quality, access for the handicapped, etc. Each of these areas has an impact on the design of HVAC systems.

Particular codes are sufficiently diverse in their adoption that it is unwise for this book to list any specifics. The HVAC system designer should simply know that life is not without constraint; that systems will conform to codes, or else a permit to build and use will be denied; and that willful violation of codes by the designer is done only at great personal risk.

The recommended practice for every HVAC design assignment is to make an initial review of the locally enforced codes and regulations, to become thoroughly familiar with the applicable paragraphs, and to religiously follow the prescribed practices, even though such an approach seems to stifle creativity.

Occasionally code constraints seem to violate or interfere with the objective of a construction. At these times, it is often possible to request a variance from the authority. There is no guarantee of acceptance, but nothing ventured, nothing gained. Good preparation generates hope and understanding, and differentiates you from the unending stream of charlatans who seek to sidestep codes and regulations for personal financial gain.

Variance procedures notwithstanding, in general the best idea is to know the codes and to design within them. See Ref. 2 for further discussion of this topic.

1.5 Fluid Mechanics*

Fluid mechanics, a fundamental area of physics, has to do with the behavior of fluids, both at rest and in motion. It deals with properties of fluids, such as density and viscosity, and relates to other aspects of physics, such as thermodynamics and heat transfer, which add the issues of energy to the functions of the basic fluid flow. For this brief reminder paragraph, remember:

- The static pressure at a point in a fluid system is directly proportional to the density of the fluid and to the height of the fluid column. Static pressure is exerted equally in all directions.

---

*See also Chap. 16.

- The velocity pressure is proportional to the square of the fluid velocity; i.e., doubling the velocity quadruples the velocity pressure.

- The friction loss of a flowing fluid is proportional to the square of the velocity.

- The pumping power is proportional to the fluid density, to the volume of fluid handled, and to the pressure against which the fluid is pumped.

    Since the friction loss is proportional to the square of the flow, pumping power in a defined system is proportional overall to the cube of the relative flow rate.

For HVAC purposes, air is considered to be an incompressible fluid. For incompressible fluids, the amount of fluid in a system is constant. Any outflows must be offset by equivalent inflows, or there must be a change in the amount of fluid held in the system. This allows us to account for fluid in a process just as we count money in the bank. See Ref. 3 for further discussion of this topic.

1.6   Thermodynamics*

Thermodynamics has to do with the thermal characteristics of matter and with the natural affinity of the universe to go from a higher to a lower energy state. Thermodynamics deals with the ability of matter to accept changes in energy level (relates to specific heat as a property and to enthalpy as a scale of measurement of energy level). For this reminder paragraph, remember:

- The energy acceptance capacity of a substance is called *specific heat* with units of Btu per pound per degree Fahrenheit. Water with a specific heat of 1.0 Btu/(lb · °F) is one of the best heat-accepting media.

- The energy acceptance capacity in a change of phase is called the *latent heat of vaporization* from liquid to gas (i.e., water to steam) and *latent heat of fusion* from liquid to solid (i.e., water to ice). Again, water with a latent heat of vaporization of approximately 1000 Btu/lb and a latent heat of fusion of 144 Btu/lb is very good at involving large quantities of energy at constant temperature in the phase change.

- Thermodynamics can be used to examine the refrigeration cycles with tools and techniques to analyze performance of equipment and systems.

---

*See also Chap. 17.

- The *first law of thermodynamics* says that "energy is conserved." For matter as for money, we can account for energy inputs, outputs, and storage. Combining thermodynamics with fluid mechanics allows us to calculate energy flows piggybacked onto fluid flows with accuracy and confidence.

- The *second law of thermodynamics* says that energy left to itself always goes from high to low, from fast to slow, from warm to cold. To make things go uphill, to go otherwise, we must expend energy. There is no such thing as a perpetual-motion machine.

- Psychrometrics is a specialty of thermodynamics involving the physics of moist air, a mixture of air and water vapor. See Ref. 4 for further discussion of this topic.

1.7 Heat Transfer*

In studying heat transfer, we study energy in motion — through a mass by conduction, from a solid to a moving liquid by convection, or from one body to another through space by radiation. Remember:

- Heat is transferred from warmer to colder — always, without exception.

- Heat transfer for conduction and for convection is directly proportional to the driving temperature differential. Double the difference to double the heat transfer rate $(T_1 - T_2)$.

- Heat transfer by radiation is proportional to the fourth power of the temperature difference $(T_1^4 - T_2^4)$. Small changes in temperature can create relatively large changes in radiation heat transfer rates.

- For heat transfer between fluids, counterflow (opposite direction) is *much* more effective than parallel flow (same direction).

- Insulation to reduce heat transfer follows a law of diminishing returns, the reciprocal of the amount of insulation used, for instance, 1, ½, ⅓, ¼, ... The first insulation is most valuable, with every succeeding increment less so. It is a design challenge to find the cost-effective happy median.

- Fouling of heat transfer surfaces is detrimental to equipment performance.

- Quantitative heat transfer is proportional to the heat transfer surface area.

---

*See also Chap. 18.

- Although it is not a classic form of heat "transfer," heat can be transported by a fluid (e.g., air in ducts and water in pipes) from one point to another. This action is better classified as a combination of fluid mechanics and thermodynamics (mixing of fluids of different thermodynamic conditions). See Ref. 5 for further discussion of this topic.

1.8 Psychrometrics*

Psychrometrics is the science of the properties of moist air, i.e., air mixed with water vapor. This subset of thermodynamics is important to the HVAC industry since air is the primary environment for all HVAC work. Whereas oxygen, nitrogen, and other components of dry air behave similarly in only a vapor phase in the HVAC temperature range, water will undergo a change of state in the same temperature range based on pressure, or in the same pressure range based on temperature. In the human comfort temperature range, the comfort of people and the quality of the environment for health, for structures, and for preservation of materials are also related to the moisture in the air. Control of the moist-air condition is a primary objective of the HVAC system. Remember the following:

- Air is considered to be *saturated* with moisture when the evaporation of water into the air at a given temperature and atmospheric pressure is offset by a concurrent condensation of water vapor to liquid. Cooling of saturated air results in dew, fog, rain, or snow. Warm air can hold more moisture than cold air.
- *Percent relative humidity* measures how much water vapor is in the air compared to how much there would be if the air were saturated at the same temperature. The adjective *relative* is appropriate because the absolute amount of water that air can hold is relative to both temperature and barometric pressure. Changes in barometric pressure related to altitude or to weather conditions affect the moisture-holding capacity of air.
- A psychrometric chart which presents properties of mixtures of moist air on a single graph is a most useful tool for quantitatively calculating and analyzing HVAC processes. Familiarity and facility with these charts are a must for the HVAC designer.
- It is impossible to remove moisture from air in a heat exchange cooling process without bringing the air near to the saturation line.

---

*See also Chap. 19.

Moisture may be removed by desiccants without approaching saturation.

- Optimum conditions for human health and comfort range from 70 to 75°F and 40 to 50 percent relative humidity. In terms of perceived comfort, a little higher relative humidity can offset a little lower ambient temperature.

- Moist air in cold climates is a problem and a liability for building designers. Since the inside environment usually is moister than the outside air, insulation and vapor barriers are required to prevent condensation in the structural cavities. Failure to respect this liability may lead to early deterioration of a building. Swimming pools and humidified buildings (hospitals, etc.) are particularly vulnerable. See Ref. 6 and Chap. 19 for further discussion of this topic.

1.9   Sound and Vibration*

Sound and vibration have become a topic of interest for the HVAC designer, not that they are part of the primary heating, cooling, and air conditioning functions but because they are secondary factors which, if not properly handled, can destroy an otherwise successful HVAC installation.

All sounds and vibrations are forms of kinetic energy, and in the HVAC world they are usually derived from moving equipment, moving air, or other moving fluid. A problem arises when an HVAC system component generates noise or vibration within, or adjacent to, a habited or process-sensitive space. If the generated sound or vibration level exceeds the local tolerance level, the HVAC system is deemed unacceptable.

For an HVAC system to be acceptable in terms of sound and vibration, an occupant or a process in a served space must be essentially unaware of, or at least not impaired by, the active functions of the HVAC system. Airborne sound in an office or theater must not draw attention to itself. The space must seem quiet when all is still, and allow conversation or music to go on without intrusion. The same is true for vibration. Operation of the HVAC system should not, often must not, be apparent to building occupants in the sense of a vibrating floor or desk, or visibly moving structural components like a light fixture. Recognize that in less sophisticated spaces like shops or equipment rooms, some sound and vibration is expected and tolerated at higher levels, so the HVAC designer must understand first the origins,

---

*See also Chap. 20.

then the level of acceptable performance, and finally the mechanisms of control of sound and vibration to achieve an acceptable level of service.

"Sound" is a generic term for airborne vibrations transmitted to the ear or equivalent acoustic sensing device. When sound offends, it is called "noise." Sound power levels are measured in watts, and with 10^{-12} W being a threshold of hearing, this is defined as being 0 decibels (dB). Sound is usually measured within and for each octave band, where the frequency of each successive octave band is twice that of the previous. A vibration frequency of 31.5 hertz or cycles per second (Hz) defines the midpoint of the first octave band. Middle C is in the middle of the fifth octave band.

Sound or noise is generated by something in motion which sets up airborne vibration. The sound "radiates" from the point of origin to the point of detection. Sound power levels in open air diminish with the square of the distance, but in a smaller confined space, with high reflectance, the sound power level may be relatively constant. Sound may be eliminated by absorption or confinement. Dense fibrous mats and accoustical duct liner are examples of absorptive materials. Masonry or concrete structure, and lead fabrics around a noise generator are examples of confinement (containment). Combinations of confinement and absorption are often used to control or eliminate system-generated sound.

Vibration becomes a problem when the effects of the cyclical motion of a piece of equipment are carried into and through the structure to a point where it is sensed, by definition, as an irritation or offense. Any rotating piece of equipment that is slightly out of balance can set up a vibration. Vibrations are often carried in structures to remote points. A person sitting at a desk may sense the floor or the desk vibrating. A pan of liquid on a laboratory table may exhibit waveforms on the surface form structural vibrations. Even as windstorms can cause buildings to shake and tremble, so can air moving in ducts or fluid moving in pipes give rise to vibrations and noise.

Vibration is best controlled by balance and isolation. Balance of moving components eliminates vibration at its origin. For large elements such as fans, compressors, and engines, isolation with springs or resilient pads is common practice. Lower-frequency vibrations require more initial spring deflection to be effective. The vibrating element must be relieved of any hard physical connection to its environment. Hence all pipe and conduit or duct connections must be flexibly connected. This is also helpful in making sure that the equipment does not carry the weight of the attached piping.

See Chap. 20 and Ref. 7 for further discussion of this issue.

1.10 Energy Conservation

A constriction in the international energy supply pipeline in 1973 to 1974 created worldwide awareness of the vital role of energy in the economic well-being of developed countries. The ability of common folk in the industrial nations to live more opulent lives than the kings and queens of yesteryear is directly related to the energy servant. Transportation, communication, agriculture, housing, manufacturing, health care services, and the like are all facilitated by the readily available supplies of energy, both fuel and power.

At the same time, there are recognizable limits to available energy sources. Despite new discoveries, coal, oil, gas, and other fossil or organic fuels can be estimated in total volume, with less this year than there was last. There is a finite limit to the potential development of renewable energies, all derived from the sun, such as wood, wind, hydropower, etc. In terms of magnitude, nuclear power (the fusion concept, at that) is the only known energy source that offers hope of being able to industrialize the third world and to sustain a growing world population over centuries rather than decades. Yet the use of nuclear power is mired in political quicksand in the United States.

The conclusion for the designer of HVAC systems, all of which by definition are energy-managing and energy-consuming, is that there is an international mandate as well as a moral and economic imperative to design systems which are modest in their use of energy. Although the United States lacks a well-defined national energy policy, local and regional energy codes give some direction to the HVAC systems designer. These codes encourage the construction of buildings which have lower inherent energy requirements, lighting systems which derive more illumination from fewer watts, and air-handling systems which move more air and water with less fan, pump, and compressor power. Most of these codes are based on the American Society of Heating, Refrigerating, and Air-Conditioning Engineers (ASHRAE) Standard 90.1.

In a time when much HVAC design involves the renovation and retrofit of older buildings and systems, there is good opportunity for substitution of components and system concepts which will provide similar or improved comfort by using less energy. Thus we suggest the five T's of energy conservation in both new and retrofit construction:

1. *Turn it off!* There is no substitute for the off switch. Provide a mechanism to turn off energy-using systems when they are not needed.

2. *Turn it down!* If it has to run, design it to run at the lowest level which will still meet the duty. Try to provide modulating control for all energy consumers.

3. *Tune it up!*   To operators: Keep things in good operating condition. To designers: Design for reliability and for maintainability.

4. *Turn it around!*   For retrofit designers: If you find a system which consumes disproportionate amounts of energy, improve it.

5. *Throw it away!*   If a system is an energy hog and does not lend itself to rehabilitation, be willing to take it out. The retrofit design market for the 1990s on into the next decades is a major industry market.

One good thing about energy conservation is that it nearly always pays for itself. But sometimes a bit of teaching is needed to get the message to the person controlling the purse strings.

A word of caution. Energy conservation is important in HVAC design, but it is not the purpose or function of the HVAC system. HVAC systems are intended to provide comfort, or a controlled environment. If we conserve energy to the point that we lose sight of the system's function, then we have failed in our duty. There is no glory in owning a building that drives tenants away with its energy-conserving but uncomfortable HVAC systems. Nor is there gratitude to an energy manager in an electronics plant where the production yield drops for lack of proper air quality even though energy costs are low. "Waste not, want not" is the energy motto. "Use what you need, but need what you use" is a corollary. See Ref. 8 for more on this topic.

1.11   Summary

A first-hand knowledge of engineering fundamentals is a must for HVAC designers. All problems are helped toward solution by an ability to think clearly, to move from problem definition to identification of alternatives, to evaluation of alternatives, to selection and implementation of preferred action. Knowledge of political constraints (codes and regulations) as well as technical and ethical factors is an important part of the designer's tool kit. Knowledge of fundamental engineering physics in fluid mechanics, thermodynamics, heat transfer, and psychrometrics is absolutely essential to the HVAC designer. Energy conservation in system design has become a moral imperative. An understanding of these fundamentals will make all other issues treated in this book easier to understand.

References

1. Dell'Isolo, Alphonse, *Value Engineering in the Construction Industry,* Construction Publishing Company, Inc., New York, 1973.

2. American Society of Heating, Refrigeration, and Air-Conditioning Engineers ASH-RAE Handbook, *1995 HVAC Applications,* Chap. 51, "Codes and Standards," Atlanta, GA.
3. ASHRAE Handbook, *1995 Fundamentals,* Chap. 2, "Fluid Flow."
4. Ibid., Chap. 1, "Thermodynamics and Refrigeration Cycles."
5. Ibid., Chap. 3, "Heat Transfer."
6. Ibid., Chap. 6, "Psychrometrics."
7. Ibid., Chap. 7, "Sound and Vibration."
8. ASHRAE Handbook, *1995 HVAC Applications,* Chap. 32, "Energy Management."

2

HVAC Engineering Fundamentals: Part 2

2.1 Introduction

A heating, ventilating, and air conditioning (HVAC) system is designed to satisfy the environmental requirements of comfort or a process, in a specific building or portion of a building and in a particular geographic locale. Designers must understand a great deal beyond basic HVAC system design and the outdoor climate. They must also understand the process or the comfort requirements.

In addition, designers must understand how the building is (or will be) constructed and whether that construction is suitable for the stipulated use of the space. For example, a warehouse cannot be used as a clean room without a great many modifications and improvements; to attempt to air-condition it to clean-room specifications without making the necessary architectural and structural revisions would be futile. While most are more subtle than this, many HVAC systems do not perform satisfactorily because the basic building is unsuitable. Designers must be able to recognize this.

It is also necessary to understand the use of the building and in most buildings the use of each part. How does this use affect occupancy, activity level, humidity, temperature, and ventilation requirements? Designers must have answers to these and many other questions before they can design a suitable HVAC system.

2.2 Comfort

Comfort is a highly subjective word, reflecting sensations which vary greatly from one individual to another. Comfort can be defined only in general or statistical terms. Research by ASHRAE over many years has identified the major factors contributing to comfort: temperature, relative humidity, air movement, and radiant effects. Attempts have been made to combine these factors to obtain a single-number index called the *effective temperature*[1] or comfort index. The type and quantity of clothing and the level of activity also affect comfort, but these factors vary so greatly that they cannot be quantified.

Typically, the air system is designed to control temperature always and humidity sometimes. Good control of air movement (from supply grilles and diffusers) is also needed to improve the comfort index. Most people do not like a draft (local air velocity over about 100 ft/min), but in some hot industrial environments, a high rate of airflow at the workstation may be necessary. Radiant effects are often beyond the control of the HVAC designer but should be taken into account in the HVAC design. Typical radiant effects are caused by windows (cold in winter, hot when exposed to the sun), by lighting fixtures (especially incandescent), and by cross-radiation among people in large groups. Radiant effects can sometimes be offset by providing radiant panels which are cooled or heated as needed.[1] There is an interesting graph in the ASHRAE Handbook which plots the percentage of people who are dissatisfied at any given temperature. It can be interpreted to read a range of general comfort between 70 and 75°F, but it also shows that there is *no* temperature at which everybody will be satisfied.[2]

2.3 Use of Psychrometrics

Psychrometrics is the study of the physical properties of air-water vapor mixtures, commonly called *moist air*. Because air is the final energy transport medium in most air conditioning processes, it is an essential tool in HVAC design. Chapter 19 discusses psychrometric principles, and a more detailed discussion can be found in the ASHRAE Handbook *Fundamentals*.[3] Many different psychrometric charts are available from various sources. This book uses the ASHRAE charts and tables.

2.4 HVAC Cycles

Figure 2.1 is a schematic representation of an elementary mechanical cooling cycle. While dehumidification is not an essential part of the cooling cycle, it usually occurs where the cooling medium is colder than the dew point temperature of the air. The cooling load Q_C in the conditioned space is a combination of internal and external loads (e.g., people, lights, solar) and is usually removed by circulating air through the space, with the entering air having a lower temperature and hu-

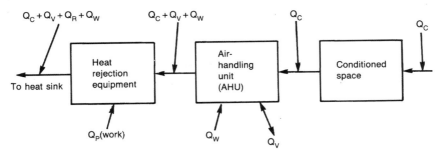

Figure 2.1 Mechanical cooling cycle.

midity than the desired space condition. To offset the cooling load, the temperature and humidity of the air are increased to equal those of the space, and then air is returned to the air-handling unit (AHU), where it is recooled and dehumidified. Most spaces require some (outside) ventilation air, which is mixed with the return air at the AHU, thereby imposing an additional cooling load Q_V. If the outside-air enthalpy is less than the space enthalpy, then Q_V will be negative and some "free cooling" will be obtained. Work energy is required to circulate the air—usually a fan driven by an electric motor—and this work Q_W becomes a part of the cooling load. The total load represented by $Q_C + Q_V + Q_W$ must be removed by the heat rejection equipment, usually a refrigeration system. Some additional work is done here by pumping fluids and driving refrigeration machines and condenser or cooling-tower fans. Ultimately all this heat energy is dumped to a heat sink—sometimes water, most often atmospheric air.

Note that the work portions of this cycle are parasitical. They contribute additional heat which must be removed, reducing the overall system efficiency.

Figure 2.2 shows schematically an elementary heating cycle, again using an air-handling unit. In this cycle, the ventilation load is usually

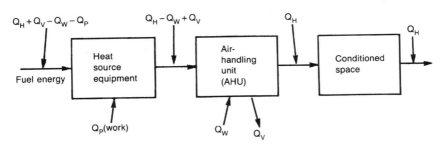

Figure 2.2 Mechanical heating cycle.

negative, requiring additional heating, but the work factors contribute to the available heat. Thus, most heating systems are more efficient overall than most cooling systems, when efficiency is defined as the heating or cooling done, divided by the energy input.

In Figs. 2.1 and 2.2, convective losses from piping, ductwork, and equipment have been neglected for simplicity. These factors may become important, particularly in large systems.

There are many different ways to accomplish the functions indicated in these diagrams. In later chapters we describe some of these methods.

2.5   Control Strategies

No HVAC system can be designed without a thorough knowledge of how it is to be controlled. Psychrometrics is helpful in control system design, since the psychrometric chart which describes the HVAC cycles and provides data for equipment selection also indicates control points and conditions to be expected. Controls are discussed in Chap. 8.

2.6   Architectural, Structural, and Electrical Considerations

The architectural design of the building provides the basis for those heating and cooling loads relating to the building envelope and to floor areas. In particular, the orientation, amount, and type of glass—together with any shading materials—are crucial to the proper calculation of solar loads and daylighting effects.

Structural and architectural factors define the space and weight carrying capacity available for HVAC equipment as well as piping and ductwork.

Electrical requirements for the HVAC equipment must be carefully and completely communicated to the electrical designer. The characteristics of the electrical service (voltage, frequency, etc.) affect the HVAC specifications and design.

Detailed and careful communication and coordination among the members of the design team are required. That statement is more than a tacit acknowledgment of the obvious. It points out an absolute necessity in the design process.

2.7   Conceptual Design

The problem-solving process of Chap. 1 should be a prelude to establishing any HVAC design concept. In most cases, a formal study is not made, but the general process is followed informally. Then the final decision is heavily biased by the desires of the client and architect as well as the experience and strengths of the HVAC design engineer. It is easy, under these conditions, to do things as they have always been done and to avoid innovation. This approach is also less expensive for the designer and avoids discussion and possible argument with the client, who is more often concerned with cost than with innovation. Still, the competent designer must be conversant with all types of systems and control strategies, in the event that the client wants suggestions for something new and different. This background is only developed through experience and study. Study is simply learning from the experience of others. This book and others like it were written for that purpose. There is a fundamental rule of any design: *Keep it simple*. No matter how complex the criteria may be, look for the simple way to solve the problem. This way is most likely to work in the real world.

2.8   Environmental Criteria for Typical Buildings

A building or space within a building may be used in many different ways. For each of these applications, the HVAC designer must determine the general criteria from personal experience or study and then add the special requirements of the user to this particular facility. The ASHRAE handbooks contain chapters on many common and exotic applications. The discussions which follow are limited to some of the more common applications. In every environment there are concerns for temperature, relative humidity, sound level and character, and the general quality of breathable air. In general, the higher the standard to be met, the more expensive the system will be to install, and probably to operate.

2.8.1   Residential

Two essential residential criteria are adequate comfort and the need of occupants to be able to adjust the controller set point. In larger residences (over 2400 ft²), the use of multiple HVAC systems should be considered, to allow zoning. These criteria extend to apartments and hotel or motel guest rooms. First cost and operating cost are concerns, as are simplicity and an acceptable sound level [from 20 to 30 noise criteria (NC)].

2.8.2   Commercial offices

Two basic needs are comfort and an adequate ventilation rate (airflow in cubic feet per minute per square foot)—so that occupants will not complain of stuffy or dead air. Where smoking is allowed, the ventilation rate and the amount of outside air must be increased. The general subject of indoor air quality is discussed in Chap. 5.

Controls are often designed to be nonadjustable by occupants. Zoning must be provided to compensate for use, occupant density, and exterior exposure. Conference rooms and corner offices should be separate zones. The ideal HVAC system is flexible enough to allow for adding or rearranging of zones as use changes. Noise levels should be stable in a range of 30 to 40 NC.

2.8.3 Hotels and motels

A small residential motel must meet only residential criteria. A large resort or conference hotel has many varying needs. The building may include not only such public areas as lobbies, restaurants, health and recreation facilities (swimming pool, sauna, etc.), meeting rooms, retail and service shops, and ballrooms, but also such behind-the-scenes facilities as kitchens, warehousing and storage for food and materials, shops (carpentry, electrical, plumbing, furniture repair), offices, central plant equipment (HVAC, electrical), laundry, employee lounge and eating areas, and others. Each area has different HVAC requirements. Comfort requirements will govern in most areas, with due allowance for occupant density and activity level. In rooms with a high occupant density, the cross-radiation effect among occupants may necessitate a lower space temperature to maintain comfort. Kitchens are seldom directly cooled, although indirect cooling using transfer air from other areas is common. Such reused air must be filtered. Laundries and health club areas have high-humidity problems, and exhaust rates must be adequate to keep humidities low enough to avoid condensation on walls and ceilings, with resultant water damage. Service and work areas may or may not be cooled, depending on the outdoor climate, but adequate ventilation must be provided. Because the main function of a hotel is to provide comfort for its guests, control, zoning and avoidance of drafts are important. The background noise level generated by the HVAC system is very important and is often overlooked. Auxiliary heating at entrances is needed even in mild climates. A wide variety of systems may be used to satisfy these various requirements. Common practice is to use AHUs of various kinds with a central chiller and heating plant.

2.8.4 Educational facilities

A school is much more than classrooms and offices, although these constitute the major portion of many schools. Comfort criteria apply in classrooms and offices, including special-purpose classrooms, e.g., music, chemistry, physics, and biology rooms. There are special exhaust requirements in laboratory rooms, especially chemistry labs. Auditoriums, with or without stages, have criteria peculiar to theaters—somewhat lower temperatures because of dense occupancy, a low background noise level, and avoidance of drafts in what is typically a high-airflow-rate situation. Many elementary schools and high schools have smaller, distributed AHUs, single or multizone, with a central plant source of heating and cooling. This approach is most suitable in campus-style schools with several buildings, but there are many other ways of air-conditioning these facilities.

At the college and university level, the facility takes on an institutional character, with emphasis on higher-quality HVAC equipment with longer life and lower maintenance costs. There are many special-purpose buildings, including auditoriums and theaters, radio and TV studios, laboratory facilities of many kinds, and physical education facilities, natatoriums, sports arenas, dormitories (residence criteria), support facilities such as maintenance shops and warehouses, and central plant facilities for heating and cooling, including sometimes elaborate distribution systems.

The acoustic design of a good theater or concert hall should be such that no electronic amplification is needed. The HVAC system must not produce a noise which will interfere with the audience's enjoyment of the performance. (See the discussion of sound in Chap. 20, and note the recommendation that the background noise level in a concert hall not exceed 25 NC.) This is not easy to achieve; it requires careful design and construction of both the building and the HVAC and electrical systems.

Laboratory facilities associated with education, public health, or industry can have very complex requirements, including humidity control and high levels of cleanliness. Most laboratories require high rates of exhaust and makeup air. The HVAC designer must work with the user to determine the exact criteria for the facility. Because the user in a research lab seldom knows "exactly" what will be needed, the HVAC design must be flexible enough to satisfy a wide range of contingencies. This tends to be costly, but is necessary in order to properly utilize the lab facility. Heat reclaim systems of various kinds are especially helpful here.

2.8.5 Hospitals

Hospitals are always interesting for the HVAC designer because of the wide variety of environmental conditions required in various depart-

ments. For example, the operating suite req
of about 65 to 70°F, with the relative humidity
50 and 60 percent, and a high percentage of outside a
is in use. To achieve the clean conditions needed, the supply a
be filtered through high-efficiency filters, preferably at or near the
discharge into the room, and a high airflow rate is needed. These requirements all have a rational basis in research; 50 to 55 percent
relative humidity (RH) is the value at which bacteria propagate least
readily. Public health authorities prescribe most of these criteria.
Some authorities are allowing reduction of airflow and some recirculation when the operating rooms are not in use. Nurseries do not require high airflow rates but do require about 55 percent RH. Offices,
public areas, cafeterias, shops, and other support areas have similar
criteria to other types of buildings. Air is often not recirculated from
patients' rooms, so individual fan-coil units are common, combined
with a small central ventilation system which provides makeup air for
exhaust. Laboratories have criteria similar to those described in Sec.
2.8.4.

2.8.6 Manufacturing facilities

Process environments dominate the manufacturing area of HVAC applications. Typical requirements include very close control of temperature (plus or minus 1°F or less is not unusual) and humidity (plus or
minus 3 to 5 percent RH is typical). These criteria can be met only by
the use of carefully designed HVAC systems and very high-quality
control devices. Clean rooms often require high airflow rates but have
normal or low heating and cooling loads. One design solution is to use
a small system to provide heating and cooling with supplemental fans
and filters to provide the necessary airflow. Research laboratory facilities are also part of the manufacturing complex. Electroplating operations require high rates of exhaust and makeup air, with a means
of removing contaminants from the exhaust. Machining operations
usually generate an oil mist which is carried in the air and deposited
everywhere. The HVAC system can be designed to control this problem
to some degree. Painting processes also involve large quantities of exhaust and makeup air, control of space pressure, and removal of contaminants from the exhaust air. Some processes generate heat and
combustion products. Many processes offer opportunities for heat reclaim. Manufacturing provides many opportunities for innovative
HVAC design. Industrial hygiene criteria complement HVAC criteria
in these environments.

2.9   Designing for Operation and Maintenance

Over the life of the HVAC system, the operating and maintenance costs of installed systems will greatly exceed the initial cost. The system design can have a substantial effect on both energy and labor costs. A system that is difficult to maintain will likely not be properly maintained, with a consequent increase in energy costs or loss of system function. The HVAC designer should observe the following basic criteria.

1. *Keep it simple.*   Complexity breeds misunderstanding and is a seedbed for dissatisfaction, even failure.

2. *Provide adequate space and accessibility for equipment.*   This includes ease of access, space for maintenance and repair, and access for removal and replacement of large equipment items. Because this involves the cooperation of the owner and the architect, the HVAC designer must be prepared to justify the needs on a long-term economic basis that takes into account the life-cycle cost.

3. *In the specifications, require written maintenance and operation procedures.*   A collection of manufacturers' descriptive and maintenance bulletins is useful as a reference but is not a procedure. A well-written operating or maintenance procedure should be simple and straightforward, not longer than one or two pages, and easy for personnel to understand and even memorize. Anything more complex or lengthy simply will not be used. The schematic flow and control diagrams from the contract drawings become primary reference material for these procedures.

4. *Require the contractor to provide basic training for operators.*   List the items to be covered and the training procedures. In this connection, the HVAC designer should remind the building owner or manager of the need for continued training and retraining of operating and maintenance personnel. The manager has an economic stake in doing this, because inadequate maintenance will result in higher operating costs. Reference 4 provides more information on this area of design.

2.10   Summary

In this chapter we discussed the fundamentals and philosophy of designing HVAC systems in very broad terms to provide a background for the technical details in succeeding chapters. While all the details are necessary to develop a functional system, the designer must never lose sight of the objective of the HVAC system: to provide a suitable environment for comfort or for a process.

References

1. ASHRAE Handbook, *1997 Fundamentals,* p. 8.13.
2. Ibid., p. 8.18.
3. Ibid., Chap. 6.
4. ASHRAE Handbook, *1995 HVAC Applications,* Chaps. 32 to 39.

Design Procedures: Part 1

Load Calculations

3.1 Introduction

All solutions to engineering problems start with a calculation or estimation of the duty which must be met (i.e., quantifying the problem). The purpose of heating and cooling load calculations, then, is to quantify the heating and/or cooling loads in the space(s) to be conditioned. Rough estimates of load may be made during the concept design phase. During design development and final design, it is essential to make orderly, detailed, and well-documented load calculations, because these form the basis for equipment selection, duct and piping design, and psychrometric analysis. Today's energy and building codes also require detailed documentation to prove compliance.

The necessity for order and documentation cannot be overemphasized. While it may sometimes seem unnecessary to list all criteria and assumptions, these data are invaluable when changes or questions arise, sometimes months or years after the design is completed.

This chapter refers to a great many data tables from the ASHRAE Handbook. Many of these tables require several pages in the 8½-in by 11-in format of the Handbook and are presented here in abstract form. For the complete tables refer to the Handbook.

3.2 Use of Computers

Current practice is to use computers for load calculations. Many load calculation programs exist, with varying degrees of complexity and accuracy. Some can be run on small personal computers while others require large computer systems which are usually provided by computer service centers. There are several important things to remember when a computer is used.

1. The input must be carefully checked for accuracy. This is not a simple task since the complete input can be voluminous and complex. In fact, it takes at least as long to properly input and check the data as it does to manually calculate the loads.

2. The output must be checked for reasonableness. Many people look on a computer printout as perfect and final. This is seldom true in HVAC work. The old rule of "garbage in, garbage out" (GIGO) is never more applicable than in HVAC calculations.

3. Different load calculation programs may yield different results for the same input data. In part, this is due to the way the programs handle solar effect and building dynamics. The differences may be significant. When using a new program, the designer is advised to manually spot-check the results.

There are also many computer programs for estimating energy consumption. Many include subroutines for calculating heating and cooling loads. These calculations are seldom suitable for design, because they tend to be "block loads" or have other limitations.

Computer calculation has one great advantage over manual calculation. With manual calculation a specific time (or times) of day must be used, with separate calculations made for each time needed. The computer can calculate the loads at 12 or more different hours from one set of input data. This is extremely valuable in organizing zones, determining maximum overall loads, and selecting equipment.

In this book, we describe manual calculations so that the reader can develop a personal understanding of the principles of HVAC load calculation and will be better able to evaluate the input and output of computer analysis.

3.3 Rule-of-Thumb Calculations

Every HVAC designer needs some handy empirical data for use in approximating loads and equipment sizes during the early conceptual stages of the design process. These are typically square feet per ton for cooling, Btu per square foot for heating, and cubic feet per minute per square foot for air-handling equipment. The values used will vary with climate and application and are always tempered by experience. These numbers can also be used as "check figures" during the detailed calculation procedure to alert the designer to unusual conditions or computational errors. As an example only, the cooling load values in Table 3.1 are based on traditional empirical data and will not be ap-

TABLE 3.1   Rough-Estimate Values for Cooling Loads

Building type	ft^2/ton
Residence	600–700
Classroom	200–250
General office	300–350
Conference room	100–200
Bowling alley	^{a}
Clean room	100–300^b
Hospital patient room	300–350^c
Manufacturing	250–350^d
Arenas, etc.	150–200
Hotel meeting room	200–250
Data processing room	80–100^e

^{a} One ton per lane, plus additional for spectator areas, food service, etc.
^{b} Eight to 10 (ft^3/min)/ft^2 required.
^{c} Most codes do not allow recirculation of return air from patient rooms.
^{d} Special areas may have other requirements.
^{e} Mainframe computers and auxiliaries.

plicable in all cases. Energy conserving practice in envelope construction, in lighting design, and in system design has resulted in decreased loads in many cases. But increased use of personal computers and other appliances has the opposite effect of increasing the air conditioning requirements. Designers must develop their own data if the data are to be reliable.

3.4   Design Criteria

The first step in any load calculation is to establish the design criteria for the project. These data should be listed on standard forms, such as those shown in Figs. 3.1, 3.2, and 3.3, and are needed for either manual or computer calculations.

For manual calculations, some specific times of day must be assumed because it is impractical to calculate manually for every hour of occupancy. Due to solar effects, maximum loads in exterior zones depend on exposure—in a typical office building, east-facing zones peak at about 10 a.m. to noon, south-facing at noon to 2 p.m., and west- and north-facing at 3 to 6 p.m, sometimes later. Because solar factors for south-facing glass are greater in winter than in summer, a south-facing space may have a greater peak cooling load in November or December than in June or July, even though the outdoor ambient condition is cooler. Load factors described below must be determined

(COMPANY NAME)

COOLING AND HEATING LOAD ESTIMATES

Job Name_____
Location_____
Job No._____ By_____
Date_____
Sheet 1 of____

DESIGN CRITERIA

Outdoor Temp: Summer_____ °F db, _____°F wb, _____% rh   h=_____
              Winter_____°F db

Indoor Temp: Summer_____ °F db, _____°F wb, _____% rh   h=_____
              Winter_____°F db, _____°F wb, _____% rh   h=_____

Summer: ΔT=_____ F, Δh=_____   Winter: ΔT=_____

Elevation above sea level _____ft
Air Density Ratio   _____
Air Density   _____lb/cu ft^3
Air Factor   _____Btu/h/CFM-F°
Latitude   _____

Basis for design: ASHRAE Handbook, Fundamentals, 1985

Factors:

  People: Sensible_____Btu/h/person,   Latent_____Btu/h/person

  Lights: Type_____   W/ft^2_____
          Allowance factor (1)_____
          Multiplier (3.41 x Allowance factor)_____
  Appliances:_____ W/ft^2
  Power   _____ W/ft^2

  Infiltration:_____

  Occupancy Schedules:
    People:
    Lights:
    Appliances:
    Power:

  CLF (Cooling Load Factors):
    People:
    Lights:
    Appliances:
    Power:

(1) Ballast factor or equivalent correction.

Figure 3.1 Design criteria form, sheet 1.

for all these times. In addition to assumed maximum loads, all zones must be calculated for one building peak time, usually 3 p.m. for an office building. Public assembly buildings such as churches and arenas will usually peak 2 to 3 h into the occupied period. The thermal mass of the building structure creates a load leveling or flywheel effect on the instantaneous load.

(COMPANY NAME)
COOLING AND HEATING LOAD ESTIMATES

Job Name_____
Job No._____ By _____
Date_____
Sheet 2 of _____

"U" Factors

Ref: ASHRAE Handbook, Fundamentals 1985

	Construction	Code	U	Heating Mult.
Roof (1)				
Roof (2)				
Wall (1)				
Wall (2)				
Glass (1)				
Glass (2)				
Partition (1)				
Partition (2)				
Floor (1)				
Floor (2)				

SHGF & SC

Ref: ASHRAE Handbook, Fundamentals 1985, Chapter 27

			SHGF				SC			
	Facing	Time ⟶								
Glass (1)										
Glass (1)										
Glass (1)										
Glass (1)										
Glass (2)										
Glass (2)										
Glass (2)										
Glass (2)										

Figure 3.2   Design criteria form, sheet 2.

3.4.1   Name of project, location, job ID, date, name of designer

That a job notebook should include the project name, location, job ID, date, name of designer, etc., is obvious. What's not so obvious is the need to show job ID, date, and designer's initials on *every page* of the calculations. Location defines latitude, longitude, altitude, and weather conditions. Latitude is important when dealing with solar heat gains. Altitude is important because it defines standard local air density, which affects airflow rates and equipment performance. Lon-

(COMPANY NAME)
COOLING AND HEATING LOAD ESTIMATES

Job   Name _____

Job   No. _____ By _____

Date _____

Sheet 3 of   _____

CLTD, CLF & ΔT

Ref: ASHRAE Handbook, Fundamentals 1985

Time	CLTD				CLF				ΔT
Roof (1) Roof (2)									
Wall Wall Wall Wall									
Wall Wall Wall Wall									
Glass Glass Glass Glass									
Glass Glass Glass Glass									
Partition (1) Partition (2) Floor (1) Floor (2)									
People Lights Appliances Power									

Figure 3.3 Design criteria form, sheet 3.

gitude places the job in a time zone, which may have an almost 1-h effect on correlation between local and solar time.

3.4.2 Outdoor and indoor design temperature and humidity

Indoor design conditions are determined by comfort or process requirements (see Sec. 1.9). For comfort cooling, conditions of 75°F and

40 to 50 percent maximum relative humidity are usually recommended, although some energy codes may require higher summer temperatures. For comfort heating, an indoor design temperature of 70 to 72°F is usually satisfactory. Many people will try to operate the systems at lower or higher temperatures than design, and this will be possible most of the time. Most HVAC systems tend and need to be oversized for various reasons, some of which will be pointed out later.

Outside design conditions are determined from published data for the specific location, based on weather bureau records. Table 3.2 is a list of data for a few selected sites. The ASHRAE Handbook *1997 Fundamentals*[1] lists data for over 1000 sites in North America and throughout the world. For comfort cooling, use of the 2.5 percent values is recommended; for comfort heating, use the 99.0 percent values, except use a median of annual extremes for certain critical heating applications. Note that the maximum wet-bulb (wb) temperature seldom occurs at the same time as the design dry-bulb (db) temperature. For sites not listed, data may be obtained by interpolation, but this should be done only by an experienced meteorologist.

The design temperature and humidity conditions should be plotted on a psychrometric chart. Then the relative humidity (RH) and enthalpy (h) can be read as well as the indoor wet-bulb temperature. (See Chap. 19 for a discussion of psychrometrics.)

3.4.3 Elevation (above sea level)

Up to about 2000 ft the altitude related change in air density has less than a 7 percent effect (see Table 3.3). With higher elevations, the decreasing air density has an increasingly significant negative effect on air-handling system performance. Heat exchanger (coil) capacities are reduced. Fans still move the same volume of air, but the heating/cooling capacity of the air is reduced because the air volume has less mass. Evaporative condenser and cooling tower capacities are slightly—but not entirely—proportionately reduced. The psychrometric chart changes are described in Chap. 19. The air factor is also affected by elevation because it includes an air density effect. The formula defining the air factor (AF) is

$$AF = \text{air density} \times SH \times 60 \text{ min/h} \tag{3.1}$$

where AF = Density = air factor for determining airflow rate, Btu/h/[(ft^3/min) · °F]
    Density = air density at design elevation and temperature (for air conditioning, 60°F is used), lb/ft^3

TABLE 3.2 Climatic Conditions in the United States

City, state	Latitude, deg	Elevation, ft	Winter, °F Design, db temp.		Summer, °F Design db mean coincident wb			Mean daily range	Design wb temp.			Median of annual extremes, –°F	
			99%	97.5%	1%	2.5%	5%		1%	2.5%	5%	Max.	Min.
Phoenix, AZ	33	1112	31	34	109/71	107/71	105/71	27	76	75	75	113	27
Los Angeles, CA	34	97	41	43	83/68	80/68	77/67	15	70	69	68	—	—
Denver, CO	40	5283	–5	1	93/59	91/59	89/59	28	64	63	62	97	–10
Washington, DC	39	14	14	17	93/75	91/74	89/74	18	78	77	76	98	7
Miami, FL	26	7	44	47	91/77	90/77	89/77	15	79	79	78	93	39
Atlanta, GA	33	1010	17	22	94/74	92/74	90/73	19	77	76	75	96	12
Chicago, IL	42	607	–5	0	94/74	91/73	88/72	20	77	75	74	—	–1
Boston, MA	42	15	6	9	91/73	88/71	85/70	16	75	74	72	96	–1
Minneapolis, MN	45	834	–16	–12	92/75	89/73	86/71	22	77	75	73	97	–22
St. Louis, MO	39	535	2	6	97/75	94/75	91/74	21	78	77	76	—	—
Albuquerque, NM	35	5311	12	16	96/61	94/61	92/61	27	66	65	64	98	5
Buffalo, NY	43	705	2	6	88/71	85/70	83/69	21	74	73	72	90	–3
Cincinnati, OH	39	758	1	6	92/73	90/72	88/72	21	77	75	74	97	0
Philadelphia, PA	40	7	10	14	93/75	90/74	87/72	21	77	76	75	96	6
Dallas, TX	33	481	18	22	102/75	100/75	97/75	20	78	78	77	—	—
Houston, TX	30	108	28	33	97/77	95/77	93/77	18	80	79	79	99	23
Seattle, WA	47	20	22	27	85/68	82/66	78/65	19	69	67	65	90	22

SOURCE: Abridged from the ASHRAE Handbook, *1993 Fundamentals*, Chap. 24, Table 1 (subsequent editions provide more extensive data). Used by permission.

TABLE 3.3 Air Factor Change with Altitude (Approximate Values at 60°F)

Altitude above sea level, ft	Atmospheric pressure, lb/in²	Density ratio	Density, lb/ft³	Air factor, Btu/h/[(ft³/min) · °F]
0	14.7	1.000	0.075	1.08
1,000	14.2	0.966	0.072	1.04
2,000	13.7	0.932	0.070	1.01
3,000	13.2	0.898	0.067	0.97
4,000	12.7	0.864	0.065	0.93
5,000	12.2	0.830	0.062	0.90
6,000	11.8	0.803	0.060	0.87
7,000	11.3	0.769	0.058	0.83
8,000	10.9	0.741	0.056	0.80
9,000	10.5	0.714	0.054	0.77
10,000	10.1	0.687	0.052	0.74

> SH = specific heat of air at design temperature and pressure, Btu/lb-°F (SH for dry air is approximately 0.24 Btu/lb-°F)

For sea level (standard air density) this becomes

$$AF = 0.075 \text{ lb/ft}^3 \times 0.24 \text{ Btu/lb} \times 60 \text{ min/h}$$

$$= 1.08 \text{ Btu/h/[(ft}^3/\text{min)} \cdot °F]$$

Some designers and handbooks use 1.10 Btu/h/[(ft³/min) · °F] (obtained by rounding off 1.08).

The air factor (AF) at altitude is obtained by multiplying the sea level air factor (1.08) by the project altitude density ratio (DR).

3.5 Factors for Load Components

3.5.1 Internal heat gains

Internal heat gains are due to people, lights, appliances, and processes. Heat gain from people is a function of the level of activity (see Table 3.4).

Heat gain from lights is a function of wattage, at a rate of 3.413 Btu/h/W. For fluorescent lighting and industrial-type fixtures, a ballast (transformer) factor must be used. A typical multiplier for fluorescents is 1.2, resulting in a heat gain of 4.1 Btu/h per nominal lighting watt. General lighting for offices requires from 2 to 3 W/ft² of floor area. The actual lighting layout should be used whenever possible. With ceiling mounted lights (recessed) some of the heat may go

TABLE 3.4 Rates of Heat Gain from Occupants of Conditioned Spaces, Btu/h

Degree of activity	Typical application	Total heat for adults, male*	Total heat adjusted†	Sensible heat	Latent heat
Seated at rest	Theater, movie	390	330	225	105
Seated, very light work writing	Offices, hotels, apartments	450	400	245	155
Seated, eating	Restaurant‡	490	550‡	275	105
Standing, light work or walking slowly	Retail store, bank	550	450	250	200
Light bench work	Factory	800	750	330	420
Walking, 3 mi/h, light machine work	Factory	1000	1000	375	105
Bowling§	Bowling alley	1500	1450	580	800
Moderate dancing	Dance hall	900	850	305	545
Heavy work, heavy machine work, lifting	Factory	1600	1600	635	965
Heavy work, athletics	Gymnasium	2000	1800	710	1090

*Tabulated values are based on 78°F room dry-bulb temperature. For 80°F room dry-bulb temperature, the total heat remains the same, but the sensible heat value should be decreased by approximately 8% and the latent heat values increased accordingly.

†Adjusted total heat gain is based on normal percentage of women, men, and children for the application listed, with the postulate that the gain from an adult female is 85% of that for an adult male, and that the gain from a child is 75% of that for an adult male.

‡Adjusted total heat value for eating in a restaurant, includes 60 Btu/h for food per individual (30 Btu/h sensible and 30 Btu/h latent).

§For bowling, figure one person per alley actually bowling, and all others as sitting (400 Btu/h) or standing and walking slowly (790 Btu/h).

SOURCE: Reprinted by permission from ASHRAE Handbook, 1997 Fundamentals, Chap. 28, Table 3.

to the ceiling plenum without being a cooling load in the room. Lighting manufacturers' literature may treat this condition.

Task lighting and appliance loads are difficult to predict. The extensive use of computer terminals and electric typewriters has made this a significant factor. The typical allowance for task lighting and appliances is 0.75 to 1.0 average W/ft^2, although localized loads may be as much as 3 W/ft^2. Some large computer components may impose 10 W/ft^2 in the vicinity of the installation.

Table 3.5 lists possible heat gains from some miscellaneous appliances. Kitchen appliances, cookers, stoves, ovens, etc., can provide large amounts of heat gain. These loads should be confirmed prior to final design effort.

Heat gains from manufacturing processes must be estimated from the energy input to the process.

3.5.2 Cooling load versus instantaneous heat gain

The internal heat gains discussed above are often greater than the actual cooling load due to those gains. This is a result of heat storage in the building and furnishings — anything that has mass. The effect is shown in Fig. 3.4. The longer the heat gain persists, the more nearly the instantaneous cooling load will approach the actual cooling load. *Cooling load factors* (CLFs) for various elements of heat gain are shown in Tables 3.6 through 3.13. The lighting-related load is particularly affected by the type of fixture and ventilation rate of the air conditioning system, as indicated in Table 3.6.

The load factor criteria pages should include schedules of use and occupancy, together with cooling load factors to be applied. See Ref. 2 for additional discussion of this topic.

3.5.3 Transmission through the building envelope

Chapter 18 discusses heat transfer and the determination of *"U"* factors — overall heat transmission coefficients — for the elements of the envelope. The criteria pages must include a description of each wall, roof, partition, and floor section which forms a boundary between conditioned and nonconditioned space. From the description a *"U"* factor is determined; note that the direction of heat flow (up, horizontal, or down) makes a difference. The units of the *"U"* factor are Btu per hour per square foot of area per degree Fahrenheit of temperature difference from inside to outside air.

For calculating the cooling load due to heat gain by conduction through opaque walls and the roof, the *sol-air* temperature con-

TABLE 3.5 Estimated Rate of Heat Release from Cooking and Miscellaneous Appliances

Appliance and capacity	Overall dim., in W × D × H	Manufacturer's input rating, Btu/h	Probable max. input, Btu/h	Estimated rate of heat release Btu/h			
				Without hood			With hood
				Sensible	Latent	Total	All sensible
Cooking, gas-burning, counter type							
Broiler-griddle	31 × 20 × 18	36,000	18,000	11,700	6,300	18,000	3,600
Coffee urn, 8-gal twin	25 in wide	20,000	10,000	7,000	3,000	10,000	2,000
Steam table, per square feet of top		2,500	1,250	750	500	1,250	250
Gas-burning, floor-mounted type							
Range, heavy-duty							
Top section	32 W × 39 D	64,000	32,000	Hood required			6,400
Oven	25 × 28 × 15	40,000	20,000	Hood required			4,000
Range, restaurant type							
Per two-burner section	12 W × 28 D	24,000	12,000	Hood required			2,400
Per oven	24 × 22 × 14	30,000	15,000	Hood required			3,000
Per broiler-griddle	24 W × 26 D	35,000	17,500	Hood required			3,500
Cooking, electric, counter type							
Coffee brewer, 240 cs/h	27 × 21 × 22	17,000	8,500	6,500	2,000	8,500	1,700
Deep-fat fryer, 14 lb	13 × 22 × 10	18,750	9,400	2,800	6,600	9,400	3,000
Toaster, cont., 360 slices/h	15 × 15 × 28	7,500	3,700	1,960	1,740	3,700	1,200
Cooking, steam-heated							
Steam table, per square feet of top		1,650	825	500	325	825	260
Steam kettle, per gallon cap.		2,000	1,000	600	400	1,000	320
Miscellaneous							
Hair dryer, helmet type		2,400		1,870	330		2,200
Instrument sterilizer		3,750		650	1,200		1,850
Bunsen burner		3,000		1,680	420		2,100

SOURCE: Abstracted from ASHRAE Handbook, *1985 Fundamentals*, Chap. 26, Tables 20 and 21. Used by permission. Subsequent editions provide more extensive data.

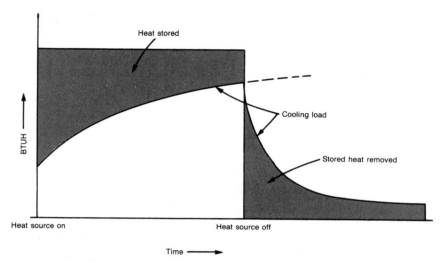

Figure 3.4 Thermal storage effect. (*Reprinted by permission from ASHRAE Handbook, 1993 Fundamentals, p. 26.7.*)

cept may be used. For a complete discussion of this concept, see the ASHRAE Handbook 1997 *Fundamentals*.[3]

Figure 3.5 illustrates the energy transfers which give rise to the sol-air concept in a wall. Both direct and diffuse solar radiation have a heating effect on the exterior surface of the wall. The surface temperature will usually be greater than the outside air temperature, which then has a cooling effect. When the exterior surface temperature is greater than the internal temperature of the wall, heat transfer into the wall will take place. Some of this heat will be stored, increasing the internal temperature of the wall. Some heat will be transferred by conduction to the cooler interior surface and then to the room, as heat gain. The process is dynamic because the exterior surface temperature is constantly changing as the angle of the sun changes. At certain times of the day and night, some of the stored heat will be transferred back to the exterior surface. Only part of the heat that enters the wall becomes cooling load, and this is delayed by storage effects. The greater the mass of the wall, the greater will be the delay.

The sol-air temperature derives an equivalent outside temperature which is a function of time of day and orientation. This value is then adjusted for the storage effect and the time delay caused by the mass of the wall or roof, the daily temperature range, which has an effect on the storage; the color of the outside surface, which affects the solar heat absorption rate; and the latitude and month. Tables 3.14 and 3.15 describe the sol-air data. When these data are combined with the in-

TABLE 3.6 Design Values of the Alpha (α) Coefficient Based on Features of
Room Furnishings, Light Fixtures, and Ventilation Arrangements

α	Furnishings	Air supply and return	Type of light fixture
0.45	Heavyweight, simple furnishings, no carpet	Low rate; supply and return below ceiling ($V \leq 0.5$)*	Recessed, not vented
0.55	Ordinary furniture, no carpet	Medium to high ventilation rate; supply and return below ceiling or through ceiling grill and space ($V \geq 0.5$)*	Recessed, not vented
0.65	Ordinary furniture, with or without carpet	Medium to high ventilation rate or fan coil or induction-type air conditioning terminal unit; supply through ceiling or wall diffuser; return around light fixtures and through ceiling space. ($V \geq 0.5$)*	Vented
0.75 or greater	Any type of furniture	Ducted returns through light fixtures	Vented or free-hanging in air stream with ducted returns

*V is room air supply rate in ft^3/(min · ft^2) of floor area.
SOURCE: Reprinted by permission from ASHRAE Handbook, *1985 Fundamentals*, Chap. 26, Table 15, p. 26.19. (Subsequent editions provide more extensive data.)

side design temperature, a *cooling load temperature difference* (CLTD) is obtained. Then the cooling load is

$$q = U \times A \times CLTD \qquad (3.2)$$

where q = cooling load, Btu/h, for the given section U is the heat transfer coefficient for the given construction, and A = area, ft^2, of the given section. Tables 3.16 through 3.20 provide data for calculating the CLTD for various orientations and solar times.

3.5.4 Conduction and solar heat gain through fenestration

Fenestration is defined as any light-transmitting opening in the exterior skin of a building. When light is transmitted, so is solar energy.

TABLE 3.7 The Beta (β) Classification Values Calculated for Different Envelope Constructions and Room Air Circulation Rates

Room envelope construction[*] (mass of floor area, lb/ft²)	Room air circulation and type of supply and return[†]			
	Low	Medium	High	Very high
2-in wood floor (10)	B	A	A	A
3-in concrete floor (40)	B	B	B	A
6-in concrete floor (75)	C	C	C	B
8-in concrete floor (120)	D	D	C	C
12-in concrete floor (160)	D	D	D	D

[*] Floor covered with carpet and rubber pad; for a floor covered only with floor tile take next classification to the right in the same row.

[†] Low: Low ventilation rate—minimum required to cope with cooling load from lights and occupants in the interior zone. Supply through floor, wall or ceiling diffuser. Ceiling space not vented and $h = 0.4$ Btu/(h · ft² · °F) (where h = inside surface convection coefficient used in calculation of b classification).

Medium: Medium ventilation rate, supply through floor, wall or ceiling diffuser. Ceiling space not vented and $h = 0.6$ Btu/(h · ft² · °F).

High: Room air circulation induced by primary air of induction unit or by fan coil unit. Return through ceiling space and $h = 0.8$ Btu/(h · ft² · °F).

Very high: High room air circulation used to minimize temperature gradients in a room. Return through ceiling space and $h = 1.2$ Btu/(h · ft² · °F).

SOURCE: Reprinted by permission from ASHRAE Handbook, *1985 Fundamentals*. (Subsequent editions provide more extensive data.)

Up to the end of World War II, fenestrations almost always used clear glass with outside shading by awnings or overhangs and inside shading by roller shades, venetian blinds or draperies. With increased use of air conditioning it was realized that solar heat gains through this type of fenestration were as much as 25 to 30 percent of the total peak air conditioning load, and efforts were made to reduce the effects. Reducing the amount of glass has a claustrophobic effect on people, so much of the effort centered on reducing the transmission through the glazing material. There are now available a multitude of materials, including heat-absorbing and heat-reflective glass.

The mechanism of solar transmission through glazing is shown in Fig. 3.6. When direct or diffuse radiation falls on the glazing, some is reflected. Some radiation is absorbed, heating the glazing material and escaping as convective or radiant heat. Some radiation passes through the glazing after which it is absorbed by materials in the room, causing a heating effect and thus a cooling load (after some time delay). If exterior shading is used, only the diffuse solar component is effective. If interior shading is used, some additional reflective and absorptive factors come into play, and the mechanism becomes even more complex. As indicated, there is also conduction through the glazing due to the temperature difference between inside and outside. At certain times of the year, conduction may represent a heat loss.

TABLE 3.8 Cooling Load Factors When Lights Are on for 8 h

a Coef-ficents	b Class-ification	0	1	2	3	4	5	6	7	8	9	10	11	12	13	14	15	16	17	18	19	20	21	22	23
																Number of hours after lights are turned on									
0.45	A	0.02	0.46	0.57	0.65	0.72	0.77	0.82	0.85	0.88	0.46	0.37	0.30	0.24	0.19	0.15	0.12	0.10	0.08	0.06	0.05	0.04	0.03	0.03	0.02
	B	0.07	0.51	0.56	0.61	0.65	0.68	0.71	0.74	0.77	0.34	0.31	0.28	0.25	0.22	0.20	0.18	0.16	0.15	0.13	0.12	0.11	0.10	0.09	0.08
	C	0.11	0.55	0.58	0.60	0.63	0.65	0.67	0.69	0.71	0.28	0.26	0.25	0.23	0.22	0.20	0.19	0.18	0.17	0.16	0.15	0.14	0.13	0.12	0.12
	D	0.14	0.58	0.60	0.61	0.62	0.63	0.64	0.65	0.66	0.22	0.22	0.21	0.20	0.20	0.19	0.19	0.18	0.18	0.17	0.16	0.16	0.16	0.15	0.15
0.55	A	0.01	0.56	0.65	0.72	0.77	0.82	0.85	0.88	0.90	0.37	0.30	0.24	0.19	0.16	0.13	0.10	0.08	0.07	0.05	0.04	0.03	0.03	0.02	0.02
	B	0.06	0.60	0.64	0.68	0.71	0.74	0.76	0.79	0.81	0.28	0.25	0.23	0.20	0.18	0.16	0.15	0.13	0.12	0.11	0.10	0.09	0.08	0.07	0.06
	C	0.09	0.63	0.66	0.68	0.70	0.71	0.73	0.75	0.76	0.23	0.21	0.20	0.19	0.18	0.17	0.16	0.15	0.14	0.13	0.12	0.11	0.11	0.10	0.10
	D	0.11	0.66	0.67	0.68	0.69	0.70	0.71	0.72	0.72	0.18	0.18	0.17	0.17	0.16	0.16	0.15	0.15	0.14	0.14	0.13	0.13	0.13	0.12	0.12
0.65	A	0.01	0.66	0.73	0.78	0.82	0.86	0.88	0.91	0.93	0.29	0.23	0.19	0.15	0.12	0.10	0.08	0.06	0.05	0.04	0.03	0.03	0.02	0.02	0.01
	B	0.04	0.69	0.72	0.75	0.77	0.80	0.82	0.84	0.85	0.22	0.19	0.18	0.16	0.14	0.13	0.12	0.10	0.09	0.08	0.08	0.07	0.06	0.06	0.05
	C	0.07	0.72	0.73	0.75	0.76	0.78	0.79	0.80	0.82	0.18	0.17	0.16	0.15	0.14	0.13	0.12	0.11	0.11	0.10	0.10	0.09	0.08	0.08	0.07
	D	0.09	0.73	0.74	0.75	0.76	0.77	0.77	0.78	0.79	0.14	0.14	0.13	0.13	0.13	0.12	0.12	0.11	0.11	0.11	0.10	0.10	0.10	0.10	0.09
0.75	A	0.01	0.76	0.80	0.84	0.87	0.90	0.92	0.93	0.95	0.21	0.17	0.13	0.11	0.09	0.07	0.06	0.05	0.04	0.03	0.02	0.02	0.02	0.01	0.01
	B	0.03	0.78	0.80	0.82	0.84	0.85	0.87	0.88	0.89	0.15	0.14	0.13	0.11	0.10	0.09	0.08	0.07	0.07	0.06	0.05	0.05	0.04	0.04	0.04
	C	0.05	0.80	0.81	0.82	0.83	0.84	0.85	0.86	0.87	0.13	0.12	0.11	0.11	0.10	0.09	0.09	0.08	0.08	0.07	0.07	0.06	0.06	0.06	0.05
	D	0.06	0.81	0.82	0.82	0.83	0.83	0.84	0.84	0.85	0.10	0.10	0.10	0.09	0.09	0.09	0.08	0.08	0.08	0.08	0.07	0.07	0.07	0.07	0.07

SOURCE: Abstracted by permission from ASHRAE Handbook, 1985 Fundamentals. (Subsequent editions provide more extensive data.)

TABLE 3.9 Cooling Load Factors When Lights Are on 10 h

a Coef-ficents	b Class-ification	0	1	2	3	4	5	6	7	8	9	10	11	12	13	14	15	16	17	18	19	20	21	22	23
												Number of hours after lights are turned on													
0.45	A	0.03	0.47	0.58	0.66	0.73	0.78	0.82	0.86	0.88	0.91	0.93	0.49	0.39	0.32	0.26	0.21	0.17	0.13	0.11	0.09	0.07	0.06	0.05	0.04
	B	0.10	0.54	0.59	0.63	0.66	0.70	0.73	0.76	0.78	0.80	0.82	0.39	0.35	0.32	0.28	0.26	0.23	0.21	0.19	0.17	0.15	0.14	0.12	0.11
	C	0.15	0.59	0.61	0.64	0.66	0.68	0.70	0.72	0.73	0.75	0.76	0.33	0.31	0.29	0.27	0.26	0.24	0.23	0.21	0.20	0.19	0.18	0.17	0.16
	D	0.18	0.62	0.63	0.64	0.66	0.67	0.68	0.69	0.69	0.70	0.71	0.27	0.26	0.26	0.25	0.24	0.23	0.23	0.22	0.21	0.21	0.20	0.19	0.19
0.55	A	0.02	0.57	0.65	0.72	0.78	0.85	0.85	0.88	0.91	0.92	0.94	0.40	0.32	0.26	0.21	0.17	0.14	0.11	0.09	0.07	0.06	0.05	0.04	0.03
	B	0.08	0.62	0.66	0.69	0.73	0.75	0.78	0.80	0.82	0.84	0.85	0.32	0.29	0.26	0.23	0.21	0.19	0.17	0.15	0.14	0.12	0.11	0.10	0.09
	C	0.12	0.66	0.68	0.70	0.72	0.74	0.75	0.77	0.78	0.79	0.81	0.27	0.25	0.24	0.22	0.21	0.20	0.19	0.17	0.16	0.15	0.14	0.14	0.13
	D	0.15	0.69	0.70	0.71	0.72	0.73	0.73	0.74	0.75	0.76	0.76	0.22	0.22	0.21	0.20	0.20	0.19	0.18	0.18	0.17	0.17	0.16	0.16	0.15
0.65	A	0.02	0.66	0.73	0.78	0.83	0.86	0.89	0.89	0.93	0.94	0.95	0.31	0.25	0.20	0.16	0.13	0.11	0.08	0.07	0.05	0.04	0.04	0.03	0.02
	B	0.06	0.71	0.74	0.76	0.79	0.81	0.83	0.84	0.86	0.87	0.89	0.25	0.22	0.20	0.18	0.16	0.15	0.13	0.12	0.11	0.10	0.09	0.08	0.07
	C	0.09	0.74	0.75	0.77	0.78	0.80	0.81	0.82	0.83	0.84	0.85	0.21	0.20	0.18	0.17	0.16	0.15	0.14	0.14	0.13	0.12	0.11	0.11	0.10
	D	0.11	0.76	0.77	0.77	0.78	0.79	0.79	0.81	0.81	0.81	0.82	0.17	0.17	0.16	0.16	0.15	0.15	0.14	0.14	0.14	0.13	0.13	0.12	0.12
0.75	A	0.01	0.76	0.81	0.84	0.88	0.90	0.92	0.93	0.95	0.96	0.97	0.22	0.18	0.14	0.12	0.09	0.08	0.06	0.05	0.04	0.03	0.03	0.02	0.03
	B	0.04	0.79	0.81	0.83	0.85	0.86	0.88	0.89	0.90	0.91	0.92	0.18	0.16	0.14	0.13	0.12	0.10	0.09	0.08	0.08	0.07	0.06	0.06	0.05
	C	0.07	0.81	0.82	0.83	0.84	0.85	0.86	0.87	0.88	0.89	0.89	0.15	0.14	0.13	0.12	0.12	0.11	0.10	0.10	0.09	0.09	0.08	0.08	0.07
	D	0.08	0.83	0.83	0.84	0.84	0.85	0.85	0.86	0.86	0.87	0.87	0.12	0.12	0.12	0.11	0.11	0.11	0.10	0.10	0.10	0.09	0.09	0.09	0.09

SOURCE: Abstracted by permission from ASHRAE Handbook, 1985 Fundamentals. (Subsequent ditions provide more extensive data.)

TABLE 3.10 Cooling Load Factors When Lights Are on for 12 h

a Coefficients	b Classification	Number of hours after lights are turned on																							
		0	1	2	3	4	5	6	7	8	9	10	11	12	13	14	15	16	17	18	19	20	21	22	23
0.45	A	0.05	0.49	0.59	0.67	0.73	0.78	0.83	0.86	0.89	0.91	0.93	0.94	0.95	0.51	0.41	0.33	0.27	0.22	0.17	0.14	0.11	0.09	0.07	0.06
	B	0.13	0.57	0.61	0.65	0.69	0.72	0.75	0.77	0.79	0.82	0.83	0.85	0.87	0.43	0.39	0.35	0.31	0.28	0.25	0.23	0.21	0.18	0.17	0.15
	C	0.19	0.63	0.65	0.67	0.69	0.71	0.73	0.74	0.76	0.77	0.79	0.80	0.81	0.37	0.35	0.33	0.31	0.29	0.27	0.26	0.24	0.23	0.21	0.20
	D	0.22	0.66	0.67	0.68	0.69	0.70	0.71	0.72	0.73	0.74	0.74	0.75	0.76	0.32	0.31	0.30	0.29	0.28	0.27	0.26	0.26	0.25	0.24	0.23
0.55	A	0.04	0.58	0.66	0.73	0.78	0.82	0.86	0.89	0.91	0.93	0.94	0.95	0.96	0.42	0.34	0.27	0.22	0.18	0.14	0.11	0.09	0.07	0.06	0.05
	B	0.11	0.65	0.68	0.72	0.74	0.77	0.79	0.81	0.83	0.85	0.86	0.88	0.89	0.35	0.32	0.28	0.26	0.23	0.21	0.19	0.17	0.15	0.14	0.12
	C	0.15	0.69	0.71	0.73	0.75	0.76	0.78	0.79	0.80	0.81	0.83	0.84	0.85	0.30	0.29	0.27	0.25	0.24	0.22	0.21	0.20	0.19	0.17	0.16
	D	0.18	0.72	0.73	0.74	0.75	0.76	0.76	0.77	0.78	0.78	0.79	0.80	0.80	0.26	0.25	0.24	0.24	0.23	0.22	0.22	0.21	0.20	0.20	0.19
0.65	A	0.03	0.67	0.74	0.79	0.83	0.86	0.89	0.91	0.93	0.94	0.95	0.96	0.97	0.33	0.26	0.21	0.17	0.14	0.11	0.09	0.07	0.06	0.05	0.04
	B	0.09	0.73	0.75	0.78	0.80	0.82	0.84	0.85	0.87	0.88	0.90	0.90	0.91	0.27	0.25	0.22	0.20	0.18	0.16	0.15	0.13	0.12	0.11	0.10
	C	0.12	0.76	0.78	0.79	0.80	0.81	0.83	0.84	0.85	0.86	0.86	0.87	0.88	0.24	0.22	0.21	0.20	0.19	0.17	0.15	0.15	0.14	0.14	0.13
	D	0.14	0.79	0.79	0.80	0.80	0.81	0.82	0.82	0.83	0.83	0.84	0.84	0.85	0.20	0.20	0.19	0.18	0.18	0.17	0.17	0.16	0.16	0.15	0.15
0.75	A	0.02	0.77	0.81	0.85	0.88	0.90	0.92	0.94	0.95	0.96	0.97	0.97	0.98	0.23	0.19	0.15	0.12	0.10	0.08	0.06	0.05	0.04	0.03	0.03
	B	0.06	0.81	0.82	0.84	0.86	0.87	0.88	0.90	0.91	0.92	0.92	0.93	0.94	0.19	0.18	0.16	0.14	0.13	0.12	0.10	0.09	0.08	0.08	0.07
	C	0.09	0.83	0.84	0.85	0.86	0.87	0.88	0.88	0.89	0.90	0.90	0.91	0.91	0.17	0.16	0.15	0.14	0.13	0.12	0.12	0.11	0.10	0.10	0.09
	D	0.10	0.85	0.85	0.86	0.86	0.86	0.87	0.87	0.88	0.88	0.88	0.89	0.89	0.14	0.14	0.14	0.13	0.13	0.12	0.12	0.12	0.11	0.11	0.11

SOURCE: Abstracted by permission from ASHRAE Handbook, *1985 Fundamentals*. (Subsequent editions provide more extensive data.)

TABLE 3.11 Sensible Heat CLFs for People

Total hours in space	Hours after each entry into space																							
	1	2	3	4	5	6	7	8	9	10	11	12	13	14	15	16	17	18	19	20	21	22	23	24
2	0.49	0.58	0.17	0.13	0.10	0.08	0.07	0.06	0.05	0.04	0.04	0.03	0.03	0.02	0.02	0.02	0.02	0.01	0.01	0.01	0.01	0.01	0.01	0.01
4	0.49	0.59	0.66	0.71	0.27	0.21	0.16	0.14	0.11	0.10	0.08	0.07	0.06	0.06	0.05	0.04	0.04	0.03	0.03	0.03	0.02	0.02	0.02	0.01
6	0.50	0.60	0.67	0.72	0.76	0.79	0.34	0.26	0.21	0.18	0.15	0.13	0.11	0.10	0.08	0.07	0.06	0.06	0.05	0.04	0.04	0.03	0.03	0.03
8	0.51	0.61	0.67	0.72	0.76	0.80	0.82	0.84	0.38	0.30	0.25	0.21	0.18	0.15	0.13	0.12	0.10	0.09	0.08	0.07	0.06	0.05	0.05	0.04
10	0.53	0.62	0.69	0.74	0.77	0.80	0.83	0.85	0.87	0.89	0.42	0.34	0.28	0.23	0.20	0.17	0.15	0.13	0.11	0.10	0.09	0.08	0.07	0.06
12	0.55	0.64	0.70	0.75	0.79	0.81	0.84	0.86	0.88	0.89	0.91	0.92	0.45	0.36	0.30	0.25	0.21	0.19	0.16	0.14	0.12	0.11	0.09	0.08
14	0.58	0.66	0.72	0.77	0.80	0.83	0.85	0.87	0.88	0.89	0.91	0.92	0.93	0.94	0.47	0.38	0.31	0.26	0.23	0.20	0.17	0.15	0.13	0.11
16	0.62	0.70	0.75	0.79	0.82	0.85	0.87	0.88	0.90	0.91	0.92	0.93	0.94	0.95	0.95	0.96	0.49	0.39	0.33	0.28	0.24	0.20	0.18	0.16
18	0.66	0.74	0.79	0.82	0.85	0.87	0.89	0.90	0.92	0.93	0.94	0.94	0.95	0.96	0.96	0.97	0.97	0.97	0.50	0.40	0.33	0.28	0.24	0.21

SOURCE: Abstracted by permission from ASHRAE Handbook, *1985 Fundamentals*. (Subsequent editions provide more extensive data.)

41

TABLE 3.12 Sensible-Heat CLFs for Appliances—Hooded

Total operational hours	Hours after appliances are on																							
	1	2	3	4	5	6	7	8	9	10	11	12	13	14	15	16	17	18	19	20	21	22	23	24
2	0.27	0.40	0.25	0.18	0.14	0.11	0.09	0.08	0.07	0.06	0.05	0.04	0.04	0.03	0.03	0.03	0.02	0.02	0.02	0.02	0.01	0.01	0.01	0.01
4	0.28	0.41	0.51	0.59	0.39	0.30	0.24	0.19	0.16	0.14	0.12	0.10	0.09	0.08	0.07	0.06	0.05	0.05	0.04	0.04	0.03	0.03	0.02	0.02
6	0.29	0.42	0.52	0.59	0.65	0.70	0.48	0.37	0.30	0.25	0.21	0.18	0.16	0.14	0.11	0.11	0.09	0.08	0.07	0.06	0.05	0.05	0.04	0.04
8	0.31	0.44	0.54	0.61	0.66	0.71	0.75	0.78	0.55	0.43	0.35	0.30	0.25	0.22	0.19	0.16	0.14	0.13	0.11	0.10	0.08	0.07	0.06	0.06
10	0.33	0.46	0.55	0.62	0.68	0.72	0.76	0.79	0.81	0.84	0.60	0.48	0.39	0.33	0.28	0.24	0.21	0.18	0.16	0.14	0.12	0.11	0.09	0.08
12	0.36	0.49	0.58	0.64	0.69	0.74	0.77	0.80	0.82	0.85	0.87	0.88	0.64	0.51	0.42	0.36	0.31	0.26	0.23	0.20	0.18	0.15	0.13	0.12
14	0.40	0.52	0.61	0.67	0.72	0.76	0.79	0.82	0.84	0.86	0.88	0.89	0.91	0.92	0.67	0.54	0.45	0.38	0.32	0.28	0.24	0.21	0.19	0.16
16	0.45	0.57	0.65	0.70	0.75	0.78	0.81	0.84	0.86	0.87	0.88	0.90	0.92	0.93	0.94	0.94	0.69	0.56	0.46	0.39	0.34	0.29	0.25	0.22
18	0.52	0.63	0.70	0.75	0.79	0.82	0.84	0.86	0.88	0.89	0.91	0.92	0.93	0.94	0.96	0.95	0.96	0.96	0.71	0.58	0.48	0.41	0.35	0.30

SOURCE: Abstracted by permission from ASHRAE Handbook, *1985 Fundamentals*. (Subsequent editions provide more extensive data.)

TABLE 3.13 Sensible-Heat CLFs for Appliances—Unhooded

Total operational hours	Hours after appliances are on																							
	1	2	3	4	5	6	7	8	9	10	11	12	13	14	15	16	17	18	19	20	21	22	23	24
2	0.56	0.64	0.15	0.11	0.08	0.07	0.06	0.05	0.04	0.04	0.03	0.03	0.02	0.02	0.02	0.02	0.01	0.01	0.01	0.01	0.01	0.01	0.01	0.01
4	0.57	0.65	0.71	0.75	0.23	0.18	0.14	0.12	0.10	0.08	0.07	0.06	0.05	0.05	0.04	0.04	0.03	0.03	0.02	0.02	0.02	0.02	0.01	0.01
6	0.57	0.65	0.71	0.76	0.79	0.82	0.29	0.22	0.18	0.15	0.13	0.11	0.10	0.08	0.07	0.06	0.06	0.05	0.04	0.04	0.03	0.03	0.03	0.02
8	0.58	0.66	0.72	0.76	0.80	0.82	0.85	0.87	0.33	0.26	0.21	0.18	0.15	0.13	0.11	0.10	0.09	0.08	0.07	0.06	0.05	0.04	0.04	0.03
10	0.60	0.68	0.73	0.77	0.81	0.83	0.85	0.87	0.89	0.90	0.36	0.29	0.24	0.20	0.17	0.15	0.13	0.11	0.10	0.08	0.07	0.07	0.06	0.05
12	0.62	0.69	0.75	0.79	0.82	0.84	0.86	0.88	0.89	0.91	0.92	0.93	0.38	0.31	0.25	0.21	0.18	0.16	0.14	0.12	0.11	0.09	0.08	0.07
14	0.64	0.71	0.76	0.80	0.83	0.85	0.87	0.89	0.90	0.92	0.93	0.93	0.94	0.95	0.40	0.32	0.27	0.23	0.19	0.17	0.15	0.13	0.11	0.10
16	0.67	0.74	0.79	0.82	0.85	0.87	0.89	0.90	0.91	0.92	0.93	0.94	0.95	0.96	0.96	0.97	0.42	0.34	0.28	0.24	0.20	0.18	0.15	0.13
18	0.71	0.78	0.82	0.85	0.87	0.89	0.90	0.92	0.93	0.94	0.94	0.95	0.96	0.96	0.97	0.97	0.97	0.98	0.43	0.35	0.29	0.24	0.21	0.18

SOURCE: Abstracted by permission from ASHRAE Handbook, *1985 Fundamentals*. (Subsequent editions provide more extensive data.)

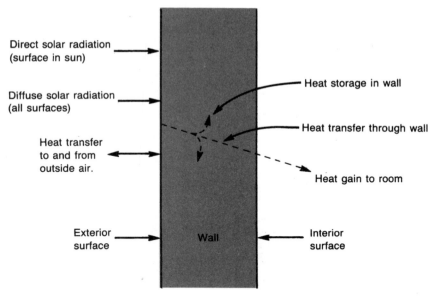

Figure 3.5 The sol-air temperature concept.

Many types of glass are treated to increase the reflective and/or absorptive components. A highly absorptive glazing can become very hot; then thermal expansion of the glass can create serious problems unless sufficient flexibility is provided in the support system. Partial shading of a pane creates thermal stress along the shadow line. In the past, there were many incidents of glass cracking, falling out, or blowing out in a high wind, or else glass leaked during a rainstorm. Most of these problems have been recognized and solved.

The cooling load due to solar radiation through fenestration is calculated from

$$q = A \times SC \times MSHGF \times CLF \qquad (3.3)$$

where q = cooling load, Btu/h
        A = net glazing area, ft^2
        SC = shading coefficient
    MSHGF = maximum solar heat gain factor
      CLF = cooling load factor

The shading coefficient is determined by using a combination of glazing material and indoor shading methods. Because almost an infinite number of combinations are available, considerable judgment is required.

TABLE 3.14 Sol-Air Temperatures for July 21, 40°N Latitude

Time	Air temp., °F	N	NE	E	SE	S	SW	W	NW	HOR
					$\alpha/h_o = 0.15$					
0100	76	76	76	76	76	76	76	76	76	69
0200	76	76	76	76	76	76	76	76	76	69
0300	75	75	75	75	75	75	75	75	75	68
0400	74	74	74	74	74	74	74	74	74	67
0500	74	74	74	74	74	74	74	74	74	67
0600	74	82	95	97	86	75	75	75	75	74
0700	75	82	103	109	97	78	78	78	78	85
0800	77	82	103	114	105	83	81	81	81	96
0900	80	85	101	114	110	92	85	85	85	106
1000	83	89	96	110	112	100	89	89	89	115
1100	87	93	94	104	111	108	96	93	93	123
1200	90	96	96	97	107	112	107	97	96	127
1300	93	99	99	99	102	114	117	110	100	129
1400	94	100	100	100	100	111	123	121	107	126
1500	95	100	100	100	100	107	125	129	116	121
1600	94	99	98	98	98	100	122	131	120	113
1700	93	100	96	96	96	96	115	127	121	103
1800	91	99	92	92	92	92	103	114	112	91
1900	87	87	87	87	87	87	87	87	87	80
2000	85	85	85	85	85	85	85	85	85	78
2100	83	83	83	83	83	83	83	83	83	76
2200	81	81	81	81	81	81	81	81	81	74
2300	79	79	79	79	79	79	79	79	79	72
2400	77	77	77	77	77	77	77	77	77	70
Avg.	83	86	89	91	90	89	90	91	89	91
					$\alpha/h_o = 0.30$					
0100	76	76	76	76	76	76	76	76	76	69
0200	76	76	76	76	76	76	76	76	76	69
0300	75	75	75	75	75	75	75	75	75	68
0400	74	74	74	74	74	74	74	74	74	67
0500	74	74	74	74	74	74	74	74	74	67
0600	74	90	117	121	99	77	77	77	77	81
0700	75	90	131	144	120	82	82	82	82	102
0800	77	87	130	151	134	89	86	86	86	122
0900	80	91	122	148	141	105	91	91	91	140
1000	83	95	109	137	141	118	96	95	95	155
1100	87	100	101	122	136	129	105	100	100	166
1200	90	103	103	104	125	134	125	104	103	172
1300	93	106	106	106	111	135	142	128	107	172
1400	94	106	106	106	107	129	152	148	120	166
1500	95	106	106	106	106	120	156	163	137	155
1600	94	104	103	103	103	106	151	168	147	139
1700	93	108	100	100	100	100	138	162	149	120
1800	91	107	94	94	94	94	116	138	134	98
1900	87	87	87	87	87	87	87	87	87	80
2000	85	85	85	85	85	85	85	85	85	78

TABLE 3.14 (Continued)

Time	Air temp., °F	N	NE	E	SE	S	SW	W	NW	HOR
				$\alpha/h_o = 0.30$						
2100	83	83	83	83	83	83	83	83	83	76
2200	81	81	81	81	81	81	81	81	81	74
2300	79	79	79	79	79	79	79	79	79	72
2400	77	77	77	77	77	77	77	77	77	70
Avg.	83	89	95	100	99	95	99	100	95	107

The hourly air temperatures in Column 2, Table 2 are for a location with a design temperature of 95°F and a range of 21°F. To compute corresponding temperatures for other locations, select a suitable design temperature from Column 6, Table 1 of Chapter 24 and note the outdoor daily range (Column 7). For each hour, take the percentage of the daily range indicated in Table 3 of this chapter and subtract from the design temperature.

Example 1: Calculate the summer dry-bulb temperature at 1200 hours for Carson City, Nevada.

Solution: From Table 1, Chapter 24, the daily range is 42°F and the 1% design dry-bulb temperature is 93°F. From Table 3, the percentage of the daily range at 1200 hours is 23%. Thus: Dry-Bulb Temperature at 1200 hours = Design Dry-Bulb − Daily Range · Percentage = 93.0 − [42(23/100)] = 83.3°F.

SOURCE: Abstracted by permission from ASHRAE Handbook, *1997 Fundamentals*, Chap. 28, Table 1, p. 28.6.

The ASHRAE Handbook[4] provides tables covering a number of combinations. These are reproduced here as Tables 3.21 through 3.25. Figure 3.7 illustrates the various drapery combinations relating to Table 3.25. Additional information is available from the glazing manufacturers. These data are best obtained by means of a solar calorimeter, in which the glazing-shading combination is measured against unrestricted solar transmission. The value of the SC must be matched against the need to see through the glazing and the use of daylighting to minimize lighting energy use.

The maximum solar heat gain factor (MSHGF) is a function of orientation, time of day, and latitude. Table 3.26 provides values of MSHGF for north latitudes from 0 to 64°.

TABLE 3.15   Percentage of the Daily Range

Time, h	%	Time, h	%	Time, h	%	Time, h	%
1	87	7	93	13	11	19	34
2	92	8	84	14	3	20	47
3	96	9	71	15	0	21	58
4	99	10	56	16	3	22	68
5	100	11	39	17	10	23	76
6	98	12	23	18	21	24	82

SOURCE: Reprinted by permission from ASHRAE Handbook, *1993 Fundamentals*, Chap. 28, Table 2, p. 28.6.

TABLE 3.16   Roof Construction Code

Roof no.	Description	Code numbers of layers
1	Steel sheet with 1-in insulation	A0, E2, E3, B5, A3, E0
2	1-in wood with 1-in insulation	A0, E2, E3, B5, B7, E0
3	4-in lightweight concrete	A0, E2, E3, C14, E0
4	2-in heavyweight concrete with 1-in insulation	A0, E2, E3, B5, C12, E0
5	1-in wood with 2-in insulation	A0, E2, E3, B6, B7, E0
6	6 in lightweight concrete	A0, E2, E3, C15, E0
7	2.5-in wood with 1-in insulation	A0, E2, E3, B5, B8, E0
8	8-in lightweight concrete	A0, E2, E3, C16, E0
9	4-in heavyweight concrete with 1-in insulation	A0, E2, E3, B5, C5, E0
10	2.5-in wood with 2-in insulation	A0, E2, E3, B6, B8, E0
11	Roof terrace system	A0, C12, B1, B6, E2, E3, C5, E0
12	6-in heavyweight concrete with 1-in insulation	A0, E2, E3, B5, C13, E0
13	4-in wood with 1-in insulation	A0, E2, E3, B5, B9, E0

SOURCE: Reprinted by permission from ASHRAE Handbook, *1985 Fundamentals,* Table 4, p. 26.7. (Subsequent editions provide more extensive data.)

The cooling-load factor (CLF) depends on the building construction and the absence or presence of internal shading, as shown in Tables 3.27 and 3.28. This factor takes into account the heat storage properties of the building and shading and the resultant time lag.

The conduction heat transfer through the glazing is equal to

$$q = A \times U \times (t_o - t_i) \tag{3.4}$$

where t_o and t_i are the outdoor and indoor temperatures. Typical U factors for glazing are shown in Tables 3.29 and 3.30.

Because of the effects of solar radiation and storage in the glass, the conduction heat gain must be treated in a manner similar to that used for walls and roofs. The design temperature difference must be corrected as shown in Table 3.31. Then the cooling load from conduction (and convection) can be calculated from

$$q = U \times A \times \text{CLTD} \tag{3.5}$$

This is added to the direct and diffuse solar load above to yield the total cooling load for fenestration.

The criteria sheets must include the necessary data for each orientation, glass-shading combination, and time of day to be used in the calculations.

3.5.5   Infiltration

The heat gain or loss due to infiltration is impossible to calculate with any assurance of accuracy. The inaccuracy results from three factors which are outside the control of the HVAC designer: building construction, chimney effects, and wind direction and velocity.

TABLE 3.17 Cooling Load Temperature Differences for Calculating Cooling Load from Flat Roofs

Description of construction	Weight, lb/ft²	U value, Btu/(h·ft²·°F)	Solar time, h													
			7	8	9	10	11	12	13	14	15	16	17	18	19	20
Without Suspended Ceiling																
1-in wood with 1-in insulation	8	0.170	-2	4	14	27	39	52	62	70	74	74	70	62	51	38
4-in lightweight concrete	18	0.213	-3	1	9	20	32	44	55	64	70	73	71	66	57	45
4-in heavyweight concrete with 1-in (or 2-in) insulation	52	0.200	8	8	10	14	20	26	33	40	46	50	53	53	52	48
4-in wood with 1-in (or 2-in) insulation	(52) 17 (18)	(0.120) 0.106 (0.078)	22	20	18	17	16	17	18	21	24	28	32	36	39	41
With Suspended Ceiling																
1-in wood with 1-in insulation	10	0.115	2	3	7	13	21	30	40	48	55	60	62	61	58	51
4-in lightweight concrete	20	0.134	0	0	4	10	19	29	39	48	56	62	65	64	61	54
4-in heavyweight concrete with 1-in (or 2-in) insulation	53	0.128	21	20	20	21	22	24	27	29	32	34	36	38	38	38
4-in wood with 1-in (or 2-in) insulation	(54) 19 (20)	(0.090) 0.082 (0.064)	27	26	24	23	22	21	22	22	24	25	27	30	32	34

SOURCE: Abstracted from ASHRAE Handbook, *1985 Fundamentals*, Chap. 26, Table 5, p. 26.8. Used by permission.
For limitations and adjustments see notes in the Handbook table. (Subsequent editions provide more extensive data.)

TABLE 3.18 Wall Construction Group Description

Group no.	Description of construction	Weight, lb/ft²	U value, Btu/(h · ft² · °F)	Code numbers of layers (see Table 8)
4-in Face brick + (brick)				
C	Airspace + 4-in face brick	83	0.358	A0, A2, B1, A2, E0
D	4-in Common brick	90	0.415	A0, A2, C4, E1, E0
C	1-in Insulation or airspace +4-in common brick	90	0.174–0.301	A0, A2, C4, B1/B2, E1, E0
B	2-in Insulation + 4-in common brick	88	0.111	A0, A2, B3, C4, E1, E0
B	8-in Common brick	130	0.302	A0, A2, C9, E1, E0
A	Insulation or airspace + 8-in common brick	130	0.154–0.243	A0, A2, C9, B1/B2, E1, E0
4-in Face brick + (heavyweight concrete)				
C	Airspace + 2-in concrete	94	0.350	A0, A2, B1, C5, E1, E0
B	2-in Insulation + 4-in concrete	97	0.116	A0, A2, B3, C5, E1, E0
A	Airspace or insulation + 8-in or more concrete	143–190	0.110–0.112	A0, A2, B1, C10/11, E1, E0
4-in Face brick + (lightweight or heavyweight concrete block)				
E	4-in Block	62	0.319	A0, A2, C2, E1, E0
D	Airspace or insulation + 4-in block	62	0.153–0.246	A0, A2, C2, B1/B2, E1, E0
D	8-in Block	70	0.274	A0, A2, C7, A6, E0
C	Airspace or 1-in insulation + 6-in or 8-in block	73–89	0.221–0.275	A0, A2, B1, C7/C8, E1, E0
B	2-in Insulation + 8-in block	89	0.096–0.107	A0, A2, B3, C7/C8, E1, E0
4-in Face brick + (clay tile)				
D	4-in tile	71	0.381	A0, A2, C1, E1, E0
D	Airspace + 4-in tile	71	0.281	A0, A2, C1, B1, E1, E0
C	Insulation + 4-in tile	71	0.169	A0, A2, C1, B2, E1, E0
C	8-in Tile	96	0.275	A0, A2, C6, E1, E0
B	Airspace or 1-in insulation + 8-in tile	96	0.142–0.221	A0, A2, C6, B1/B2, E1, E0
A	2-in Insulation + 8-in tile	97	0.097	A0, A2, B3, C6, E1, E0

Wall description	Weight	U	Codes
Heavyweight concrete wall + (finish)			
E 4-in Concrete	63	0.585	A0, A1, C5, E1, E0
D 4-in Concrete + 1-in or 2-in insulation	63	0.119–0.200	A0, A1, C5, B2/B3, E1, E0
C 2-in Insulation + 4-in concrete	63	0.119	A0, A1, B6, C5, E1, E0
C 8-in Concrete	109	0.490	A0, A1, C10, E1, E0
B 8-in Concrete + 1-in or 2-in insulation	110	0.115–0.187	A0, A1, C10, B5/B6, E1, E0
A 2-in Insulation + 8-in concrete	110	0.115	A0, A1, B3, C10, E1, E0
B 12-in Concrete	156	0.421	A0, A1, C11, E1, E0
A 12-in Concrete + insulation	156	0.113	A0, C11, B6, A6, E0
Lightweight and heavyweight concrete block + (finish)			
F 4-in Block + airspace/insulation	29	0.161–0.263	A0, A1, C2, B1/B2, E1, E0
E 2-in Insulation + 4-in block	29–37	0.105–0.114	A0, A1, B3, C2/C3, E1, E0
E 8-in Block	47–51	0.294–0.402	A0, A1, C7/C8, E1, E0
D 8-in Block + airspace/insulation	41–57	0.149–0.173	A0, A1, C7/C8, B1/B2, E1, E0
Clay tile + (finish)			
F 4-in Tile	39	0.419	A0, A1, C1, E1, E0
F 4-in Tile + airspace	39	0.303	A0, A1, C1, B1, E1, E0
E 4-in Tile + 1-in insulation	39	0.175	A0, A1, C1, B2, E1, E0
D 2-in Insulation + 4-in tile	40	0.110	A0, A1, B3, C1, E1, E0
D 8-in Tile	63	0.296	A0, A1, C6, B1/B2, E1, E0
C 8-in Tile + airspace/1-in insulation	63	0.151–0.231	A0, A1, C6, B1/B2, E1, E0
B 2-in Insulation + 8-in tile	63	0.099	A0, A1, B3, C6, E1, E0
Metal curtain wall			
G With/without airspace + 1-in/ 2-in/3-in insulation	5–6	0.091–0.230	A0, A3, B5/B6/B12, A3, E0
Frame wall			
G 1- to 3-in insulation	16	0.081–0.178	A0, A1, B1, B2/B3/B4, E1, E0

SOURCE: Abstracted by permission from ASHRAE Handbook, *1989 Fundamentals*, Table 30, p. 26.35. (Subsequent editions provide more extensive data.)

TABLE 3.19 Cooling Load Temperature Differences for Calculating Cooling Load from Sunlit Walls

North latitude wall facing	0100	0200	0300	0400	0500	0600	0700	0800	0900	1000	1100	1200	1300	1400	1500	1600	1700	1800	1900	2000	2100	2200	2300	2400	Max. CLTD, h	Min. CLTD	Max. CLTD	Difference CLTD
Group A Walls																												
N	14	14	14	13	13	13	12	12	11	11	10	10	10	10	10	10	11	11	12	12	13	13	14	14	2	10	14	4
NE	19	19	18	18	17	17	16	15	15	15	15	15	16	16	17	18	18	19	19	19	20	20	20	20	22	15	20	5
E	24	24	23	22	22	21	20	19	18	18	19	19	20	21	22	23	24	24	25	25	25	25	25	25	22	18	25	6
SE	24	23	23	22	21	20	20	19	18	18	18	19	19	19	19	20	22	23	23	23	24	24	25	24	22	18	24	6
S	20	20	19	19	18	18	17	16	15	15	14	14	14	14	14	15	16	17	18	19	19	20	20	20	23	14	20	6
SW	25	25	25	24	24	23	22	21	20	19	19	18	18	17	17	17	18	19	20	22	23	24	25	25	24	17	25	8
W	27	27	26	26	25	24	24	23	22	21	20	19	19	18	18	18	18	19	20	22	23	25	26	26	1	18	27	9
NW	21	21	21	20	20	19	19	18	17	16	16	15	15	14	14	14	15	15	16	17	18	19	20	21	1	14	21	7
Group B Walls																												
N	15	14	14	13	12	11	11	10	9	9	9	8	9	9	9	10	12	12	13	14	14	15	15	15	24	8	15	7
NE	19	18	17	16	15	14	12	12	12	13	14	15	16	17	18	19	20	20	20	21	21	21	20	20	21	12	21	9
E	23	22	21	20	18	17	16	15	15	15	17	19	21	22	24	25	26	26	27	27	26	26	25	24	20	15	27	12
SE	23	22	21	20	18	17	16	15	15	14	15	18	18	20	21	23	25	26	26	26	26	26	25	24	21	14	26	12
S	21	20	19	18	17	15	14	13	12	11	11	11	11	12	12	13	17	18	20	21	22	22	22	22	23	11	22	11
SW	24	26	25	24	22	20	19	18	16	15	14	14	13	13	14	15	17	19	22	25	27	28	28	28	24	13	28	15
W	29	28	27	25	24	23	21	19	18	17	16	15	14	14	14	15	17	19	23	25	27	29	30	30	24	14	30	16
NW	23	22	21	20	19	18	17	15	14	13	12	12	12	11	12	12	13	15	17	19	21	22	23	23	24	11	23	9
Group C Walls																												
N	15	14	13	12	11	10	9	8	8	7	6	7	8	9	10	12	13	14	15	16	17	17	17	16	22	7	17	10
NE	19	17	16	14	13	11	10	10	11	13	17	20	19	20	21	22	23	23	23	23	23	22	21	20	20	10	23	13
E	22	21	19	18	17	16	14	12	11	13	19	27	25	27	26	29	30	30	30	29	28	27	24	24	18	11	30	18
SE	22	21	19	18	15	14	12	12	12	13	17	19	22	24	24	28	29	29	29	29	28	27	26	24	19	12	29	17
S	21	19	18	16	15	13	12	10	9	8	8	9	11	14	17	20	24	25	26	29	26	25	24	22	20	8	29	21
SW	29	27	25	22	20	18	16	15	13	12	8	9	11	13	15	18	22	26	29	32	35	33	35	33	22	9	35	27
W	31	29	27	24	22	20	18	16	14	13	13	12	12	11	14	18	24	30	32	36	38	35	35	34	22	11	38	27
NW	25	23	21	18	16	14	13	11	10	9	7	8	10	11	12	13	15	18	22	25	27	27	27	26	22	10	27	17
Group D Walls																												
N	15	13	12	11	10	9	7	6	6	6	6	7	8	10	10	13	15	17	18	19	19	19	18	16	21	6	19	13
NE	17	15	13	12	11	8	7	8	10	14	17	20	22	23	23	24	24	25	25	24	23	22	20	18	19	7	25	18
E	19	17	15	13	11	9	8	9	12	17	26	33	33	30	31	32	33	31	31	30	28	25	24	22	16	8	33	25
SE	20	17	15	13	11	10	8	9	12	17	22	27	29	31	32	34	36	34	33	31	30	28	26	24	17	8	32	24
S	19	17	15	14	13	11	9	8	7	6	9	9	11	16	20	24	27	29	29	29	27	24	22	20	19	6	29	23
SW	25	22	20	18	16	14	12	11	10	9	8	9	11	14	18	24	31	36	38	38	36	32	29	27	21	8	38	30
W	25	22	20	18	16	14	12	10	9	9	13	12	12	11	14	20	27	36	41	44	45	40	36	31	21	9	41	32
NW	20	18	16	14	13	11	9	8	7	7	6	8	10	11	13	16	20	27	32	34	33	30	27	24	22	7	34	25
Group E Walls																												
N	12	11	11	10	9	8	7	6	5	4	6	7	9	11	13	15	17	19	20	21	22	22	20	14	20	3	22	19
NE	13	13	13	12	12	11	9	9	15	20	24	25	23	23	26	26	26	26	28	24	22	20	17	15	16	4	25	22
E	14	12	11	10	10	10	11	16	26	33	38	36	37	37	29	24	20	20	18	17	17	18	19	17	13	5	38	33
SE	15	13	12	11	11	11	11	15	18	26	31	36	37	37	37	32	29	27	24	20	18	17	17	17	15	5	37	32
S	15	13	12	11	10	10	9	8	4	5	9	9	19	24	29	29	24	18	15	14	13	13	14	17	17	3	29	26
SW	22	18	16	15	14	12	10	9	5	6	7	9	13	18	24	32	34	34	35	44	34	35	34	26	19	5	34	29
W	25	21	18	16	14	12	10	8	7	7	7	8	9	11	16	26	34	38	43	44	45	40	38	29	20	6	45	40
NW	20	17	14	12	11	10	8	6	5	5	6	8	10	13	16	20	26	32	36	38	36	32	28	24	22	5	38	33

Group F Walls

Orientation	1	2	3	4	5	6	7	8	9	10	11	12	13	14	15	16	17	18	19	20	21	22	23	24
N	8	6	5	3	2	1	2	4	6	7	9	11	14	17	19	21	22	23	24	23	20	16	13	11
NE	9	7	5	3	2	1	5	14	23	28	30	29	28	27	27	27	27	26	24	22	19	16	13	11
E	10	7	6	4	3	2	6	17	28	38	44	45	43	39	36	34	32	30	27	24	21	17	15	12
SE	10	7	6	4	3	2	6	10	28	28	36	41	43	42	39	36	34	31	28	25	21	18	15	12
S	10	8	6	4	3	3	4	1	19	7	13	20	27	34	38	39	38	35	31	26	22	18	15	12
SW	15	11	9	6	5	3	3	2	3	5	8	11	17	26	34	44	50	53	52	45	37	28	23	18
W	17	13	10	7	5	4	3	3	4	6	8	11	14	20	28	39	49	57	60	54	43	34	27	21
NW	14	10	8	6	4	3	2	2	3	5	8	10	13	15	21	27	35	42	46	43	35	28	22	18

Group G Walls

Orientation	1	2	3	4	5	6	7	8	9	10	11	12	13	14	15	16	17	18	19	20	21	22	23	24
N	3	2	1	0	-1	2	7	8	9	12	15	18	21	23	24	25	26	24	22	20	18	15	11	9
NE	2	2	1	0	-1	9	27	36	39	35	30	27	26	26	27	26	25	18	14	11	9	9	7	5
E	4	2	1	0	-1	11	31	47	54	50	40	33	31	30	29	29	27	19	10	8	6	6	6	5
SE	4	2	1	0	-1	5	18	32	42	51	49	42	36	36	32	30	24	19	11	10	8	6	6	6
S	4	2	1	0	-1	1	5	5	12	22	31	39	45	46	43	37	25	20	14	10	8	8	8	6
SW	5	4	3	1	1	2	5	5	8	11	16	26	38	50	59	63	61	52	43	31	23	18	13	10
W	6	5	3	2	1	2	5	5	8	11	15	29	37	56	72	67	61	48	29	20	15	11	10	8
NW	5	3	2	1	0	2	5	5	8	11	15	18	21	27	37	47	55	41	27	18	15	13	11	7

1. *Direct application of the table without adjustments:*
Values in the table were calculated using the same conditions for walls as outlined for the roof CLTD table, Table 5. These values may be used for all normal air-conditioning estimates usually without correction (except as noted below) when the load is calculated for the hottest weather. For totally shaded walls use the north orientation values.

2. *Adjustments to table values:*
The following equation makes adjustments for conditions other than those listed in note 1.

$$\text{CLTD}_{corr} = (\text{CLTD} + \text{LM})K + (78 - T_R) + (T_o - 85)$$

where CLTD is from Table 7 at the wall orientation.
LM is the latitude-month correction from Table 9.
K is a color adjustment factor applied after first making month-latitude adjustment
    $K = 1.0$ if dark colored or light in an industrial area
    $K = 0.83$ if permanently medium-colored (rural area)
    $K = 0.65$ if permanently light-colored (rural area)
Credit should not be taken for wall color other than dark except where permanence of color is established by experience, as in rural areas or where there is little smoke.
SOURCE: Abstracted by permission from ASHRAE Handbook, *1989 Fundamentals*, Table 31, p. 26.36. (Subsequent editions provide more extensive data.)

TABLE 3.20 CLTD Correction for Latitude and Month Applied to Walls and Roofs, North Latitudes

Lat.	Month	N	NNE NNW	NE NW	ENE WNW	E W	ESE WSW	SE SW	SSE SSW	S	HOR
0	Dec	−3	−5	−5	−5	−2	0	3	6	9	−1
	Jan/Nov	−3	−5	−4	−4	−1	0	2	4	7	−1
	Feb/Oct	−3	−2	−2	−2	−1	−1	0	−1	0	0
	Mar/Sept	−3	0	1	−1	−1	−3	−3	−5	−8	0
	Apr/Aug	5	4	3	0	−2	−5	−6	−8	−8	−2
	May/Jul	10	7	5	0	−3	−7	−8	−9	−8	−4
	Jun	12	9	5	0	−3	−7	−9	−10	−8	−5
8	Dec	−4	−6	−6	−6	−3	0	4	8	12	−5
	Jan/Nov	−3	−5	−6	−5	−2	0	3	6	10	−4
	Feb/Oct	−3	−4	−3	−3	−1	−1	1	2	4	−1
	Mar/Sept	−3	−2	−1	−1	−1	−2	−2	−3	−4	0
	Apr/Aug	2	2	2	0	−1	−4	−5	−7	−7	−1
	May/Jul	7	5	4	0	−2	−5	−7	−9	−7	−2
	Jun	9	6	4	0	−2	−6	−8	−9	−7	−2
16	Dec	−4	−6	−8	−8	−4	−1	4	9	13	−9
	Jan/Nov	−4	−6	−7	−7	−4	−1	4	8	12	−7
	Feb/Oct	−3	−5	−5	−4	−2	0	2	5	7	−4
	Mar/Sept	−3	−3	−2	−2	−1	−1	0	0	0	−1
	Apr/Aug	−1	0	−1	−1	−1	−3	−3	−5	−6	0
	May/Jul	4	3	3	0	−1	−4	−5	−7	−7	0
	Jun	6	4	4	1	−1	−4	−6	−8	−7	0
24	Dec	−5	−7	−9	−10	−7	−3	3	9	13	−13
	Jan/Nov	−4	−6	−8	−9	−6	−3	3	9	13	−11
	Feb/Oct	−4	−5	−6	−6	−3	−1	3	7	10	−7
	Mar/Sept	−3	−4	−3	−3	−1	−1	1	2	4	−3
	Apr/Aug	−2	−1	0	−1	−1	−2	−1	−2	−3	0
	May/Jul	1	2	2	0	0	−3	−3	−5	−6	1
	Jun	3	3	3	1	0	−3	−4	−6	−6	1
32	Dec	−5	−7	−10	−11	−8	−5	2	9	12	−17
	Jan/Nov	−5	−7	−9	−11	−8	−4	2	9	12	−15
	Feb/Oct	−4	−6	−7	−8	−4	−2	4	8	11	−10
	Mar/Sept	−3	−4	−4	−4	−2	−1	3	5	7	−5
	Apr/Aug	−2	−2	−1	−2	0	−1	0	1	1	−1
	May/Jul	1	1	1	0	0	−1	−1	−3	−3	1
	Jun	1	2	2	1	0	−2	−2	−4	−4	2
40	Dec	−6	−8	−10	−13	−10	−7	0	7	10	−21
	Jan/Nov	−5	−7	−10	−12	−9	−6	1	8	11	−19
	Feb/Oct	−5	−7	−8	−9	−6	−3	3	8	12	−14
	Mar/Sept	−4	−5	−5	−6	−3	−1	4	7	10	−8
	Apr/Aug	−2	−3	−2	−2	0	0	2	3	4	−3
	May/Jul	0	0	0	0	0	0	0	0	1	1
	Jun	1	1	1	0	1	0	0	−1	−1	2
48	Dec	−6	−8	−11	−14	−13	−10	−3	2	6	−25
	Jan/Nov	−6	−8	−11	−13	−11	−8	−1	5	8	−24
	Feb/Oct	−5	−7	−10	−11	−8	−5	1	8	11	−18
	Mar/Sept	−4	−6	−6	−7	−4	−1	4	8	11	−11
	Apr/Aug	−3	−3	−3	−3	−1	0	4	6	7	−5
	May/Jul	0	−1	0	0	1	1	3	3	4	0
	Jun	1	1	2	1	2	1	2	2	3	2

No building is tight. Windows, even when fixed in place, leak air around the frames and gaskets. Doors leak air even when closed, and traffic increases air movement. Walls are porous. Airflows have been measured even through masonry walls, and joints in metal panel walls are never airtight. Vertical air movement in multistory buildings takes place through elevator shafts, stairwells, utility chases, ducts, and numerous construction openings.

TABLE 3.20 *(Continued)*

Lat.	Month	N	NNE NNW	NE NW	ENE WNW	E W	ESE WSW	SE SW	SSE SSW	S	HOR
56	Dec	−7	−9	−12	−16	−16	−14	−9	−5	−3	−28
	Jan/Nov	−6	−8	−11	−15	−14	−12	−6	−1	2	−27
	Feb/Oct	−6	−8	−10	−12	−10	−7	0	6	9	−22
	Mar/Sept	−5	−6	−7	−8	−5	−2	4	8	12	−15
	Apr/Aug	−3	−4	−4	−4	−1	1	5	7	9	−8
	May/Jul	0	0	0	0	2	2	5	6	7	−2
	Jun	2	1	2	1	3	3	4	5	6	1
64	Dec	−7	−9	−12	−16	−17	−18	−16	−14	−12	−30
	Jan/Nov	−7	−9	−12	−16	−16	−16	−13	−10	−8	−29
	Feb/Oct	−6	−8	−11	−14	−13	10	−4	1	4	−26
	Mar/Sept	−5	−7	−9	−10	−7	−4	2	7	11	−20
	Apr/Aug	−3	−4	−4	−4	−1	1	5	9	11	−11
	May/Jul	1	0	1	0	3	4	6	8	10	−3
	Jun	2	2	2	2	4	4	6	7	9	0

(1) Corrections in this table are in °F. The correction is applied directly to the CLTD for a wall or roof as given in Tables 5A and 7A.

(2) The CLTD correction given in this table is *not* applicable to Table 10, Cooling Load Temperature Differences for Conduction through Glass.

(3) For south latitudes, replace Jan. through Dec. by July through June.

SOURCE: Abstracted by permission from ASHRAE Handbook, *1989 Fundamentals,* Chap. 26, Table 32, p. 26.37.

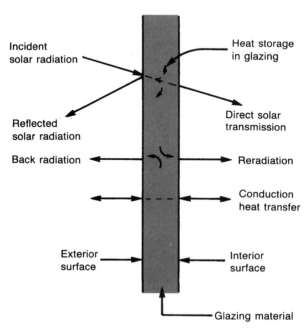

Figure 3.6 Heat gain through fenestration.

TABLE 3.21 Shading Coefficients for Single Glass and Insulating Glass*

Type of glass	Nominal thickness,[‡] in	Solar trans.[†]	Shading coefficient	
			$h_o = 4.0$	$h_o = 3.0$
A. Single Glass				
Clear	1/8	0.86	1.00	1.00
	1/4	0.78	0.94	0.95
	3/8	0.72	0.90	0.92
	1/2	0.67	0.87	0.88
Heat-absorbing	1/8	0.64	0.83	0.85
	1/4	0.46	0.69	0.73
	3/8	0.33	0.60	0.64
	1/2	0.24	0.53	0.58
B. Insulating Glass				
Clear out, clear in	1/8 ‡	0.71[¶]	0.88	0.88
Clear out, clear in	1/4	0.61	0.81	0.82
Heat-absorbing[§] out, clear in	1/4	0.36	0.55	0.58

*Refers to factory-fabricated units with 3/16, 1/4, or 1/2-in airspace or to prime windows plus storm sash.
[†]Refer to manufacturer's literature for values.
[‡]Thickness of each pane of glass, not thickness of assembled unit.
[§]Refers to gray, bronze, and green tinted heat-absorbing float glass.
[¶]Combined transmittance for assembled unit.
SOURCE: Reprinted by permission from ASHRAE Handbook, *1993 Fundamentals,* Chap. 27, Table 11, p. 27.19. (The 1997 edition has similar but more extensive data.)

Wind creates a positive pressure on the windward side of a building and a negative pressure on the leeward side. These pressures change with changes in wind direction and velocity. Wind effects make it difficult or impossible to obtain consistent measurement of interior or exterior pressures for the purposes of control.

The chimney effect in a multistory building (or even in a single-story building) is related to variations in the air density due to temperature and height and is aggravated by wind. The effect is minor during warm weather but significant in winter. The buoyancy of the warm air inside the building makes the air rise, creating a pressure gradient, as shown in Fig. 3.8. The lower floors are negative with respect to the outside, while the upper floors are positive. The neutral point will vary depending on the building construction and height but can be observed by a ride up the elevator that stops at every floor. The effect is to cause infiltration on the lower floors and exfiltration on the upper floors. This effect is also a driving force for smoke spread in a fire. Even when the fire is on a low floor, most deaths due to smoke occur on upper floors. The supply and exhaust systems can have a positive or negative influence on the chimney effect.

TABLE 3.22 Shading Coefficients for Single Glass with Indoor Shading by Venetian Blinds or Roller Shades

			Type of shading				
			Venetian blinds		Roller shade		
					Opaque		Translucent
	Nominal thickness,* in	Solar trans.[†]	Medium	Light	Dark	White	Light
Clear	3/32 – 1/4	0.87–0.80					
Clear	1/4 – 1/2	0.80–0.71					
Clear pattern	1/8 – 1/2	0.87–0.79	0.64	0.55	0.59	0.25	0.39
Heat-absorbing pattern	1/8	—					
Tinted	3/16, 7/32	0.74, 0.71					
Heat-absorbing[‡]	3/16, 1/4	0.46					
Heat-absorbing pattern	3/16, 1/4	—	0.57	0.53	0.45	0.30	0.36
Tinted	1/8, 7/32	0.59, 0.45					
Heat-absorbing or pattern	—	0.44–0.30	0.54	0.52	0.40	0.28	0.32
Heat-absorbing[‡]	3/8	0.34					
Heat-absorbing	—	0.29–0.15	0.42	0.40	0.36	0.28	0.31
Pattern	—	0.24					
Reflective coated glass[§]							
SC = 0.30			0.25	0.23			
0.40			0.33	0.29			
0.50			0.42	0.38			
0.60			0.50	0.44			

* Refer to manufacturer's literature for values.
[†] For vertical blinds with opaque white and beige louvers in the tightly closed position, SC is 0.25 and 0.29 when used with glass of 0.71 to 0.80 transmittance.
[‡] Refers to gray, bronze, and green tinted heat-absorbing glass.
[§] SC for glass with no shading device.
SOURCE: Reprinted by permission from ASHRAE Handbook, *1993 Fundamentals*, Chap. 27, Table 28, p. 27.36. (Data in the 1997 edition, Chap. 29, Table 29, is somewhat more extensive.)

TABLE 3.23 Shading Coefficients for Insulating Glass* with Indoor Shading by Venetian Blinds or Roller Shades

Type of glass	Nominal thickness, each light, in	Solar trans.†		Type of shading				
				Venetian blinds‡		Roller shade		
		Outer pane	Inner pane			Opaque		Translucent
				Medium	Light	Dark	White	Light
Clear out	3/32,1/8	0.87	0.87	0.57	0.51	0.60	0.25	0.37
Clear in								
Clear out	1/4	0.80	0.80					
Clear in								
Heat-absorbing§ out, clear in	1/4	0.46	0.80	0.39	0.36	0.40	0.22	0.30
Reflective coated glass¶								
SC = 0.20				0.19	0.18			
0.30				0.27	0.26			
0.40				0.34	0.33			

* Refers to factory-fabricated units with 3/16, 1/4, or 1/2-in airspace, or to prime windows plus storm windows.

† Refer to manufacturer's literature for exact values.

‡ For vertical blinds with opaque white or beige louvers, tightly closed, SC is approximately the same as for opaque white roller shades.

§ Refers to bronze, or green tinted, heat-absorbing glass.

¶ SC for glass with no shading device.

SOURCE: Reprinted by permission from ASHRAE Handbook, *1993 Fundamentals*, Chap. 27, Table 29, p. 27.36. (Data in the 1997 edition, Chap. 29, Table 29, is somewhat more extensive.)

TABLE 3.24 Shading Coefficients for Double Glazing with Between-Glass Shading

Type of glass	Nominal each pane, in	Solar trans.[*] Outer pane	Solar trans.[*] Inner pane	Description of airspace	Type of shading — Venetian blinds Light	Venetian blinds Medium	Louvered sun screen
Clear out, clear in	3/32,1/8	0.87	0.87	Shade in contact with glass or shade separated from glass by airspace.	0.33	0.36	0.43
Clear out, clear in	1/4	0.80	0.80	Shade in contact with glass voids filled with plastic.	—	—	0.49
Heat-absorbing[†] out, clear in	1/4	0.46	0.80	Shade in contact with glass or shade separated from glass by airspace.	0.28	0.30	0.37
				Shade in contact with glass voids filled with plastic.	—	—	0.41

[*] Refer to manufacturer's literature for exact values.

[†] Refers to gray, bronze, and green tinted heat-absorbing glass.

SOURCE: Reprinted by permission from ASHRAE Handbook, *1993 Fundamentals*, Chap. 27, Table 30, p. 27.36. (Data in the 1997 edition, Chap. 29, Table 30, is somewhat more extensive.)

TABLE 3.25 Shading Coefficients for Single and Insulating Glass with Draperies

Glazing	Glass trans.	Glass SC	SC for index letters in Fig. 3.7									
			A	B	C	D	E	F	G	H	I	J
Single glass												
¼ in Clear	0.80	0.95	0.80	0.75	0.70	0.65	0.60	0.55	0.50	0.45	0.40	0.35
½ in Clear	0.71	0.88	0.74	0.70	0.66	0.61	0.56	0.52	0.48	0.43	0.39	0.35
¼ in heat-absorbing	0.46	0.67	0.57	0.54	0.52	0.49	0.46	0.44	0.41	0.38	0.36	0.33
½ in heat-absorbing	0.24	0.50	0.43	0.42	0.40	0.39	0.38	0.36	0.34	0.33	0.32	0.30
Reflective coated	—	0.60	0.57	0.54	0.51	0.49	0.46	0.43	0.41	0.38	0.36	0.33
(see manufacturers' literature	—	0.50	0.46	0.44	0.42	0.41	0.39	0.38	0.36	0.34	0.33	0.31
for exact values)	—	0.40	0.36	0.35	0.34	0.33	0.32	0.30	0.29	0.28	0.27	0.26
	—	0.30	0.25	0.24	0.24	0.23	0.23	0.23	0.22	0.21	0.21	0.20
Insulating glass ½-in airspace clear out and clear in	0.64	0.83	0.66	0.62	0.58	0.56	0.52	0.48	0.45	0.42	0.37	0.35
Heat-absorbing out and clear in	0.37	0.55	0.49	0.47	0.45	0.43	0.41	0.39	0.37	0.35	0.33	0.32
Reflective coated	—	0.40	0.38	0.37	0.37	0.36	0.34	0.32	0.31	0.29	0.28	0.28
(see manufacturers' literature	—	0.30	0.29	0.28	0.27	0.27	0.26	0.26	0.25	0.25	0.24	0.24
for exact values)	—	0.20	0.19	0.19	0.18	0.18	0.17	0.17	0.16	0.16	0.15	0.15

SOURCE: Reprinted by permission from ASHRAE Handbook, *1997 Fundamentals*, Chap. 29, Table 29, p. 27.40.

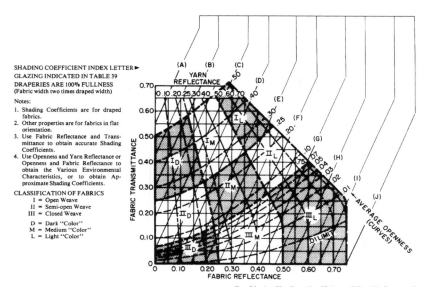

SHADING COEFFICIENT INDEX LETTER ►
GLAZING INDICATED IN TABLE 39
DRAPERIES ARE 100% FULLNESS
(Fabric width two times draped width)

Notes:

1. Shading Coefficients are for draped fabrics.
2. Other properties are for fabrics in flat orientation.
3. Use Fabric Reflectance and Transmittance to obtain accurate Shading Coefficients.
4. Use Openness and Yarn Reflectance or Openness and Fabric Reflectance to obtain the Various Environmental Characteristics, or to obtain Approximate Shading Coefficients.

CLASSIFICATION OF FABRICS

    I = Open Weave
    II = Semi-open Weave
    III = Closed Weave

    D = Dark "Color"
    M = Medium "Color"
    L = Light "Color"

To Obtain Fabric Designator (III_L, I_M, etc.): Using coordinates (1) Fabric Transmittance and Fabric Reflectance, or (2) Openness and Yarn Reflectance, find point on chart and note designator for that area. If properties are not known, classification may be approximated by eye using Fig. 24, Note 2. See Table 40 for application classifications.

To Obtain Shading Coefficient (SC): (1) Locate drapery fabric as a point using its known properties, or approximate using its fabric classification designator. For accuracy, use Fabric Transmittance and Fabric Reflectance. (2) Follow diagonal SC lines to lettered columns in Table 39. Find SC on line with glazing used. *Example:* SC of point "X" is 0.45 with 0.25 in. Clear Single Glass (Column H).

Note: SC are for 45 deg incident angle. For 30 deg or less, add 5% to number found in Table 39.

Figure 3.7 Indoor shading properties of drapery fabrics. (*Reprinted by permission from ASHRAE Handbook,* 1997 Fundamentals, *part of Table 29, p. 27.40.*)

The combination of wind and chimney effect can create serious problems at ground floor entrances during the heating season; this is a major concern of the HVAC designer. Even when vestibules or revolving doors are provided, additional heating is needed to offset the entrance infiltration. An approximate value for a 30-story office building with an inside-outside temperature difference of 75°F and a vestibule entrance is about 4500 ft³/min per 3-ft by 7-ft door, based on a traffic load of 500 persons per hour.[5]

Residential infiltration rates are often estimated at 0.5 to 1.5 air changes per hour, depending on the tightness of construction. The higher value is preferable since too low an infiltration/ventilation rate can result in excessive buildup of toxic gases from carpets, wall finishes, and other elements within the building.

3.6 Load Calculations

Once the criteria and multipliers are developed as described above, the load calculations are simply a matter of accurately determining

TABLE 3.26   Maximum Solar Heat Gain Factors, Btu/(h · ft²), for Sunlit Glass, North Latitudes

	N	NNE/ NNW	NE/ NW	ENE/ WNW	E/ W	ESE/ WSW	SE/ SW	SSE/ SSW	S	HOR
					0°N Lat.					
Jan.	34	34	88	177	234	254	235	182	118	296
Feb.	36	39	132	205	245	247	210	141	67	306
Mar.	38	87	170	223	242	223	170	87	38	303
Apr.	71	134	193	224	221	184	118	38	37	284
May	113	164	203	218	201	154	80	37	37	265
June	129	173	206	212	191	140	66	37	37	255
July	115	164	201	213	195	149	77	38	38	260
Aug.	75	134	187	216	212	175	112	39	38	276
Sept.	40	84	163	213	231	213	163	84	40	293
Oct.	37	40	129	199	236	238	202	135	66	299
Nov.	35	35	88	175	230	250	230	179	117	293
Dec.	34	34	71	164	226	253	240	196	138	288

	N	NNE/ NNW	NE/ NW	ENE/ WNW	E/ W	ESE/ WSW	SE/ SW	SSE/ SSW	S	HOR
					4°N Lat.					
Jan.	33	33	79	170	229	252	237	193	141	286
Feb.	35	35	123	199	242	248	215	152	88	301
Mar.	38	77	163	219	242	227	177	96	43	302
Apr.	55	125	189	223	223	190	126	43	38	287
May	93	154	200	220	206	161	89	38	38	272
June	110	164	202	215	196	147	73	38	38	263
July	96	154	197	215	200	156	85	39	38	267
Aug.	59	124	184	215	214	181	120	42	40	279
Sept.	39	75	156	209	231	216	170	93	44	293
Oct.	36	36	120	193	234	239	207	148	86	294
Nov.	34	34	79	168	226	248	232	190	139	284
Dec.	33	33	62	157	221	250	242	206	160	277

	N	NNE/ NNW	NE/ NW	ENE/ WNW	E/ W	ESE/ WSW	SE/ SW	SSE/ SSW	S	HOR
					8°N Lat.					
Jan.	32	32	71	163	224	250	242	203	162	275
Feb.	34	34	114	193	239	248	219	165	110	294
Mar.	37	67	156	215	241	230	184	110	55	300
Apr.	44	117	184	221	225	195	134	53	39	289
May	74	146	198	220	209	167	97	39	38	277
June	90	155	200	217	200	141	82	39	39	269
July	77	145	195	215	204	162	93	40	39	272
Aug.	47	117	179	214	216	186	128	51	41	282
Sept.	38	66	149	205	230	219	176	107	56	290
Oct.	35	35	112	187	231	239	211	160	108	288

areas, internal load densities such as people and watts per square foot, special process loads, and any unusual conditions. Where more than one room or zone has the same size, exposure, and internal loads, a "typical" calculation can be done. Corner zones should always be calculated separately. East-facing zone loads will normally peak from 10 to 12 a.m., while most building loads will peak from 3 to 5 p.m. South-facing zones are similar but will peak usually from noon to 2 p.m. and

TABLE 3.26 *(Continued)*

	N	NNE/ NNW	NE/ NW	ENE/ WNW	E/ W	ESE/ WSW	SE/ SW	SSE/ SSW	S	HOR
					8°N Lat.					
Nov.	33	33	71	161	220	245	233	200	160	273
Dec.	31	31	55	149	215	246	247	215	179	265

	N	NNE/ NNW	NE/ NW	ENE/ WNW	E/ W	ESE/ WSW	SE/ SW	SSE/ SSW	S	HOR
					12°N Lat.					
Jan.	31	31	63	155	217	246	247	212	182	262
Feb.	34	34	105	186	235	248	226	177	133	286
Mar.	36	58	148	210	240	233	190	124	73	297
Apr.	40	108	178	219	227	200	142	64	40	290
May	60	139	194	220	212	173	106	40	40	280
June	75	149	198	217	204	161	90	40	40	274
July	63	139	191	215	207	168	102	41	41	275
Aug.	42	109	174	212	218	191	135	62	142	282
Sept.	37	57	142	201	229	222	182	121	73	287
Oct.	34	34	103	180	227	238	219	172	130	280
Nov.	32	32	63	153	214	241	243	209	179	260
Dec.	30	30	47	141	207	242	251	223	197	250

	N	NNE/ NNW	NE/ NW	ENE/ WNW	E/ W	ESE/ WSW	SE/ SW	SSE/ SSW	S	HOR
					16°N Lat.					
Jan.	30	30	55	147	210	244	251	223	199	248
Feb.	33	33	96	180	231	247	233	188	154	275
Mar.	35	53	140	205	239	235	197	138	93	291
Apr.	39	99	172	215	227	204	150	77	45	289
May	52	132	189	218	215	179	115	45	41	282
June	66	142	194	217	207	167	99	41	41	277
July	55	132	187	214	210	174	111	44	42	277
Aug.	41	100	168	209	219	196	143	74	46	282
Sept.	36	50	134	196	227	224	191	134	93	282
Oct.	33	33	95	174	223	237	225	183	150	270
Nov.	30	30	55	145	206	241	247	220	196	246
Dec.	29	29	41	132	198	241	254	233	212	234

	N	NNE/ NNW	NE/ NW	ENE/ WNW	E/ W	ESE/ WSW	SE/ SW	SSE/ SSW	S	HOR
					20°N Lat.					
Jan.	29	29	48	138	201	243	253	233	214	232
Feb.	31	31	88	173	226	244	238	201	174	263

may peak in winter. All zones should be calculated at both zone peak (for sizing air-handling equipment) and building peak (for sizing central equipment). Buildings such as churches and restaurants will usually have peaks at times within an hour or two after maximum occupancy occurs. Judgment and experience must be applied.

Calculations should be done on "standard" forms. There are many such forms. Designers should use whatever form is found most satis-

TABLE 3.26 *(Continued)*

		NNE/ NNW	NE/ NW	ENE/ WNW	E/ W	ESE/ WSW	SE/ SW	SSE/ SSW	S	HOR
	N									

20°N Lat. *(Continued)*

	N	NNE/ NNW	NE/ NW	ENE/ WNW	E/ W	ESE/ WSW	SE/ SW	SSE/ SSW	S	HOR
Mar.	34	49	132	200	237	236	206	152	115	284
Apr.	38	92	166	213	228	208	158	91	58	287
May	47	123	184	217	217	184	124	54	42	283
June	59	135	189	216	210	173	108	45	42	279
July	48	124	182	213	212	179	119	53	43	278
Aug.	40	91	162	206	220	200	152	88	57	280
Sept.	36	46	127	191	225	225	199	148	114	275
Oct.	32	32	87	167	217	236	231	196	170	258
Nov.	29	29	48	136	197	239	249	229	211	230
Dec.	27	27	35	122	187	238	254	241	226	217

24°N Lat.

	N	NNE/ NNW	NE/ NW	ENE/ WNW	E/ W	ESE/ WSW	SE/ SW	SSE/ SSW	S	HOR
Jan.	27	27	41	128	190	240	253	241	227	214
Feb.	30	30	80	165	220	244	243	213	192	249
Mar.	34	45	124	195	234	237	214	168	137	275
Apr.	37	88	159	209	228	212	169	107	75	283
May	43	117	178	214	218	190	132	67	46	282
June	55	127	184	214	212	179	117	55	43	279
July	45	116	176	210	213	185	129	65	46	278
Aug.	38	87	156	203	220	204	162	103	72	277
Sept.	35	42	119	185	222	225	206	163	134	266
Oct.	31	31	79	159	211	237	235	207	187	244
Nov.	27	27	42	126	187	236	249	237	224	213
Dec.	26	26	29	112	180	234	247	247	237	199

28°N Lat.

	N (Shade)	NNE/ NNW	NE/ NW	ENE/ WNW	E/ W	ESE/ WSW	SE/ SW	SSE/ SSW	S	HOR
Jan.	25	25	35	117	183	235	251	247	238	196
Feb.	29	29	72	157	213	244	246	224	207	234
Mar.	33	41	116	189	231	237	221	182	157	265
Apr.	36	84	151	205	228	216	178	124	94	278
May	40	115	172	211	219	195	144	83	58	280
June	51	125	178	211	213	184	128	68	49	278
July	41	114	170	208	215	190	140	80	57	276
Aug.	38	83	149	199	220	207	172	120	91	272
Sept.	34	38	111	179	219	226	213	177	154	256
Oct.	30	30	71	151	204	236	238	217	202	229
Nov.	26	26	35	115	181	232	247	243	235	195
Dec.	24	24	24	99	172	227	248	251	246	179

factory, or is required. Figure 3.9 is a form which has been used satisfactorily. Zone and room calculations must then be summarized, by grouping rooms and zones in the way in which air-handling systems will be applied. A typical summary form is shown in Fig. 3.10.

The summary includes a column for listing the design airflow in cubic feet per minute, denoted by CFM, based on the cooling load and a design temperature difference between the entering air and the space temperature.

TABLE 3.26 (*Continued*)

32°N Lat.

	N (Shade)	NNE/ NNW	NE/ NW	ENE/ WNW	E/ W	ESE/ WSW	SE/ SW	SSE/ SSW	S	HOR
Jan.	24	24	29	105	175	229	249	250	246	176
Feb.	27	27	65	149	205	242	248	232	221	217
Mar.	32	37	107	183	227	237	227	195	176	252
Apr.	36	80	146	200	227	219	187	141	115	271
May	38	111	170	208	220	199	155	99	74	277
June	44	122	176	208	214	189	139	83	60	276
July	40	111	167	204	215	194	150	96	72	273
Aug.	37	79	141	195	219	210	181	136	111	265
Sept.	33	35	103	173	215	227	218	189	171	244
Oct.	28	28	63	143	195	234	239	225	215	213
Nov.	24	24	29	103	173	225	245	246	243	175
Dec.	22	22	22	84	162	218	246	252	252	158

36°N Lat.

	N (Shade)	NNE/ NNW	NE/ NW	ENE/ WNW	E/ W	ESE/ WSW	SE/ SW	SSE/ SSW	S	HOR
Jan.	22	22	24	90	166	219	247	252	252	155
Feb.	26	26	57	139	195	239	248	239	232	199
Mar.	30	33	99	176	223	238	232	206	192	238
Apr.	35	76	144	196	225	221	196	156	135	262
May	38	107	168	204	220	204	165	116	93	272
June	47	118	175	205	215	194	150	99	77	273
July	39	107	165	201	216	199	161	113	90	268
Aug.	36	75	138	190	218	212	189	151	131	257
Sept.	31	31	95	167	210	228	223	200	187	230
Oct.	27	27	56	133	187	230	239	231	225	195
Nov.	22	22	24	87	163	215	243	248	248	154
Dec.	20	20	20	69	151	204	241	253	254	136

40°N Lat.

	N (Shade)	NNE/ NNW	NE/ NW	ENE/ WNW	E/ W	ESE/ WSW	SE/ SW	SSE/ SSW	S	HOR
Jan.	20	20	20	74	154	205	241	252	254	133
Feb.	24	24	50	129	186	234	246	244	241	180
Mar.	29	29	93	169	218	238	236	216	206	223
Apr.	34	71	140	190	224	223	203	170	154	252
May	37	102	165	202	220	208	175	133	113	265
June	48	113	172	205	216	199	161	116	95	267
July	38	102	163	198	216	203	170	129	109	262
Aug.	35	71	135	185	216	214	196	165	149	247
Sept.	30	30	87	160	203	227	226	209	200	215
Oct.	25	25	49	123	180	225	238	236	234	177
Nov.	20	20	20	73	151	201	237	248	250	132
Dec.	18	18	18	60	135	188	232	249	253	113

TABLE 3.26   *(Continued)*

	N (Shade)	NNE/ NNW	NE/ NW	ENE/ WNW	E/ W	ESE/ WSW	SE/ SW	SSE/ SSW	S	HOR
					44°N Lat.					
Jan.	17	17	17	64	138	189	232	248	252	109
Feb.	22	22	43	117	178	227	246	248	247	160
Mar.	27	27	87	162	211	236	238	224	218	206
Apr.	33	66	136	185	221	224	210	183	171	240
May	36	96	162	201	219	211	183	148	132	257
June	47	108	169	205	215	203	171	132	115	261
July	37	96	159	198	215	206	179	144	128	254
Aug.	34	66	132	180	214	215	202	177	165	236
Sept.	28	28	80	152	198	226	227	216	211	199
Oct.	23	23	42	111	171	217	237	240	239	157
Nov.	18	18	18	64	135	186	227	244	248	109
Dec.	15	15	15	49	115	175	217	240	246	89

	N (Shade)	NNE/ NNW	NE/ NW	ENE/ WNW	E/ W	ESE/ WSW	SE/ SW	SSE/ SSW	S	HOR
					48°N Lat.					
Jan.	15	15	15	53	118	175	216	239	245	85
Feb.	20	20	36	103	168	216	242	249	250	138
Mar.	26	26	80	154	204	234	239	232	228	188
Apr.	31	61	132	180	219	225	215	194	186	226
May	35	97	158	200	218	214	192	163	150	247
June	46	110	165	204	215	206	180	148	134	252
July	37	96	156	196	214	209	187	158	146	244
Aug.	33	61	128	174	211	216	208	188	180	223
Sept.	27	27	72	144	191	223	228	223	220	182
Oct.	21	21	35	96	161	207	233	241	242	136
Nov.	15	15	15	52	115	172	212	234	240	85
Dec.	13	13	13	36	91	156	195	225	233	65

	N (Shade)	NNE/ NNW	NE/ NW	ENE/ WNW	E/ W	ESE/ WSW	SE/ SW	SSE/ SSW	S	HOR
					52°N Lat.					
Jan.	13	13	13	39	92	155	193	222	230	62
Feb.	18	18	29	85	156	202	235	247	250	115
Mar.	24	24	73	145	196	230	239	238	236	169
Apr.	30	56	128	177	215	224	220	204	199	211
May	34	98	154	198	217	217	199	175	167	235
June	45	111	161	202	214	210	188	162	152	242
July	36	97	152	194	213	212	195	171	163	233
Aug.	32	56	124	169	208	216	212	197	193	208
Sept.	25	25	65	136	182	218	228	228	227	163
Oct.	19	19	28	80	148	192	225	238	240	114
Nov.	13	13	13	39	90	152	189	217	225	62
Dec.	10	10	10	19	73	127	172	199	209	42

TABLE 3.26 *(Continued)*

	56°N Lat.									
	N (Shade)	NNE/ NNW	NE/ NW	ENE/ WNW	E/ W	ESE/ WSW	SE/ SW	SSE/ SSW	S	HOR
Jan.	10	10	10	21	74	126	169	194	205	40
Feb.	16	16	21	71	139	184	223	239	244	91
Mar.	22	22	65	136	185	224	238	241	241	149
Apr.	28	58	123	173	211	223	223	213	210	195
May	36	99	149	195	215	218	206	187	181	222
June	53	111	160	199	213	213	196	174	168	231
July	37	98	147	192	211	214	201	183	177	221
Aug.	30	56	119	165	203	216	215	206	203	193
Sept.	23	23	58	126	171	211	227	230	231	144
Oct.	16	16	20	68	132	176	213	229	234	91
Nov.	10	10	10	21	72	122	165	190	200	40
Dec.	7	7	7	7	47	92	135	159	171	23

	60°N Lat.									
	N (Shade)	NNE/ NNW	NE/ NW	ENE/ WNW	E/ W	ESE/ WSW	SE/ SW	SSE/ SSW	S	HOR
Jan.	7	7	7	7	46	88	130	152	164	21
Feb.	13	13	13	58	118	168	204	225	231	68
Mar.	20	20	56	125	173	215	234	241	242	128
Apr.	27	59	118	168	206	222	225	220	218	178
May	43	98	149	192	212	220	211	198	194	208
June	58	110	162	197	213	215	202	186	181	217
July	44	97	147	189	208	215	206	193	190	207
Aug.	28	57	114	161	199	214	217	213	211	176
Sept.	21	21	50	115	160	202	222	229	231	123
Oct.	14	14	14	56	111	159	193	215	221	67
Nov.	7	7	7	7	45	86	127	148	160	22
Dec.	4	4	4	4	16	51	76	100	107	9

	64°N Lat.									
	N (Shade)	NNE/ NNW	NE/ NW	ENE/ WNW	E/ W	ESE/ WSW	SE/ SW	SSE/ SSW	S	HOR
Jan.	3	3	3	3	15	45	67	89	96	8
Feb.	11	11	11	43	89	144	177	202	210	45
Mar.	18	18	47	113	159	203	226	236	239	105
Apr.	25	59	113	163	201	219	225	225	224	160
May	48	97	150	189	211	220	215	207	204	192
June	62	114	162	193	213	216	208	196	193	203
July	49	96	148	186	207	215	211	202	200	192
Aug.	27	58	109	157	193	211	217	217	217	159
Sept.	19	19	43	103	148	189	213	224	227	101
Oct.	11	11	11	40	83	135	167	191	199	46
Nov.	4	4	4	4	15	44	66	87	93	8
Dec.	0	0	0	0	1	5	11	14	15	1

SOURCE: Abstracted by permission from ASHRAE Handbook, *1997 Fundamentals,* Chap. 29, Tables 15–20, pp. 29.29–29.35.

TABLE 3.27 Cooling Load Factors for Glass without Interior Shading, North Latitudes

Fenestration facing	Room construction	Solar time, h																							
		1	2	3	4	5	6	7	8	9	10	11	12	13	14	15	16	17	18	19	20	21	22	23	24
N (Shaded)	L	0.17	0.14	0.11	0.09	0.08	0.33	0.42	0.48	0.56	0.63	0.71	0.76	0.80	0.82	0.82	0.79	0.75	0.84	0.61	0.48	0.38	0.31	0.25	0.20
	M	0.23	0.20	0.18	0.16	0.14	0.34	0.41	0.46	0.53	0.59	0.65	0.70	0.73	0.75	0.76	0.74	0.70	0.79	0.61	0.50	0.42	0.36	0.31	0.27
	H	0.25	0.23	0.21	0.20	0.19	0.38	0.45	0.49	0.55	0.60	0.65	0.69	0.72	0.72	0.72	0.70	0.70	0.75	0.57	0.46	0.39	0.34	0.31	0.28
NNE	L	0.06	0.05	0.04	0.03	0.03	0.26	0.43	0.47	0.44	0.41	0.40	0.39	0.38	0.38	0.36	0.33	0.30	0.26	0.20	0.16	0.13	0.10	0.08	0.07
	M	0.09	0.08	0.07	0.06	0.06	0.24	0.38	0.42	0.39	0.38	0.37	0.35	0.34	0.33	0.34	0.33	0.28	0.27	0.22	0.18	0.16	0.14	0.12	0.10
	H	0.11	0.10	0.09	0.09	0.08	0.26	0.39	0.42	0.44	0.36	0.35	0.34	0.34	0.33	0.32	0.31	0.28	0.25	0.21	0.18	0.16	0.14	0.13	0.12
NE	L	0.04	0.04	0.03	0.02	0.02	0.23	0.41	0.51	0.51	0.45	0.39	0.36	0.33	0.31	0.30	0.28	0.23	0.19	0.15	0.12	0.10	0.08	0.06	0.05
	M	0.07	0.06	0.06	0.05	0.04	0.21	0.36	0.45	0.45	0.40	0.36	0.33	0.31	0.30	0.28	0.26	0.23	0.21	0.17	0.15	0.11	0.09	0.09	0.08
	H	0.09	0.08	0.08	0.07	0.07	0.23	0.37	0.44	0.44	0.39	0.34	0.31	0.29	0.30	0.26	0.24	0.22	0.20	0.17	0.14	0.12	0.11	0.11	0.10
ENE	L	0.04	0.03	0.03	0.02	0.02	0.21	0.40	0.52	0.57	0.53	0.45	0.39	0.34	0.31	0.28	0.26	0.21	0.18	0.14	0.12	0.09	0.08	0.06	0.05
	M	0.07	0.06	0.05	0.05	0.04	0.20	0.35	0.45	0.49	0.47	0.41	0.36	0.33	0.30	0.28	0.26	0.23	0.20	0.17	0.14	0.11	0.11	0.09	0.08
	H	0.09	0.08	0.08	0.07	0.07	0.22	0.36	0.46	0.49	0.45	0.38	0.33	0.30	0.27	0.25	0.23	0.21	0.19	0.16	0.14	0.12	0.12	0.11	0.10
E	L	0.04	0.04	0.03	0.02	0.02	0.19	0.37	0.51	0.57	0.57	0.50	0.42	0.36	0.32	0.29	0.25	0.22	0.19	0.15	0.12	0.10	0.08	0.06	0.05
	M	0.07	0.07	0.06	0.05	0.05	0.18	0.33	0.44	0.51	0.51	0.46	0.39	0.35	0.31	0.29	0.26	0.23	0.21	0.17	0.15	0.12	0.11	0.10	0.08
	H	0.09	0.09	0.08	0.07	0.07	0.20	0.34	0.45	0.49	0.49	0.43	0.36	0.32	0.29	0.26	0.24	0.22	0.19	0.17	0.15	0.13	0.12	0.11	0.10
ESE	L	0.05	0.05	0.04	0.03	0.03	0.17	0.34	0.49	0.58	0.61	0.57	0.48	0.41	0.36	0.32	0.28	0.24	0.20	0.16	0.13	0.10	0.09	0.07	0.06
	M	0.08	0.07	0.07	0.06	0.05	0.16	0.30	0.43	0.51	0.54	0.51	0.44	0.39	0.35	0.32	0.29	0.26	0.22	0.19	0.16	0.14	0.12	0.11	0.09
	H	0.10	0.10	0.09	0.08	0.08	0.19	0.32	0.43	0.50	0.52	0.49	0.41	0.36	0.32	0.29	0.26	0.24	0.21	0.18	0.16	0.14	0.13	0.12	0.11
SE	L	0.05	0.04	0.04	0.03	0.03	0.13	0.28	0.43	0.55	0.62	0.63	0.57	0.48	0.42	0.37	0.33	0.28	0.24	0.19	0.15	0.12	0.10	0.08	0.07
	M	0.09	0.08	0.07	0.06	0.05	0.14	0.26	0.38	0.48	0.54	0.56	0.51	0.45	0.40	0.36	0.33	0.29	0.25	0.21	0.18	0.16	0.14	0.12	0.10
	H	0.11	0.10	0.10	0.09	0.08	0.17	0.28	0.40	0.49	0.53	0.53	0.48	0.41	0.36	0.33	0.30	0.27	0.24	0.20	0.18	0.16	0.14	0.13	0.12
SSE	L	0.07	0.06	0.05	0.04	0.03	0.06	0.16	0.26	0.38	0.48	0.55	0.60	0.60	0.60	0.52	0.45	0.35	0.29	0.23	0.18	0.15	0.12	0.10	0.08
	M	0.11	0.09	0.08	0.07	0.06	0.08	0.16	0.26	0.36	0.45	0.51	0.57	0.57	0.54	0.48	0.43	0.35	0.30	0.25	0.21	0.18	0.16	0.14	0.12
	H	0.12	0.11	0.11	0.10	0.09	0.12	0.19	0.29	0.38	0.46	0.54	0.55	0.55	0.51	0.44	0.39	0.35	0.31	0.27	0.23	0.20	0.18	0.16	0.15
S	L	0.08	0.07	0.06	0.06	0.05	0.06	0.09	0.11	0.14	0.19	0.34	0.48	0.59	0.65	0.59	0.50	0.43	0.36	0.28	0.22	0.18	0.15	0.12	0.10
	M	0.12	0.11	0.09	0.08	0.07	0.08	0.11	0.14	0.18	0.21	0.31	0.42	0.52	0.57	0.58	0.54	0.47	0.36	0.29	0.24	0.18	0.16	0.14	0.14
	H	0.13	0.12	0.12	0.11	0.10	0.13	0.15	0.17	0.21	0.24	0.33	0.43	0.51	0.55	0.56	0.53	0.50	0.44	0.36	0.30	0.25	0.21	0.18	0.17
SSW	L	0.10	0.09	0.07	0.06	0.06	0.06	0.09	0.11	0.12	0.14	0.16	0.27	0.39	0.52	0.62	0.67	0.65	0.58	0.43	0.36	0.27	0.22	0.18	0.15
	M	0.14	0.12	0.11	0.09	0.08	0.09	0.11	0.13	0.15	0.18	0.18	0.25	0.35	0.48	0.58	0.59	0.59	0.53	0.41	0.34	0.27	0.23	0.19	0.16
	H	0.15	0.14	0.13	0.12	0.11	0.12	0.14	0.16	0.18	0.21	0.20	0.27	0.37	0.44	0.53	0.55	0.56	0.53	0.44	0.37	0.28	0.25	0.20	0.17
SW	L	0.12	0.10	0.08	0.07	0.06	0.06	0.08	0.10	0.12	0.14	0.16	0.24	0.36	0.49	0.60	0.66	0.66	0.58	0.43	0.33	0.22	0.18	0.15	0.14
	M	0.15	0.13	0.12	0.10	0.09	0.09	0.10	0.12	0.13	0.15	0.17	0.23	0.33	0.44	0.53	0.58	0.59	0.53	0.41	0.33	0.24	0.21	0.19	0.17
	H	0.15	0.14	0.13	0.12	0.11	0.12	0.13	0.14	0.16	0.17	0.19	0.25	0.34	0.44	0.52	0.56	0.56	0.49	0.37	0.30	0.25	0.22	0.19	0.17
WSW	L	0.12	0.10	0.08	0.07	0.06	0.06	0.07	0.09	0.10	0.12	0.13	0.17	0.26	0.40	0.52	0.62	0.66	0.61	0.44	0.34	0.27	0.22	0.18	0.15
	M	0.15	0.13	0.12	0.10	0.10	0.09	0.10	0.11	0.12	0.13	0.14	0.19	0.24	0.35	0.46	0.54	0.58	0.55	0.42	0.34	0.28	0.24	0.21	0.18
	H	0.15	0.14	0.13	0.12	0.11	0.11	0.13	0.14	0.15	0.16	0.16	0.19	0.26	0.36	0.43	0.50	0.56	0.51	0.38	0.30	0.25	0.21	0.19	0.17
W	L	0.12	0.10	0.08	0.06	0.05	0.06	0.07	0.08	0.10	0.11	0.12	0.14	0.14	0.20	0.32	0.44	0.57	0.64	0.61	0.44	0.34	0.27	0.22	0.18
	M	0.15	0.13	0.12	0.10	0.09	0.09	0.10	0.11	0.12	0.13	0.14	0.15	0.16	0.24	0.35	0.47	0.58	0.56	0.41	0.33	0.27	0.23	0.20	0.17
	H	0.14	0.14	0.12	0.11	0.11	0.11	0.12	0.13	0.15	0.16	0.16	0.18	0.19	0.25	0.36	0.46	0.52	0.52	0.38	0.30	0.24	0.21	0.18	0.16
WNW	L	0.12	0.10	0.08	0.06	0.05	0.06	0.07	0.09	0.10	0.12	0.13	0.15	0.17	0.19	0.22	0.33	0.47	0.62	0.63	0.44	0.34	0.27	0.22	0.18
	M	0.15	0.13	0.12	0.10	0.09	0.09	0.10	0.11	0.12	0.13	0.14	0.15	0.17	0.24	0.35	0.47	0.58	0.55	0.41	0.33	0.27	0.23	0.20	0.17
	H	0.14	0.14	0.13	0.11	0.10	0.11	0.12	0.13	0.15	0.16	0.16	0.18	0.20	0.25	0.36	0.46	0.53	0.52	0.38	0.30	0.24	0.20	0.18	0.16
NW	L	0.11	0.09	0.07	0.06	0.05	0.06	0.06	0.08	0.10	0.12	0.14	0.16	0.17	0.19	0.21	0.33	0.42	0.51	0.39	0.33	0.26	0.21	0.17	0.14
	M	0.14	0.12	0.11	0.10	0.09	0.10	0.11	0.12	0.11	0.12	0.13	0.15	0.16	0.23	0.30	0.42	0.51	0.54	0.41	0.32	0.27	0.22	0.19	0.16
	H	0.14	0.12	0.12	0.11	0.10	0.10	0.11	0.13	0.14	0.15	0.16	0.18	0.19	0.22	0.30	0.41	0.50	0.51	0.36	0.29	0.23	0.20	0.17	0.15
NNW	L	0.12	0.11	0.09	0.08	0.06	0.07	0.07	0.09	0.11	0.14	0.16	0.18	0.20	0.22	0.24	0.33	0.44	0.57	0.44	0.33	0.26	0.22	0.17	0.14
	M	0.15	0.13	0.12	0.10	0.10	0.10	0.10	0.11	0.13	0.15	0.18	0.21	0.24	0.27	0.28	0.31	0.39	0.56	0.41	0.33	0.27	0.23	0.19	0.16
	H	0.14	0.13	0.12	0.11	0.11	0.12	0.12	0.13	0.15	0.17	0.20	0.23	0.26	0.28	0.31	0.38	0.49	0.53	0.38	0.30	0.25	0.21	0.18	0.16
HOR	L	0.11	0.09	0.07	0.06	0.05	0.07	0.14	0.24	0.36	0.48	0.58	0.66	0.72	0.74	0.73	0.67	0.59	0.47	0.37	0.29	0.24	0.19	0.16	0.13
	M	0.16	0.14	0.12	0.11	0.09	0.11	0.16	0.24	0.33	0.43	0.52	0.59	0.64	0.67	0.66	0.62	0.56	0.47	0.38	0.32	0.28	0.24	0.21	0.18
	H	0.17	0.16	0.14	0.13	0.10	0.15	0.20	0.24	0.36	0.45	0.52	0.62	0.64	0.64	0.62	0.58	0.56	0.42	0.35	0.29	0.26	0.23	0.21	0.19

L = Light construction: frame exterior wall, 2-in concrete floor slab, approximately 30 lb of material/ft² of floor area.

M = Medium construction: 4-in concrete exterior wall, 4-in concrete floor slab, approximately 70 lb of building material/ft² of floor area.

H = Heavy construction: 6-in concrete exterior wall, 6-in concrete floor slab, approximately 130 lb of building materials/ft² of floor area.

... from ASHRAE Handbook, 1989 Fundamentals, Chap. 26, Table

TABLE 3.28 Cooling Load Factors for Glass with Interior Shading, North Latitudes

Fenes-tration facing	Solar time, h																							
	0100	0200	0300	0400	0500	0600	0700	0800	0900	1000	1100	1200	1300	1400	1500	1600	1700	1800	1900	2000	2100	2200	2300	2400
N	0.08	0.07	0.06	0.06	0.07	0.73	0.66	0.65	0.73	0.80	0.86	0.89	0.89	0.86	0.82	0.75	0.78	0.91	0.24	0.18	0.15	0.13	0.11	0.10
NNE	0.03	0.03	0.02	0.02	0.03	0.64	0.77	0.62	0.42	0.37	0.37	0.37	0.36	0.35	0.32	0.28	0.23	0.17	0.08	0.07	0.06	0.05	0.04	0.04
NE	0.03	0.02	0.02	0.02	0.02	0.56	0.76	0.74	0.58	0.37	0.29	0.27	0.26	0.24	0.22	0.20	0.16	0.12	0.06	0.05	0.04	0.04	0.03	0.03
ENE	0.03	0.02	0.02	0.02	0.02	0.52	0.76	0.80	0.71	0.52	0.31	0.26	0.24	0.22	0.20	0.18	0.15	0.11	0.06	0.05	0.04	0.04	0.03	0.03
E	0.03	0.02	0.02	0.02	0.02	0.47	0.72	0.80	0.76	0.62	0.41	0.27	0.24	0.22	0.20	0.17	0.14	0.11	0.06	0.05	0.04	0.04	0.03	0.03
ESE	0.03	0.03	0.02	0.02	0.02	0.41	0.67	0.79	0.80	0.72	0.54	0.34	0.27	0.24	0.21	0.19	0.15	0.12	0.07	0.06	0.05	0.04	0.03	0.03
SE	0.03	0.03	0.02	0.02	0.02	0.30	0.57	0.74	0.81	0.79	0.68	0.49	0.33	0.28	0.25	0.22	0.18	0.13	0.08	0.07	0.06	0.05	0.04	0.04
SSE	0.04	0.03	0.03	0.03	0.02	0.12	0.31	0.54	0.72	0.81	0.81	0.71	0.54	0.38	0.32	0.27	0.22	0.16	0.09	0.08	0.07	0.06	0.05	0.04
S	0.04	0.04	0.03	0.03	0.03	0.09	0.16	0.23	0.38	0.58	0.75	0.83	0.80	0.68	0.50	0.35	0.27	0.19	0.11	0.09	0.08	0.07	0.06	0.05
SSW	0.05	0.04	0.04	0.03	0.03	0.09	0.14	0.18	0.22	0.27	0.43	0.63	0.78	0.84	0.80	0.66	0.46	0.25	0.13	0.11	0.09	0.08	0.06	0.06
SW	0.05	0.05	0.04	0.04	0.03	0.07	0.11	0.14	0.16	0.19	0.22	0.38	0.59	0.75	0.83	0.81	0.69	0.45	0.16	0.12	0.10	0.09	0.07	0.06
WSW	0.05	0.05	0.04	0.04	0.03	0.07	0.10	0.12	0.14	0.16	0.17	0.23	0.44	0.64	0.78	0.84	0.78	0.55	0.16	0.12	0.10	0.09	0.07	0.06
W	0.05	0.05	0.04	0.04	0.03	0.06	0.09	0.11	0.13	0.15	0.16	0.17	0.31	0.53	0.72	0.82	0.81	0.61	0.16	0.12	0.10	0.08	0.07	0.06
WNW	0.05	0.05	0.04	0.04	0.03	0.07	0.10	0.12	0.14	0.16	0.17	0.18	0.22	0.43	0.65	0.80	0.84	0.66	0.16	0.12	0.10	0.08	0.07	0.06
NW	0.05	0.04	0.04	0.03	0.03	0.07	0.11	0.14	0.17	0.19	0.20	0.21	0.22	0.30	0.52	0.73	0.82	0.69	0.16	0.12	0.10	0.08	0.07	0.06
NNW	0.05	0.05	0.04	0.03	0.03	0.11	0.17	0.22	0.26	0.30	0.32	0.33	0.34	0.34	0.39	0.61	0.82	0.76	0.17	0.12	0.10	0.08	0.07	0.06
HOR.	0.06	0.05	0.04	0.04	0.03	0.12	0.27	0.44	0.59	0.72	0.81	0.85	0.85	0.81	0.71	0.58	0.42	0.25	0.14	0.12	0.10	0.08	0.07	0.06

SOURCE: Abstracted by permission from ASHRAE Handbook, *1989 Fundamentals*, Chap. 26, Table 39, p. 26.43.

TABLE 3.29 Overall Coefficients of Heat Transmission (U Factor) of Windows and Sliding Patio Doors for Use in Peak Load Determination and Mechanical Equipment Sizing Only, Btu/(h · ft² · °F)

	No storm sash				Glass outdoor storm sash 1-in airspace[b] added to described product			
	No shade		Indoor shade		No shade		Indoor shade	
	Winter*	Summer**	Winter*	Summer**	Winter*	Summer**	Winter*	Summer**
Flat glass[c]								
Single glass, clear	1.10	1.04	0.83	0.81	0.50	0.50	0.44	0.49
Single glass, low emittance coating[d]								
e = 0.60	1.02	1.00	0.76	0.80	0.47	0.60	0.39	0.55
e = 0.40	0.91	0.90	0.68	0.70	0.44	0.60	0.37	0.55
e = 0.20	0.79	0.75	0.59	0.55	0.40	0.50	0.33	0.45
Insulating glass, double[e]								
3/16-in airspace[f]	0.62	0.65	0.52	0.58	0.37	0.40	0.29	0.37
1/4-in airspace[f]	0.58	0.61	0.48	0.55	0.35	0.39	0.28	0.36
1/2-in airspace[g]	0.49	0.56	0.42	0.52	0.32	0.39	0.25	0.37
1/2-in airspace, low emittance coating[h]								
e = 0.60	0.43	0.53	0.38	0.49	0.41	0.30	0.24	0.37
e = 0.40	0.38	0.47	0.36	0.43	0.27	0.39	0.22	0.35
e = 0.20	0.32	0.39	0.30	0.36	0.24	0.33	0.20	0.30
Insulating glass, triple[e]								
1/4-in airspace[f]	0.39	0.44	0.31	0.40	0.27	0.32	0.22	0.30
1/2-in airspace[i]	0.31	0.39	0.26	0.36	0.23	0.31	0.19	0.29

	Glass indoor storm sash 1-in airspace[b] added to described product				Acrylic indoor storm sash 1-in airspace[b] added to described product			
	No shade		Indoor shade		No shade		Indoor shade	
	Winter*	Summer**	Winter*	Summer**	Winter*	Summer**	Winter*	Summer**
Flat glass[c]								
Single glass, clear	0.50	0.50	0.44	0.49	0.48	0.48	0.42	0.47
Single glass, low emittance coating[d]								
e = 0.60	0.47	0.50	0.39	0.45	0.45	0.50	0.38	0.45
e = 0.40	0.42	0.45	0.36	0.40	0.41	0.45	0.35	0.40
e = 0.20	0.37	0.35	0.32	0.30	0.36	0.35	0.31	0.30

| | Glass indoor storm sash 1-in airspace[b] added to described product | | | | Acrylic indoor storm sash 1-in airspace[b] added to described product | | | |
| | No shade | | Indoor shade | | No shade | | Indoor shade | |
	Winter*	Summer**	Winter*	Summer**	Winter*	Summer**	Winter*	Summer**
Insulating glass, double[e]								
3/16-in airspace[f]	0.37	0.40	0.29	0.36	0.35	0.39	0.28	0.35
1/4-in airspace[f]	0.35	0.39	0.28	0.36	0.34	0.38	0.27	0.34
1/2-in airspace[g]	0.31	0.38	0.25	0.35	0.30	0.37	0.24	0.33
1/2-in airspace, low emittance coating[h]								
e = 0.60	0.29	0.37	0.24	0.33	0.28	0.35	0.23	0.31
e = 0.40	0.27	0.33	0.22	0.30	0.26	0.32	0.22	0.29
e = 0.20	0.25	0.29	0.20	0.26	0.24	0.28	0.20	0.25
Insulating glass, triple[e]								
1/4-in airspace[f]	0.27	0.32	0.22	0.30	0.26	0.31	0.22	0.29
1/2-in airspace[i]	0.23	0.30	0.19	0.28	0.22	0.29	0.18	0.28

[a] See Part C for appropriate adjustments for various windows and sliding patio doors. Window manufacturers should be consulted for specific data.

[b] 1/8-in. glass or acrylic as noted, 1 to 4-in. air space.

[c] Hemispherical emittance of uncoated glass surface = 0.84, coated glass surface as specified.

[d] Coating on second surface, i.e., room side of glass.

[e] Double and triple refer to number of lights of glass.

[f] 1/8-in. glass.

[g] 1/4-in. glass.

[h] Coating on either glass surface 2 or 3 for winter, and on surface 2 for summer U-factors.

[i] Window design 1/4-in. glass, 1/8-in. glass and 1/4-in. glass.

* 15 mph outdoor air velocity; 0°F outdoor air; 70°F inside air temperature, natural convections.

** 7.5 mph outdoor air velocity; 89°F outdoor air; 75°F inside air; natural convection; solar radiation 248.3 Btu/h · ft².

SOURCE: Abstracted by permission from ASHRAE Handbook, *1985 Fundamentals*, Chap. 27, Table 13, pp. 27.10–27.11.

TABLE 3.30 **Overall Coefficients of Heat Transmission (U Factor) of Exterior Horizontal Panels (Skylights) for Use in Peak Load Determination and Mechanical Equipment Sizing Only,** Btu/(h · ft^2 · °F)

Description	Winter	Summer
Flat glass		
Single glass	1.23	0.83
Insulating glass; double		
$\frac{3}{16}$-in airspace	0.70	0.57
$\frac{1}{4}$-in airspace	0.65	0.54
$\frac{1}{2}$-in airspace	0.59	0.49
$\frac{1}{2}$-in airspace, low emittance coating		
$e = 0.60$	0.56	0.46
$e = 0.40$	0.52	0.42
$e = 0.20$	0.48	0.36
Plastic domes		
Single-walled	1.15	0.80
Double-walled	0.70	0.46

SOURCE: Reprinted by permission from ASHRAE Handbook, *1985 Fundamentals,* Chap. 27, Table 13.13, p. 27.10.

The design airflow rate is

$$\text{CFM} = \frac{q_s}{\text{TD} \times \text{AF}} \qquad (3.6)$$

where CFM = design airflow rate for cooling, ft^3/min
  q_s = sensible cooling load, Btu/h
  TD = design temperature difference between space and supply air
  AF = air factor, Btu/h/[(ft^3/min) · °F] (see Table 3.3)

The selection of a cooling TD for an air-handling unit is typically routine and rather casual.[6] However, it is actually very critical because the real operating TD is determined by the physical laws governing the performance of the air system. This is discussed briefly below.

The psychrometric chart is used because this allows us to make a graphical study of the processes. To simplify the analysis, the effects of heat pickup in return air and the effects of fan work are neglected.

For comfort cooling (Fig. 3.11), assume a single-zone AHU, an inside design condition of 75°F db and 50 percent RH, an outside design condition of 95°F db and 75°F wb, and 20 percent minimum outside air. Assume also a sensible cooling load of 50,000 Btu/h and a latent load of 10,000 Btu/h. Then, from the chart, the mixed-air condition will be 79°F db and 65.3°F wb, with an enthalpy h of 30.2 Btu/lb of

TABLE 3.31 Cooling Load Temperature Differences for Conduction through Glass

Solar time, h	0100	0200	0300	0400	0500	0600	0700	0800	0900	1000	1100	1200	1300	1400	1500	1600	1700	1800	1900	2000	2100	2200	2300	2400
CLTD, °F	1	0	−1	−2	−2	−2	−2	0	2	4	7	9	12	13	14	14	13	12	10	8	6	4	3	2

Corrections: The values in the table were calculated for an inside temperature of 78°F and an outdoor maximum temperature of 95°F with an outdoor daily range of 21°F. The table remains approximately correct for other outdoor-maximums 93–102°F and other outdoor daily ranges 16–34°F, provided the outdoor daily average temperature remains approximately 85°F. If the room air temperature is different from 78°F and/or the outdoor daily average temperature is different from 85°F, the following rules apply: (a) For room air temperature less than 78°F, add the difference between 78°F and room air temperature; if greater than 78°F, subtract the difference. (b) For outdoor daily average temperature less than 85°F, subtract the difference; between 85°F and the daily average temperature; if greater than 85°F, add the difference.

SOURCE: Reprinted by permission from ASHRAE Handbook, *1989 Fundamentals*.

71

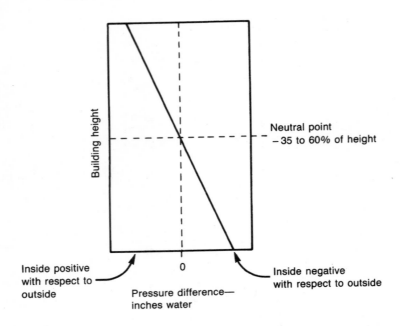

Building height

Neutral point
– 35 to 60% of height

Inside positive
with respect to
outside

0

Inside negative
with respect to outside

Pressure difference—
inches water

Note: Measured values will not necessarily follow the theoretical curve, due to
      construction effects and HVAC supply/exhaust influences.

Figure 3.8 Pressure gradient in a high-rise building (stack effect).

dry air. The room condition will include an h of 28.1 and a specific humidity w of 0.0092 lb moisture/lb dry air.

The design condition of the air supplied to the room is determined in one of two ways:

1. If the psychrometric chart includes a "protractor" (e.g., the ASH-RAE chart), a line may be drawn through the room state point with a slope equal to the sensible/total (S/T) ratio (in this case ⅚, or 0.833). In the figure this is line RS. Theoretically, the supply air point may be anywhere on this line. In practice, there are limitations, as discussed below.

2. On any chart the slope of the supply air process line may be determined by assuming a TD and calculating the resulting Δw. In this example, a TD of 20°F has been used, so that the supply point S is at 55°F. Then the design CFM will be

$$\text{CFM} = \frac{50,000 \text{ Btu/h}}{20°\text{F} \times 1.08} = 2315 \text{ ft}^3/\text{min}$$

(COMPANY NAME)
COOLING AND HEATING LOAD ESTIMATES

Job Name _____
Job No. _____ By _____
Space _____
Time of Day _____
Date _____
Sheet_____ of _____

Conduction

Surface - Facing	Dimen.	Area, Ft²	U	Heating		Cooling	
				ΔT	Btuh	ΔT/CLTD/CLF	Btuh
Roof ( )							
Wall ( ) Wall ( ) Wall ( ) Wall ( )							
Partition ( ) Floor ( )							
Glass ( ) Glass ( ) Glass ( ) Glass ( )							
			Total Heating			Subtotal	

Solar

Glass - Facing	Dimen.	Area, Ft²	SHGF	SC	CLF		Btuh
Glass ( ) Glass ( ) Glass ( ) Glass ( )							
						Subtotal	

Internal Sensible Heat

	Quantity	Factor	CLF		Btuh
Occupants Lights	people watts				
Appliances Power	Btuh watts				
				Subtotal	
				Total Sensible Cooling	

Latent

Source	Quantity	Factor	CLF	Btuh
			Total Latent Cooling	

Figure 3.9 Calculation form.

where 1.08 is the air factor for standard air. Using this CFM, we can calculate the value of Δw:

$$\Delta w = \frac{10{,}000 \text{ Btu/h}}{2315 \text{ ft}^3/\text{min} \times 60 \times 0.075 \times 1059} = 0.0009 \text{ lbw/lba}$$

Job Name _____ By _____
Job No. _____
Date _____
Sheet _____ of _____

(COMPANY NAME)
COOLING AND HEATING LOAD ESTIMATES
SUMMARY — ZONE _____

| Space | | Latent | Outside Air | | Sensible Cooling Load | | | | CFM at ___ΔT | CFM/ Ft² | Heat Loss | Adj. | NewΔT | |
No.	Name	Area	Btu/h	cfm	Btu/h	Space Max	Time	Bldg. Max	Time			Btu/h	cfm	Htg	Clg

Figure 3.10 Summary form.

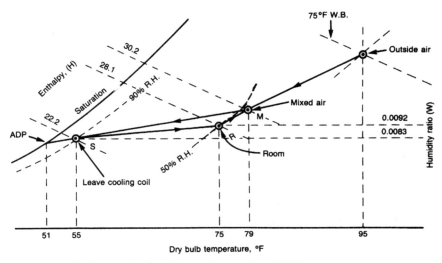

Figure 3.11 Psychrometric chart for comfort cooling (CHW).

where 60 = min/h
  0.075 = air density, lb/ft³ (standard air)
  1059 = latent heat of vaporization at 60°F, Btu/lb

The Δw of 0.0009 subtracted from the room w of 0.0092 equals 0.0083, the needed w of the supply air at point S. Then, from the chart, the supply air properties are 55°F db, 53.3°F wb, $h = 22.2$, $w = 0.0083$, and RH = 90 percent. By projection to the saturation line, the *apparatus dew point* (ADP) is 51°F.

The figure of 90 percent RH presents a problem because it implies a coil bypass factor of about 14 percent (see the coil discussion in Chap. 9). A present-day cooling coil, even at four rows deep, will do much better, with a bypass factor as low as 5 percent. It follows that the design condition will not be obtained in practice and, if the supply air temperature is controlled at 55°F, the resulting room condition will be at a somewhat higher humidity than the design value. In this example, the error is probably not serious, but the design is, in fact, flawed. While the 20°F TD is not too far off, a TD of 15 or 16°F would be unrealistic (unless reheat were used—a no-no in these energy-conscious days except for humidity control).

The ADP of 51°F will require a supply water temperature of about 45°F. With a DX coil, the ADP will tend to be between 40 and 45°F (see Fig. 3.12), which will pull the room humidity downward, increasing the load due to dehumidification of outside air. It will also lower

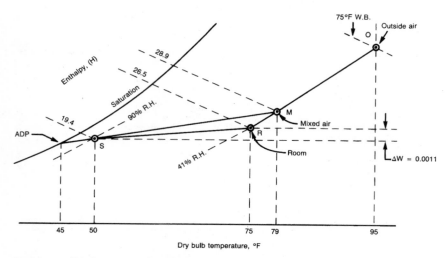

Figure 3.12 Psychrometric chart for comfort cooling (DX).

the supply air temperature so that the TD will be 25°F or more. Then the airflow rate should be 1850 ft³/min. From experience this would be expected to result in cold drafts and rapid two-position response with a tendency to short-cycle unless a wide differential is used, with resulting discomfort. The control system gets the blame, but actually there is a design deficiency.

At part-load conditions, which prevail most of the time, the DX system cycles even more often, because the supply air temperature does not modulate but varies between the design condition when the compressor is running and the return-air condition when the compressor is off. With chilled water, the throttling of water flow by the control valve allows the bypass factor to increase and the supply air temperature to modulate. Better control can be obtained at the expense of a slight increase in room humidity.

With *variable air volume* (VAV), accurate control becomes more difficult because of the system gains (see Chap. 8). There is a limited tendency for the entire process to move upward on the psychrometric chart, with a resulting increase in room humidity. This will be more noticeable if the supply air temperature is reset upward with decreasing air volume, as frequently recommended.

There is a tendency among designers to accept the "standard" CFM of a package AHU and to use whatever TD results. This can be unrealistic in terms of the CFM and coil bypass factor, and almost always it will result in poor control and wide temperature swings. Package unit CFMs are nearly always adjustable and should be specified at an appropriate value.

This mistake is even easier to make when you are remodeling and rearranging zones with existing AHUs. The "CFM is there, why not use it," philosophy can be hazardous to comfort.

While this discussion concerns comfort cooling, it should be apparent that these phenomena become even more important in systems for process cooling, especially those requiring close control of temperature and humidity. Then reheat becomes a necessity, but the need for energy conservation requires a careful look at the coil TD.

When the cooling CFM is compared to the heat loss, the temperature difference for heating will be found to vary from room to room or from zone to zone. If the variation is small, this will not be a problem; if large, it may be necessary to provide supplemental heating in some rooms or zones. The tabulated heat loss is a gross heat loss at design conditions, but with no credit for the internal heat gains which occur when the room is occupied. Because of all these factors, the tabulated "heating temperature difference" is not really meaningful except in an unoccupied building at about 6 a.m.

The CFM per square foot tabulation in the summary is a very useful check item for both manual and computer calculations. Values below about 0.7 in an occupied space are suspect and will usually result in inadequate ventilation rates with complaints of "stuffiness" and high humidity. For VAV a minimum design rate should be about 0.75 to 1.0 $(ft^3/min)/ft^2$, which will result in a building average actual circulation rate of 0.5 to 0.6 $(ft^3/min)/ft^2$. Values above 3.0 $(ft^3/min)/ft^2$ can create distribution problems, with high-velocity drafts on the occupants. If a recheck of the calculations shows that these high airflows are necessary, then special attention should be paid to the air distribution technique. Clean rooms and large computer rooms typically need airflow rates of 8 to 10 $(ft^3/min)/ft^2$. See Chap. 5 for methods of dealing with these rates.

Experience has shown that small adjustments of high rates down to 3.0 and low rates up to 0.75 $(ft^3/min)/ft^2$ will cause fewer problems than using the calculated rates. Again, judgment and common sense are needed.

The above comments imply that the system concept must precede the calculations. That is, the types of HVAC systems to be used, zoning, location of equipment, and control strategies must be at least approximated before the summaries are made. While this is not an absolute requirement, it will make the calculation process more efficient. The concept should be flexible, so as to accommodate changes required by unforeseen complications and last-minute alterations in the use and occupancy of certain building spaces.

3.7 Dynamic versus Static Load Calculations

All manual load calculations and many of the computer programs assume that a static or steady-state condition exists. But steady-state conditions do not exist in an air conditioning situation. If the HVAC systems and controls are functioning properly, then the indoor environment will vary only slightly. However, the internal and external loads are constantly changing. The function of the cooling load factor (CLF) in the calculation is to approximate the effect of these transient factors so that the static load calculation will yield results more like the "real" dynamic load.

The research which led to these factors resulted from the widely recognized condition that older calculation methods invariably led to oversizing of HVAC systems and equipment. The increases in energy and equipment costs during the 1970s led to a broad acceptance of the new methods because, in general, overall operating efficiency decreases if equipment is oversized. Even so, the factors in the tables are conservative, and some oversizing will normally result.

3.8 Ventilation Loads

Infiltration has already been discussed. However, most building codes require positive ventilation in public buildings, with a fixed minimum rate which relates to occupancy. Typically the required code amounts are in the range of 5 to 10 ft^3/min per person. The new and still evolving ASHRAE Standard 62-1989 increases this to 15 to 20 ft^3/min per person, which values will probably be adopted by most local code authorities. In addition, many processes require large amounts of exhaust, for which makeup air is required. Outside air for ventilation and makeup must be introduced through an air-handling unit, where it can be filtered and tempered (brought to design condition for heating or cooling). This whole matter is further complicated by the knowledge that outdoor air quality may not be acceptable in many indoor environments, so that special treatment to remove contaminants may be necessary (see Chap. 5).

Thus, while the ventilation load is not a part of the space load, it is reflected in the air-handling unit and central plant capacity. For heating, it is the minimum outside air quantity multiplied by the design temperature difference and the proper air factor. Thus,

$$q_h = \text{CFM}_{oa} \times 1.08 \times (t_i - t_o) \tag{3.7}$$

where q_h = ventilation load for heating, Btu/h
t_i = design inside temperature, heating

t_o = design outside temperature, heating
CFM_{oa} = outside-air quantity, ft^3/min
1.08 = air factor, $Btu/h/[(ft^3/min) \cdot °F]$ (for standard air—must be adjusted for higher elevations)

For cooling, the ventilation load is the minimum outside air quantity multiplied by the design enthalpy difference. Thus,

$$q_t = CFM_{oa} \times (h_o - h_i) \times 0.075 \times 60 \qquad (3.8)$$

where q_t = total cooling load for ventilation, Btu/h
h_o = enthalpy at design outside conditions, Btu/lb
h_i = enthalpy at design inside conditions, Btu/lb
0.075 = air density, lb/ft^3 (for standard air—must be adjusted for higher elevations)
60 = min/h

The sensible cooling load is calculated from the design temperature difference. Thus,

$$q_s = CFM_{oa} \times (t_o - t_i) \times 1.08 \qquad (3.9)$$

where q_s = sensible cooling for ventilation, Btu/h
t_o = outside design temperature, cooling
t_i = inside design temperature, cooling

If the AHU capacity is calculated from a psychrometric chart analysis, as shown in Chap. 4, the ventilation load is automatically included.

3.9   Other Loads

There are some additional factors which contribute to the cooling and heating loads. Among these are fan and pump work as well as duct and piping losses. These are discussed in subsequent chapters.

3.10   Summary

In this chapter, cooling and heating loads have been discussed, with emphasis on manual procedures and the elements of the loads. The principal source of information on this subject is the ASHRAE Handbook *Fundamentals,* to which the interested reader is referred for a more detailed discussion. All load calculations, whether manual or computerized, should be carefully checked for consistency and reasonableness. This requires the application of judgment, common sense, and experience.

References

1. ASHRAE Handbook, *1997 Fundamentals,* Chap. 26, "Climatic Design Information."
2. Ibid., Chap. 28, "Non Residential Air Conditioning, Cooling, and Heating Load."
3. Ibid., Chap. 28, "Non Residential Air Conditioning, Cooling, and Heating Load."
4. Ibid., Chap. 29, "Fenestration."
5. T. C. Min, "Winter Infiltration through Swinging Door Entrances in Multistory Buildings," *ASHRAE Transactions,* vol. 64, 1958, p. 421.
6. R. W. Haines, "Selecting a Delta *T* for an AHU," *Heating / Piping / Air Conditioning,* Nov. 1968, p. 210.
7. ASHRAE Standard 62-1989, *Ventilation for Acceptable Indoor Air Quality,* 1989 (currently in revision).

4

Design Procedures: Part 2

General Concepts for Equipment Selection

4.1 Introduction

The purpose of this chapter is to outline the criteria used in the HVAC system and equipment selection process, to describe some of the systems and equipment available, and to develop some of the underlying philosophy and background related to system selection.

Details of specific systems and items of equipment are discussed in later chapters.

4.2 Criteria for System and Equipment Selection

The problem-solving process requires some criteria that can be applied in describing and evaluating alternatives. In the selection of HVAC systems, the following criteria (Table 4.1) are used — consciously or unconsciously — because only rarely is the problem-solving process formally applied.

1. *Requirements of comfort or process.* These requirements include temperature, always; humidity, ventilation, and pressurization, sometimes; and zoning for better control, if needed. In theory at least, the comfort requirement should have a high priority. In practice, this criterion is sometimes subordinated to first cost or to the desires of someone in authority. This is happening less often as building occupants become more sophisticated in their expectations. Process requirements are more difficult and require a thorough inquiry by the HVAC designer into the process and its needs. Until the process is fully under-

TABLE 4.1   Criteria for HVAC System and Equipment Selection

1. Demands of comfort or process
2. Energy conservation, code requirements
3. First cost versus life-cycle cost
4. Desires of owner, architect, and/or design office
5. Space limitations
6. Maintainability/reliability
7. Central plant versus distributed systems
8. Simplicity and controllability

stood, the designer cannot provide an adequate HVAC system. Most often, different parts of the process have different temperature, humidity, pressure, and cleanliness requirements; the most extreme of these can penalize the entire HVAC system.

2. *Energy conservation.*   This is usually a code requirement and not optional. State and local building codes almost invariably include requirements constraining the use of new, nonrenewable energy. *Nonrenewable* refers primarily to fossil-fuel sources. Renewable sources include solar power, wind, water, geothermal, waste processing, heat reclaim, and the like. The strictest codes prohibit any form of reheat (except from reclaimed or renewable sources) unless humidity control is essential. This restriction eliminates such popular systems as terminal reheat, multizone, and constant volume dual-duct systems, although the two-fan dual-duct system is still possible and the three-duct multizone system is acceptable (see Chap. 11). Most HVAC systems for process environments have opportunities for heat reclaim and other ingenious ways of conserving energy. Off-peak storage systems are becoming popular for energy cost savings, although these systems may actually consume more energy than conventional systems.[1]

3. *First cost and life-cycle cost.*   The first cost reflects only the initial price, installed and ready to operate. The first cost ignores such factors as expected life, ease of maintenance, and even, to some extent, efficiency, although most energy codes require some minimum efficiency rating. The life-cycle cost includes all cost factors (first cost, operation, maintenance, replacement, and estimated energy use) and can be used to evaluate the total cost of the system over a period of years. The usual method of comparing the life-cycle costs of two or more systems is to convert all costs to present-worth values. Typically, first cost governs in buildings being built for speculation or short-term investment. Life-cycle costs are most often used by institutional builders — schools, hospitals, government — and owners who expect to occupy the building for an indefinite extended period.

4. *Desires of owner, architect, or design office.* Very often, someone in authority lays down guidelines which must be followed by the designer. This is particularly true for institutional owners and major retailers. Here the designer's job is to follow the criteria of the employer or the client unless it is obvious that some requirements are unsuitable in an unusual environment. Examples of such environmental conditions are extremely high or low outside-air humidity, high altitude (which affects the AHU and air-cooled condenser capacity), and contaminated outside air (which may require special filtration and treatment).

5. *Space limitations.* Architects can influence the HVAC system selection by the space they make available in a new building. In retrofit situations, designers must work with existing space. Sometimes in existing buildings it is necessary to take additional space to provide a suitable HVAC system. For example, in adding air conditioning to a school, it is often necessary to convert a classroom to an equipment room. Rooftop systems are another alternative where space is limited, if the building structure will support such systems. In new buildings, if space is too restricted, it is desirable to discuss the implications of the space limitations in terms of equipment efficiency and maintainability with the architect. There are ways of providing a functional HVAC system in very little space, such as individual room units and rooftop units, but these systems often have a high life-cycle cost.

6. *Maintainability.* This criterion includes equipment quality (the mean time between failures is commonly used); ease of maintenance (are high-maintenance items readily accessible in the unit?); and accessibility (Is the unit readily accessible? Is there adequate space around it for removing and replacing items?). Rooftop units may be readily accessible if an inside stair and a roof penthouse exist; but if an outside ladder must be climbed, the adjective *readily* must be deleted. Many equipment rooms are easy to get to but are too small for adequate access or maintenance. This criterion is critical in the life-cycle cost analysis and in the long-term satisfaction of the building owner and occupants.

7. *Central plant versus distributed systems.* Central plants may include only a chilled water source, both heating and chilled water, an intermediate temperature water supply for individual room heat pumps, or even a large, central air-handling system. Many buildings have no central plant. This decision is, in part, influenced by previously cited criteria and is itself a factor in the life-cycle cost analysis. In general, central plant equipment has a longer life than packaged equipment and can be operated more efficiently. The disadvantages include the cost of pumping and piping or, for the central AHU, longer

duct systems and more fan horsepower. There is no simple answer to this choice. Each building must be evaluated separately.

8. *Simplicity and controllability.*   Although listed last, this is the most important criterion in terms of how the system will really work. There is an accepted truism that operators will soon reduce the HVAC system and controls to their level of understanding. This is not to criticize the operator, who may have had little or no instruction about the system. It is simply a fact of life. The designer who wants or needs to use a complex system must provide for adequate training — and retraining — for operators. The best rule is: Never add an unnecessary complication to the system or its controls.

4.3   Options in System and Equipment Selection

The various systems and equipment available are described in later chapters. They are briefly listed here to summarize the options available to the designer.

4.3.1   Air-handling units

Air-handling units (AHUs) include factory-assembled package units and field-erected, built-up units (see Chap. 11). The common components are a fan or fans, cooling and/or heating coils, and air filters. Most units also include a mixing chamber with outside and return air connections with dampers. The size range is from small fan-coil units with as little as 100 ft^3/min capacity to built-up systems handling over 100,000 ft^3/min. When a package unit includes a cooling source, such as a refrigeration compressor and condenser, or a heating source, such as a gas-fired heater or electric heating coil, or both, then the unit is said to be *self-contained.* This classification includes heat pumps. Many systems for rooftop mounting are self-contained, with capacities as great as 100 tons or more of cooling and a comparable amount of heating. Some room units for wall or window installation have capacities as small as 0.5 or 0.75 ton. Split-system packages are also available, with the heating and/or cooling source section matching the fan-coil section but installed outdoors. The two sections are connected by piping. Cooling coils may use chilled water, brine, or refrigerant (direct expansion). Heating coils may use steam or high- or low-temperature water; or "direct-fired" heating may be used, usually gas or electric resistance. Heat reclaim systems of various types are employed. Humidification equipment includes the steam grid, evaporative, and slinger/atomizer types. Dehumidification equipment includes the dehumidifying effect of most cooling coils as well as absorption-type de-humidifiers.

Thus, the designer has a wide range of equipment to choose from. Although generalizations are dangerous, some general rules may be applied, but the designer must also develop, through experience, an understanding of the best and worst choices. There are some criteria which are useful:

1. Packaged equipment should be tested, rated, and certified in accordance with standards of American Society of Heating, Refrigeration, and Air Conditioning Engineers (ASHRAE), Air Conditioning and Refrigeration Institute (ARI), Association of Home Appliance Manufacturers (AHAM), Air Movement and Control Association (AMCA), or others as applicable.

2. Minimum unit efficiencies should be in accordance with codes or higher.

3. In general, packaged equipment has a lower first cost and a shorter life than equipment used in built-up systems. This is not always true, and comparisons must be made for the specific application.

4. In general, packaged equipment is designed to be as small as possible for a given capacity. This may create problems of access for maintenance. Also the supplier should show that capacity ratings were determined for the package as assembled and not just for the separate components. See particularly the discussion on the effects of geometry on fan performance in Chap. 5.

5. In hotels, motels and apartments, individual room units should be used to give occupants maximum control of their environment. Where many people share the same space, central systems are preferable, with controls which cannot be reset by occupants.

6. Noise is a factor in almost any HVAC installation, yet noise is often neglected in equipment selection and installation. Noise ratings are available for all types of HVAC equipment and should be used in design and specifications (see Chap. 20).

4.3.2 Radiant and convective heating and cooling

Convector radiators, using steam or hot water, are one of the oldest heating methods and are still in common use. Modern systems are more compact than the old cast-iron radiators and depend more on natural convection than on radiation. Rating methods are standardized by the Hydronics Institute.

Radiant heating by means of floor, wall, or ceiling panels is common. Hot-water piping or electric resistance heating tape is used. Maximum temperatures of the surface must be limited, and there are some control problems, particularly in floor panels, due to the mass of the panel.

Radiant cooling by means of wall or ceiling panels may also be used. Surface temperatures must be kept above the dew point; therefore, any dehumidification required must be accomplished by other means.

In modern practice, radiant and/or convective heating or cooling is usually a supplement to the air system and is used primarily to offset exterior wall, roof, and floor heat gains or losses.

4.3.3 Refrigeration equipment

Source cooling equipment includes refrigeration compressors of reciprocating, centrifugal, and screw types; absorption chillers using steam, hot water, or direct fuel firing; water chiller heat exchangers; condensers cooled by air, water, and evaporation; cooling towers; and evaporative coolers, including spray, slinger, and drip types (see Chap. 9).

Self-contained package AHUs always use direct-expansion cooling with reciprocating or rotary compressors. Other AHUs may use direct expansion, chilled water, or brine cooling, with the cooling medium provided by a separate, centralized, refrigeration system (see Chap. 6). Evaporative cooling is used primarily in climates with low design ambient wet-bulb temperatures, although it may be used in almost any climate to achieve some cooling. Evaporative cooler efficiencies are highest for the spray type and lowest for the drip type. Centrifugal and screw-type compressors and absorption refrigeration are used almost entirely in large central-station water or brine chillers. Absorption refrigeration may be uneconomical unless there is an adequate source of waste heat or solar energy. Air-cooled condensers are less costly to purchase and maintain than cooling towers or evaporative condensers, but they result in higher peak condensing temperatures at design conditions and may result in lower overall efficiency in the cooling system.

The selection of the source cooling equipment is influenced primarily by the selection of the AHU equipment and systems. Often both are selected at the same time. The use of individual room units does not preclude the use of central-station chillers; this combination may be preferable in many situations. For off-peak cooling with storage, a central chilling plant is an essential item.

4.3.4 Heating equipment

Source heating equipment includes central plant boilers for steam and high-, medium-, and low-temperature hot water; heat pumps, both central and unitary; direct-fired heaters; solar equipment, including solar-assisted heat pumps; and geothermal and heat reclaim. Fuels include coal, oil, natural and manufactured gas, and peat. Waste products such as *refuse-derived fuel* (RDF) and sawdust are also being used in limited ways. Electricity for resistance heating is not a fuel in the combustion sense but is a heat source. Heat reclaim takes many forms, some of which are discussed in Chap. 10.

Self-contained package AHUs use direct-fired heaters—usually gas or electric—or heat pumps. For other systems, some kind of central plant equipment is needed. The type of equipment and fuel used is determined on the basis of the owner's criteria, local availability and comparative cost of fuels, and, to some extent, the expertise of the designer. Large central plants for high-pressure steam or high-temperature hot water present safety problems and require special expertise on the part of the designer, contractor, and operator. New buildings connected to existing central plants will require the use of heat exchangers, secondary pumping or condensate return pumping, and an understanding of limitations imposed by the existing plant, such as limitations on the pressure and temperature of returned water or condensate.

4.4   The Psychrometric Chart

When the system type has been selected and a summary completed, showing design CFM, temperature difference (TD), and latent load, it is time to complete the psychrometric chart. For a detailed discussion of psychrometrics, see Chap. 19. Consider a single-zone, draw-through air-handling system, as in Fig. 4.1. In summer the return air is mixed with some minimum amount of outside air; is pulled through the filter, the coils, and the fan; and is supplied to the space. Cooling is provided

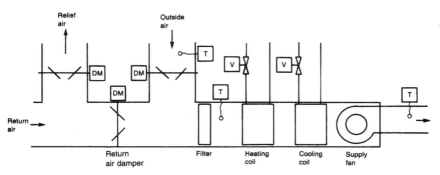

Figure 4.1 Single-zone draw-through air-handling unit.

by the cooling coil through the use of chilled water (as in this example) or by direct refrigerant expansion. On the psychrometric chart (Fig. 4.2), the designer first plots the inside and outside design conditions—say 75°F and 50 percent RH inside and 95°F dry bulb (db), 75°F wet bulb (wb) outside. The return air to the system will usually be at a higher temperature than the space due to heat gains in return-air plenums. (This does not hold for direct-return units.) This heat gain can be estimated or calculated from the geometry of the building, the wattage of recessed lighting, etc. For this example, assume a 3°F rise. Then the return air is at 78°F with the same humidity ratio w as the space. A straight line between this point and the outside air state point represents the mixing process. The mixed-air state point lies on this line at a distance from the return-air point equal to the design minimum percentage of outside air—for this example, 20 percent. Then the mixed-air condition is 81.4°F db and 66.2°F wb, with an enthalpy (h) of 30.8 (Btu/lb). The design condition of the supply air is calculated as described in Chap. 3 and for this example is assumed to be 56°F db with a humidity ratio (w) equal to 0.0086 (lbw/lba). Because this is a draw-through system, there is some heating effect due to fan work. If the fan horsepower and efficiency are known, this can be calculated. For preliminary purposes, a temperature rise of 0.5°F per inch of pressure rise across the fan can be assumed—for this example, 3-in static pressure or 1.5°F. Then the air must leave the coil at 54.5°F db, with w equal to 0.0086 as above. The resulting point has an h value of 22.5. Now the cooling (coil) process can be represented by a straight line from the mixed-air point through the "leaving coil" point and can be extended to the saturation curve. The

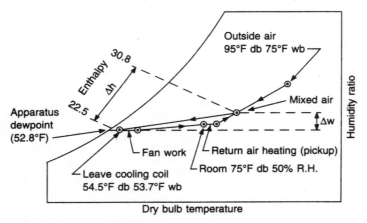

Figure 4.2 Psychrometric chart, cooling cycle example.

intersection with the saturation curve is called the *apparatus dew point* (ADP) and is the coil surface temperature required to obtain the design process (here about 53°F). The total cooling load in Btu/h is determined from the difference in h values multiplied by the total CFM, 60 min/h, and the air density in pounds per cubic foot (0.075 for standard air). Thus

$$q_c = \text{CFM} \times (h_m - h_c) \times 60 \times 0.075 \qquad (4.1)$$

where q_c = total cooling, Btu/h
$\quad h_m$ = enthalpy of mixed air entering cooling coil, Btu/lb
$\quad h_c$ = enthalpy of air leaving the cooling coil, Btu/lb

The cooling load thus calculated includes the sensible and latent space load — the total load due to outside air, fan work, and any return-air "pickup."

The cycle at winter design conditions can also be plotted, as shown in Fig. 4.3. The mixed air is controlled at the low-limit condition, say 60°F, although this may be reset upward as the outside temperature decreases for energy conservation. Return air will be about 3°F above the 72°F space temperature, or 75°F. For this example, the outside design temperature is assumed to be 32°F and 50 percent RH. Heating will be provided as required to maintain the space conditions (some design heating temperature difference will be calculated). If space humidity is uncontrolled, the cycle will automatically fall into a position such that the humidity ratio difference between supply air and space will be the same as that for cooling. This will typically result in a lower space humidity in winter.

Most designers use the psychrometric chart only for the design cooling cycle, or for both heating and cooling if humidity control is pro-

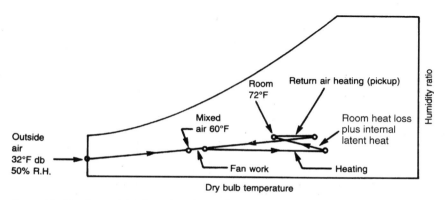

Figure 4.3 Psychrometric chart, heating cycle example.

vided. It is sometimes useful to look at intermediate conditions such as in Fig. 4.4. Here an outside temperature of 60°F is assumed, and 100 percent outside air is used by the economy cycle. The inside humidity will depend on the outside humidity, as discussed before. No mechanical cooling and little or no heating should be needed. Other intermediate conditions can be examined in similar ways.

4.5   Effects of Altitude and Temperature

Air density varies directly and linearly with temperature, and inverse/exponentially with altitude. See Fig. 4.5. Standard conditions are defined as 0.075 lb/ft³ at 59°F. All HVAC airside calculations are recognized as being inexact, where seasonal ambient temperatures may vary from 0 to 100°F, a 10 percent effect, and local barometric pressures may fluctuate plus or minus 2 percent, depending on weather. But changes in density related to altitude and related to heating or cooling processes may compound all other effects and should not be taken lightly.

4.5.1   Changes due to altitude

Atmospheric pressure and related air density decreases as altitude or elevation above sea level increases. Inversely, atmospheric pressure increases for those elevations below sea level or some other reference point. At elevations up to 6000 ft where the altitude/density variation is nearly linear, the rate of density change is approximately 3 to 4 percent per 1000 ft of elevation change. This corresponds to furnace manufacturer's counsel to derate natural draft burner equipment at 4

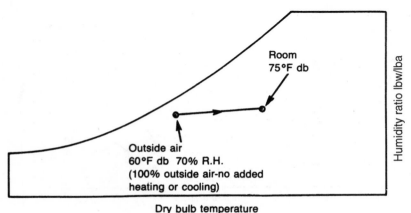

Figure 4.4 Psychrometric chart, intermediate cycle example.

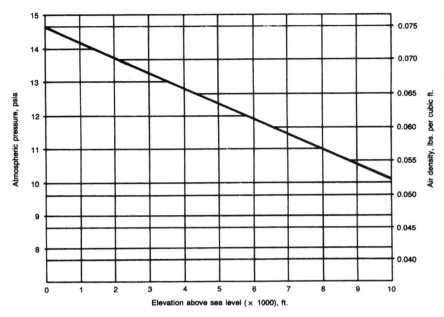

Figure 4.5 Pressure and density versus altitude.

percent per 1000 ft elevation. Altitude effects are often ignored below 2000 ft elevation where more than 90 percent of U.S. commerce is involved, but negative effects at altitudes above 2500 ft of 10 percent and more can threaten the success of an otherwise competent design.

Most of the basic effects of altitude density variation can be predicted. Reduced mass of air per unit volume results in lower air pressure drops due to reduced friction as air passes through filters, coils, and ducts. Air volume must be increased to transport the same amount of cooling or heating capacity for a given air temperature differential. For the same amount of energy transport, fan speeds must increase, but fan horsepower stays about the same. Airside heat transfer in coils is reduced for a given coil face area, while heat transfer loss can be offset with higher face velocities, but then moisture carryover from a cooling coil may be a problem.

Evaporative media and equipment performance is less affected by altitude and air density change. As air pressure goes down, the water vapor holding capacity of the air increases. The inverse can be observed as water drains from the receiver tank of an air compressor. Instruments for measuring and monitoring airflow may be affected if they are dependent on mass flow rate or velocity pressure.

There is an often overlooked effect of altitude on steam system performance. Gauge pressures seem to be the same at altitude as at sea level, but the absolute pressure is reduced, which determines the actual steam temperature. Zero lb/in² gauge steam is 212°F in Seattle or Miami, but only 202°F in Denver or Albuquerque, and less than that in the high mountain ski resorts. Steam-driven absorption chillers may be altitude-sensitive, as may some other heat-exchange devices.

4.5.2 Changes due to temperature

Air density varies linearly with absolute temperature. The change begins to be significant above 120 to 140°F. In HVAC design, commonly encountered problems are high-temperature exhaust involving temperatures of 200°F or more or low-temperature "sharp-freeze" air-cooling systems using air at 20 to 50°F below zero. Equipment for these applications must be specially designed with fan bearings and other materials suitable for the design temperatures. Motor horsepower must be adjusted to the nonstandard density. In high-temperature applications, the required horsepower will be decreased, but the motor must still be large enough for operation at normal temperatures under no-load or start-up conditions.

4.6 Use of Computers

Many manufacturers now provide computerized selection of their equipment. Some manufacturers provide working software to the design office as they would a catalog. For others, the designer provides to the sales engineer a summary of the pertinent criteria; the information is either handled locally or forwarded to the factory. Usually, a printout returns with several alternative selections. For example, the designer may need a cooling coil to provide 120,000 Btu/h total, with a sensible/total ratio of 0.8; 3210 ft³/min of entering air at 81.4°F db, 66.2°F wb; leaving air at 54.5°F db, 53.8°F wb; and 24 gal/min of chilled water supplied at 45°F. There may also be some kind of coil face velocity and air and water pressure-drop limitations, typically about 500–550 ft/min, 1 in of water (APD), and 10 to 20 ft of water (WPD). Four or five coils will be selected by the computer, with a variety of configurations, rows, and costs.

The computerized selections will always assume the maximum optimum performance of the standard-size coils checked. The design conditions specified will always be altered somewhat to compensate for this. The designer must be aware that another manufacturer's coil will not be identical, and care must be taken in equipment schedules to allow for these variations if competition is to be maintained. Similar situations arise with other types of equipment, with package air-handling systems having the greatest variations.

4.7   Summary

In this chapter the general philosophy and concept of HVAC equipment selection was discussed. The details of various equipment items are covered in subsequent chapters.

References

1. R. W. Haines, "Energy Cost versus Energy Conservation," *Heating / Piping / Air Conditioning,* July 1987, p. 90.

5

Design Procedures: Part 3

Air-Handling Systems

5.1 Introduction

In most HVAC systems, the final transport medium is moist air — a mixture of dry air and water vapor. This is conveyed through filters, heat exchange equipment, ducts, and various terminal devices to the space to be air-conditioned. The power to move the air is supplied by fans. This chapter discusses fans and duct systems, together with related subjects such as grilles, registers, diffusers, dampers, filters, and noise control.

5.2 Fans

According to Air Moving and Conditioning Association (AMCA) Standard 210,[1] "A fan is a device for moving air which utilizes a power-driven, rotating impeller." The three fan types of primary interest in HVAC systems are centrifugal (Figs. 5.1 and 5.2), axial (Fig. 5.3), and propeller. The fan motor may be directly connected to the impeller, directly connected through a gearbox, or indirectly connected by means of a belt-drive system.

5.2.1 Fan law equations

The fan law equations are used to predict the performance of a fan at some other condition than that at which it is tested and rated. The HVAC designer is particularly interested in the effect on horsepower, pressure, and volume of varying the speed of the fan in a system.

Figure 5.1 Double-width centrifugal fan. (*Courtesy of New York Blower Co.*)

The fan laws expressed in the following equations relate only to the effect of varying speed, assuming that fan size and air density remain constant.

$$CFM_2 = CFM_1 \frac{RPM_2}{RPM_1} \tag{5.1}$$

$$SP_2 = SP_1 \left(\frac{RPM_2}{RPM_1}\right)^2 \tag{5.2}$$

$$TP_2 = TP_1 \left(\frac{RPM_2}{RPM_1}\right)^2 \tag{5.3}$$

$$BHP_2 = BHP_1 \left(\frac{RPM_2}{RPM_1}\right)^3 \tag{5.4}$$

Figure 5.2 Single-width centrifugal fan. (*Courtesy of Buffalo Forge Co.*)

where CFM = airflow rate, ft³/min
      SP = static pressure
      TP = total pressure
      BHP = brake horsepower, bhp

Expressed in simple language, the fan laws say that when fan size and air density are unchanged, the airflow rate varies directly as the

Figure 5.3 Axial fan. (*Courtesy of Buffalo Forge Co.*)

change in speed, the pressure varies as the square of the change in speed, and the power required varies as the cube of the change in speed.

The complete fan laws also include terms for changes in fan size and air density. The laws are valid only when fans of different sizes (diameters) are geometrically similar.

$$\frac{\text{CFM}_2}{\text{CFM}_1} = \frac{\text{RPM}_2}{\text{RPM}_1} \times \left(\frac{D_2}{D_1}\right)^3 \tag{5.5}$$

$$\frac{\text{TP}_2}{\text{TP}_1} = \frac{\text{SP}_2}{\text{SP}_1} = \frac{\text{VP}_2}{\text{VP}_1} = \left(\frac{\text{RPM}_2}{\text{RPM}_1}\right)^2 \left(\frac{D_2}{D_1}\right)^2 \frac{d_2}{d_1} \tag{5.6}$$

$$\frac{\text{BHP}_2}{\text{BHP}_1} = \left(\frac{\text{RPM}_2}{\text{RPM}_1}\right)^3 \left(\frac{D_2}{D_1}\right)^5 \cdot \frac{d_2}{d_1} \tag{5.7}$$

where D = fan diameter and d = air density.

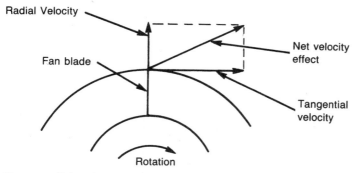

Figure 5.4 Principle of operation of centrifugal fan.

For further variations, see ASHRAE Handbook, *1996 HVAC Systems and Equipment,* Chap. 18, Table 2, p. 18.4.

5.2.2 Centrifugal fans

A centrifugal fan creates pressure and air movement by a combination of centrifugal (radial) velocity and rotating (tangential) velocity. As shown in Fig. 5.4, these two effects combine to create a net velocity vector. When the fan is enclosed in a scroll (housing) as shown in Fig. 5.5, some of the velocity pressure is converted to static pressure. The

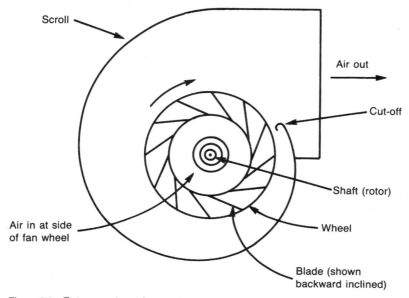

Figure 5.5 Cutaway view of centrifugal fan.

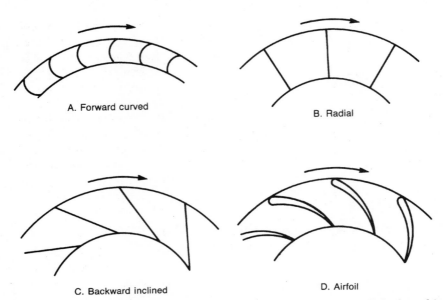

Figure 5.6 Centrifugal fan blade types. *A*. Forward-curved. *B*. Radial. *C*. Backward inclined. *D*. Airfoil.

fan characteristics can be changed by changing the shape of the blade. Typical shapes (Fig. 5.6) are forward-curved, straight radial, backward-inclined (straight or curved), and airfoil.

The geometry of the fan wheel, inlet cone, and scroll also has an effect on fan performance and efficiency. Figure 5.7 shows a typical

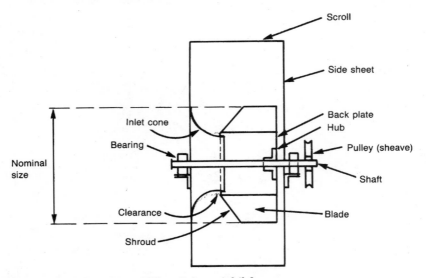

Figure 5.7 Cross section of BI, radial, or airfoil fan.

cross section for a *backward-inclined* (BI) or *airfoil* (AF) fan wheel. For a given wheel or diameter, as the blade gets narrower and longer, higher pressures can be generated but flow rates are reduced. The inlet cone is shown curved (bell-mouth) to minimize air turbulence. Straight cones are also used, at the cost of some reduction in performance. The clearance between the inlet cone and the wheel shroud must be minimized for efficiency, because some air is bypassed through this opening. The *forward-curved* (FC) wheel (Fig. 5.8) usually has a short, wide blade and a flat shroud. The inlet cone is curved or tapered and is mounted to minimize the clearance between the inlet cone and shroud. This type of fan handles large air volumes at low pressures. The illustrations show *single-width, single-inlet* (SWSI) fans. *Double-width, double-inlet* (DWDI) fans are also made.

5.2.3 Fan testing procedures

Fans for HVAC applications should be tested for performance rating in accordance with AMCA Standard 210,[1] promulgated by the Air Moving and Conditioning Association. Also, ASHRAE Standard 51 prescribes the test setup and data-gathering procedures in fan testing. For a line of several sizes of geometrically similar fans, only the smallest fan in the line is actually tested. Performance of all other sizes is calculated, by using formulas based on the fan laws. The testing setup and procedures are designed for ideal inlet and outlet conditions, with a minimum of turbulence. Later in this chapter we discuss the effect of the nonideal conditions usually found in HVAC system installations.

The test procedure includes measuring the airflow and horsepower against varying pressures, for a constant fan speed. Pressure is measured in inches of water, by using an oil or water manometer. Airflow

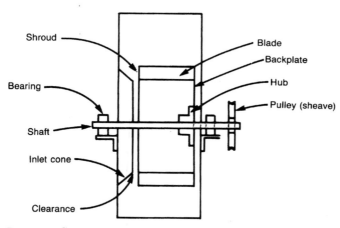

Figure 5.8 Cross section of FC fan.

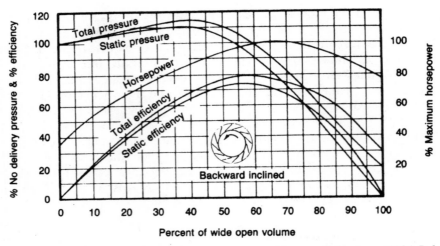

Figure 5.9 Normalized curves for a BI fan. (*Reprinted by permission from AMCA Publication 201,* Fans and Systems.)

is measured in cubic feet per minute. The data can then be plotted as a series of curves similar to Fig. 5.9. This figure contains "normalized" typical curves for a BI fan. Airfoil fan curves are similar with slightly higher efficiencies. The curves for an FC fan have a different shape, as shown in Fig. 5.10. When the fan speed is varied, the result is a family of parallel curves, as shown in Fig. 5.11.

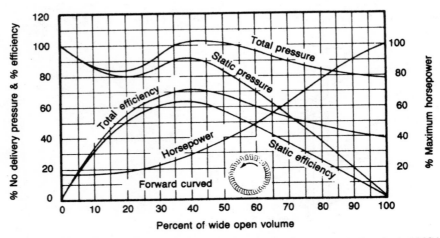

Figure 5.10 Normalized curves for an FC fan. (*Reprinted by permission from AMCA Publication 201,* Fans and Systems.)

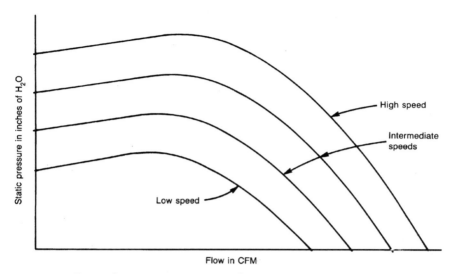

Figure 5.11 Fan performance at various speeds.

5.2.4 Fan performance data

The HVAC system in which a fan is to be installed has a system-curve characteristic relating to its geometry. In accordance with the laws of hydraulics, the system pressure loss varies as the square of the change in airflow rate. The system curve can be superimposed on a fan curve, resulting in something like Fig. 5.12. For purposes of illustration, this shows two different system curves. These two curves are the recommended limits between which the fan can be efficiently and safely used. The manufacturer's performance tables normally cover this area of the graph. Operation at conditions outside the recommended limits can result in inefficiency, noise, and instability (surge).

The point of intersection of the fan curve and the system curve determines the actual operating condition — flow rate versus pressure. This assumes that the system resistance has been accurately estimated and that the fan is installed so that inlet and outlet conditions are comparable to those used in the laboratory test. In fact, this is seldom or never the case. AMCA Publication 201,[2] *Fans and Systems,* discusses system effects in detail and includes a great deal of data on multipliers to be used for various system effects which are too voluminous to include in this book. The effect is illustrated in Fig. 5.13, which is taken from AMCA Publication 201 and used by permission. The theoretical fan selection would be at condition 1 on the calculated duct system curve. However, if the actual system curve is as shown by the dashed line, then the fan selected at condition 1 will actually perform at condition 4, with a higher pressure and lower flow than

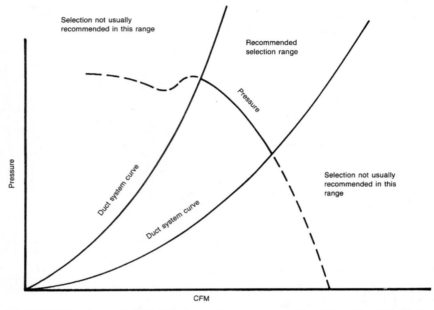

Figure 5.12 Recommended performance of a typical centrifugal fan. (*Reprinted by permission from AMCA Publication 201,* Fans and Systems.)

the design values. To get the design airflow rate, the fan will have to be speeded up to get to condition 2. This might not be possible with the original fan and horsepower selection, and a different size fan will be needed. If the problem is discovered after installation, it could be very costly to fix.

The most common installation errors relate to fan inlet and outlet conditions. The ideal in both cases is a gradual transition with no turns close to the fan. Turning vanes must be provided in inlet duct elbows. An inlet condition that creates a swirling motion in the direction of rotation will reduce the pressure-volume curve by an amount depending on the intensity of the vortex; this is the principle used by inlet vane dampers. A condition that causes a swirl opposite to the direction of rotation will cause a substantial increase in horsepower.

Installation in an intake plenum (as in most packaged HVAC systems) or discharging directly into a plenum (as in most multizone and dual-duct systems) will affect fan performance adversely.

The performance curves indicate fan classes. Classes I, II, III, and IV relate to structural considerations required to accommodate higher speeds and pressures. These include stronger frames and wheels and larger shafts and bearings.

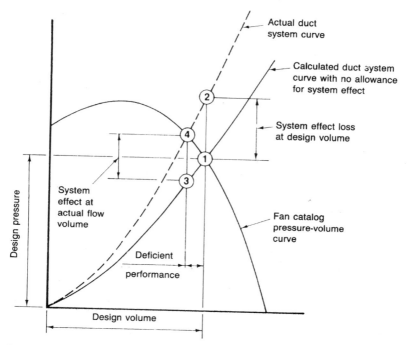

Figure 5.13 Deficient duct system performance, system effects ignored. (*Reprinted by permission from AMCA Publication 201*, Fans and Systems.)

5.2.5 Inlet vane dampers for fan volume control

A common method of fan volume control employs the inlet vane damper (Fig. 5.14). This consists of a ring of pie-shaped elements which open and close in parallel. Control may be manual or automatic. When properly installed to provide an inlet swirl in the direction of fan rotation, the damper alters the fan performance curve as shown in Fig. 5.15. The fan horsepower is also reduced, although not as much as would be predicted by the fan laws, because the damper increases the system pressure loss. Use of discharge dampers is not recommended for volume control, only for fan isolation.

5.2.6 Mechanical and structural considerations

The mounting and driving mechanism for the fan wheel entails many mechanical and structural considerations. The bearings which support the shaft come in many kinds, depending on the speed of rotation, weight of the fan wheel, power transmitted, and whether the fan

Figure 5.14   Inlet vane damper. (*Courtesy of Chicago Blower Corp.*)

wheel is overhung (Fig. 5.16) or supported between bearings (Fig. 5.17). Sometimes three bearings are used; then alignment must be precise. Bearing supports must be strong enough to support the bearings without flexing. The drive shaft must be strong and rigid enough to support the fan wheel between the bearings and to transmit the required power without undue flexing over a specified rotational speed. All shafts have a critical speed at which excessive vibration, noise, and possible failure will occur. Many shafts have two or more critical speeds. Sometimes the lower critical speed is less than the normal speed range of the fan. This is satisfactory when the fan is accelerated quickly through the critical speed. It will not be satisfactory if the fan is to be used in a speed-controlled VAV application.

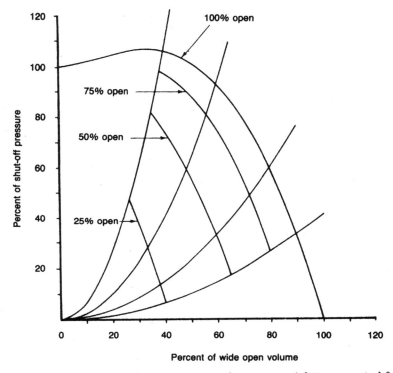

Figure 5.15 Typical normalized pressure-volume curve—inlet vane control for a BI fan. (*Reprinted by permission from AMCA Publication 201,* Fans and Systems.)

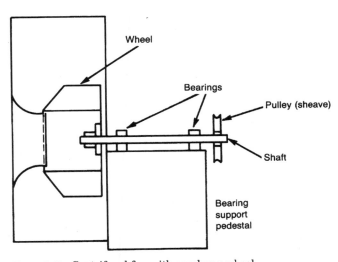

Figure 5.16 Centrifugal fan with overhung wheel.

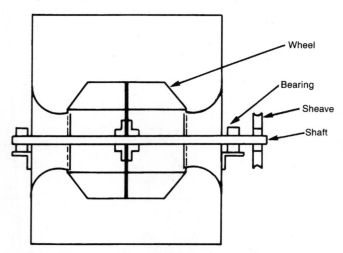

Figure 5.17 Centrifugal fan with wheel between bearings.

5.2.7 Axial fans

Axial-flow fans impart energy to the airstream by giving it a swirling motion. Straightening vanes must be provided in the tubular housing to improve flow and efficiency for use with duct systems, as shown in Fig. 5.18. Belt or direct drive may be used. Typical performance curves for an axial-flow fan are shown in Fig. 5.19. Note that the horsepower increases toward the no-flow condition. Care must be taken to avoid selection in these areas, and motors must be large enough to avoid overloading. In this connection, note that centrifugal fans are nonoverloading—there is some horsepower rating which will never be exceeded at a given speed.

Axial fan performance is governed by the blade shape and pitch, the ratio of hub diameter to tip diameter, and the number of blades. Large hub-to-tip ratios (0.6 to 0.8) relate to lower flow rates and higher pressures than small hub-to-tip ratios (0.4 to 0.5). Airfoil blades are most efficient. The fan volume can be varied by changing the speed. The more common method of volume control is to use variable-pitch blades, which allow changes in pressure and flow performance characteristics.

5.2.8 Propeller fans

The propeller fan is used primarily for moving air at low pressure, usually without ductwork. It is often used in roof and wall ventilators for makeup, exhaust, and relief. Efficiencies are low, and the horsepower increases rapidly as the no-flow condition is approached (Fig.

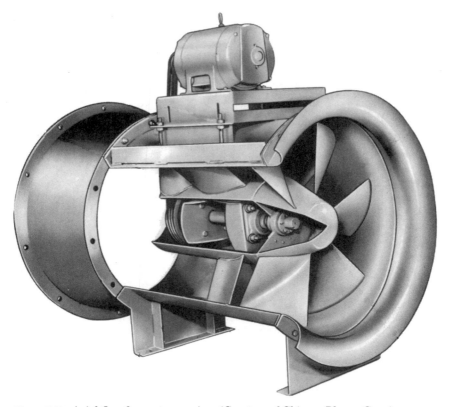

Figure 5.18   Axial-flow fan, cutaway view. (*Courtesy of Chicago Blower Corp.*)

5.20). The fan is usually mounted in a circular orifice or venturi plate, with direct or belt drive.

5.2.9   Fan noise

Fans operate most quietly in the region of highest efficiency, because energy not converted to power is usually converted to noise. In general, noise levels increase with higher outlet velocities (centrifugal fans) or tip speeds (propeller and axial fans), given equivalent efficiencies. However, higher outlet velocities or tip speeds are required at higher pressures.

5.2.10   Variable speed control of fans

Since fan capacity is directly related to fan speed for all types of fans, fan speed selection has always been a part of fan system design. Belt

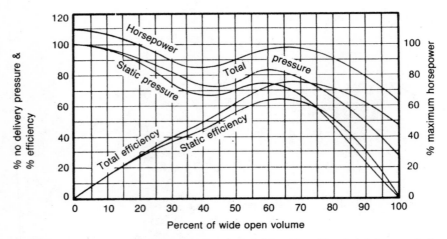

Figure 5.19 Typical normalized curve for an axial-flow fan. (*Reprinted by permission from AMCA Publication 201,* Fans and Systems.)

drives with adjustable or exchangeable sheaves have been a traditional method for adjusting a given fan to a specific system capacity requirement. Two-speed or multispeed motors have provided a means of fan speed and capacity adjustment on a high-low or high-medium-low basis. For small motors, a manually adjustable speed switch similar to a lighting dimmer switch has been available.

Speed adjustment for fans to modify capacity to exactly meet the load has great potential benefit for fan energy conservation. If volume flow is proportional to fan speed, and developed fan pressure is proportional to the square of the fan speed, fan power requirements are

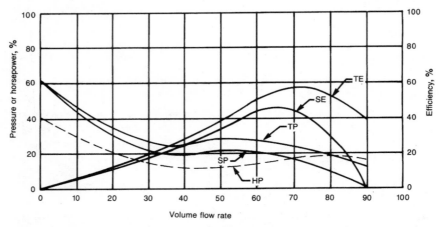

Figure 5.20 Typical normalized curve for a propeller fan. (*Reprinted by permission from the ASHRAE PDS-1 Workbook.*)

proportional to the cube of the fan speed. This means that a 50 percent reduction of fan speed results in a reduction of fan power (theoretical) to one-eighth of the original power requirement. Where air-handling systems serve loads of varying intensity, an opportunity to directly control fan volume by varying fan speed is of great potential benefit. Many variable air volume (VAV) systems run in a range of 40 to 80 percent of design capacity most of the time. Fan speed adjustment can be in addition, and a corollary, to modulation of a primary heating and cooling medium for the supply air stream.

If inlet vane dampers or adjustable belt drives have been common mechanical-type volume adjustment techniques, with industrial fluid drives as a high-priced option, recent development of competitively priced variable-frequency drives, also derived from an industrial market, are proving to be a great addition to the HVAC system configuration. These units, which vary in size from fractional up through several hundred horsepower capacity, use transistor technology to rectify alternating current in the standard 50- or 60-Hz format, and reconstitute it at any desired frequency. Since induction motor speed depends on frequency of the power supply, varying voltage frequency determines output of the fan, which can be automatically adjusted to match the connected load.

While variable or adjustable speed drives (VSD or ASD), also called variable frequency drives (VFD), are electrical in nature, they are directly involved in the mechanical duty. Selection and specification may be a joint mechanical/electrical assignment but should be abrogated by neither. Of many VFD brands available on the market, there is wide variation in character and configuration. There are many purchase options. Different brands are better or worse in terms of power-line "harmonic" generation. Nearly all can be programmed and can accept 4 to 20 mA, 0 to 5 V dc or 0 to 10 V dc input signals. Common voltage ratings are 208/230 or 460 V, three-phase. Motors must be carefully selected to be able to withstand higher winding temperatures, as motor cooling is typically reduced with lower fan speed. VFDs usually obtain a power factor of 0.9 or higher while imposing a 5 percent power consumption penalty on the motor use. (The VFD penalty is recovered, and more by the power savings in reduced motor speed.)

5.3 Air Duct Design

An *air duct* is an enclosed conduit through which air is conveyed from one location to another. The design of the duct system must take into account the space available, allowable noise levels, potential for duct leakage, effect of duct heat losses or gains on system performance, thermal and noise insulation, effect of air contaminants on duct materials (corrosion, etc.), fire and smoke control, and pressure losses due to friction and turbulence.

The principal references in air duct design are the ASHRAE handbooks[3] and the manuals published by the Sheet Metal and Air Conditioning Contractors' National Association (SMACNA).[4] The ASHRAE handbooks deal with both system design and duct construction; the SMACNA manuals deal primarily with duct construction methods to ensure adequate strength and minimize leakage. Many different duct materials are used, but principally galvanized sheet steel. Aluminum and fiberglass are also common in some applications.

5.3.1 Pressures

In the design of duct systems, three pressures are specified: the total pressure P_t, static pressure P_s, and velocity pressure P_v. The total pressure is the sum of the velocity and static pressures:

$$P_t = P_s + P_v \qquad (5.8)$$

In HVAC work, these pressures are usually measured and expressed in terms of the height of a water column supported by the pressure. Static pressure is the pressure which exists in the duct independent of the velocity. Static pressure is considered to be uniform in all directions. The velocity pressure is that due to the inertial or kinetic energy of the flowing fluid (air in the HVAC system) and is measured in the direction of flow. For standard air conditions (density of 0.075 lb/ft[3]), the equation for velocity pressure is

$$P_v = \left(\frac{V}{4005}\right)^2 \qquad (5.9)$$

where P_v = velocity pressure, inches of H_2O
     V = velocity, ft/min
     4005 = dimensional constant (call it K)

Because the dimensional constant includes the air density, the constant must be corrected for elevations above sea level. This is done by dividing the sea level constant 4005 by the square root of the density ratio at the desired elevation (see Table 3.3 for the density ratio). Thus, the constant for a 5000-ft elevation is

$$K = \frac{4005}{\sqrt{0.83}} = 4400$$

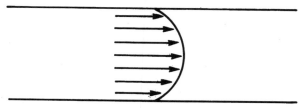

Figure 5.21 Air velocity pattern in straight duct.

5.3.2 Pressure changes in duct systems

As air flows through a duct, there is a loss of energy—measured as a total pressure reduction—due to friction and turbulence. The ASH-RAE Handbook[3] states: "Frictional losses are due to fluid viscosity, and are the result of momentum exchange between molecules in laminar flow and between particles at different velocities in turbulent flow." In simpler language, some air molecules rub against the duct wall and are slowed down; other molecules rub against the slower molecules, and so on. In a long, straight section of duct, this results in a velocity pattern something like Fig. 5.21. Friction loss in this section will be uniform and constant so long as the duct dimensions and airflow rate remain constant. Friction losses are measured in inches of water per 100 ft of duct. For calculation purposes, the mean velocity in the duct is used.

When the duct changes dimensions or direction, there is an additional dynamic loss due to turbulence such as swirls, eddies, and non-uniform velocity patterns produced by the inertial properties of the air—the tendency to keep flowing in a straight line. Thus, at an elbow the air tends to "pile up" along the heel of the elbow, creating a negative pressure and eddy in the throat (see Fig. 5.22). Similar phenomena occur at abrupt transitions (dimensional changes). Gradual transitions (Fig. 5.23) lessen or eliminate the effects of turbulence.

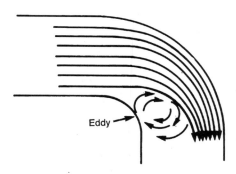

Eddy

Figure 5.22 Air velocity pattern at a duct elbow.

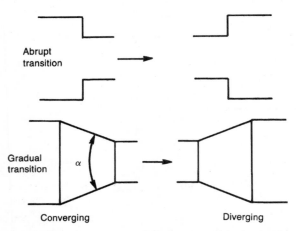

Figure 5.23 Duct transitions. *Note:* For angle of 15° or less, flow is assumed nonturbulent.

5.3.3 Friction losses

The ASHRAE Handbook *Fundamentals*[3] contains a detailed theoretical analysis of friction losses in ducts. For practical purposes, the friction loss can be determined from the flow versus friction chart (Fig. 5.24). This nomograph shows the friction loss in round ducts in inches of water per 100 ft of duct length, for a range of diameters and mean velocities. This chart is based on an absolute duct surface roughness ϵ of 0.0003, the approximate value for galvanized steel sheets. It is also based on standard air (70°F and 0.075 lb/ft³). Corrections should be made for the roughness of other duct materials and for significant differences in elevation or temperature. Table 5.1 provides information on roughness factors for several common duct materials. Figure 5.26 provides K_m factors for four degrees of roughness. The pressure loss from the friction chart is multiplied by the K_m to get the friction loss at a different roughness. Roughness correction is essential to correct design.

Corrections for the density, viscosity, and humidity of nonstandard air are quite complex and are usually neglected, although this is technically incorrect. Neglecting these factors usually results in a conservative design.

The air friction chart (Fig. 5.24) is correct for circular ducts. For rectangular and oval ducts, an equivalent diameter must be determined from a formula based on the hydraulic diameter of the duct. The hydraulic radius of a duct is the area divided by its perimeter, or A/P. The hydraulic diameter is 4 times the hydraulic radius: $D_e = 4A/P$. Table 5.2 shows circular equivalents for rectangular ducts of equivalent capacity. Note that the mean velocity of the rectangular

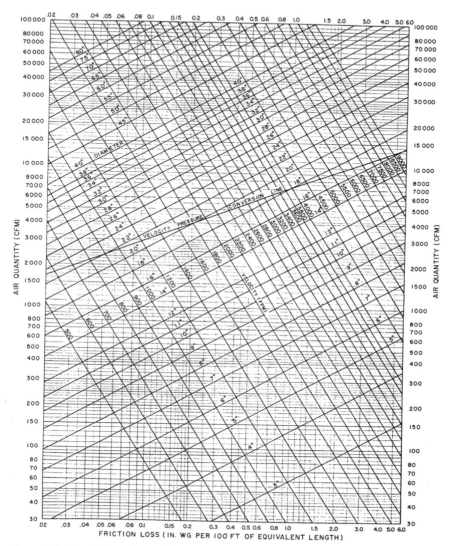

Figure 5.24 Air friction chart. (*From* Carrier-System Design Manual, *Part 2:* Air Distribution, *1960*)

TABLE 5.1 Duct Material Roughness Factors

Duct material	Roughness category	Absolute roughness ϵ, ft
Uncoated carbon steel, clean (0.00015 ft)	Smooth	.0001
PVC plastic pipe (0.00003– 0.00015 ft)		
Aluminum (0.00015–0.0002 ft)		
Galvanized steel, continuously rolled, longitudinal seams (<0.0003 ft)		
Galvanized steel, continuously rolled, spiral seams (0.0002– 0.0004 ft)	Medium-smooth	.0003
Galvanized steel, sheets, hot dipped, longitudinal seams (0.0005 ft)	Average	.0005
Fibrous glass duct, rigid	Medium-rough	.003
Fibrous glass duct liner, airside with facing material (0.005 ft)		
Fibrous glass duct liner, airside spray-coated (0.015 ft)	Rough	.01
Flexible Duct, metallic (0.004– 0.007 ft when fully extended)		
Flexible duct, all types of fabric and wire (0.0035– 0.015 ft when fully extended)		
Concrete (0.001–0.01 ft)		

SOURCE: Reprinted by permission from ASHRAE Handbook, *1997 Fundamentals,* Chap. 32, Table 1, p. 32.7.

duct will always be less than that of its circular equivalent. The friction chart and circular equivalent chart have been combined into a slide-rule type of calculator by several manufacturers; this device greatly simplifies duct sizing selections.

5.3.4 Dynamic losses

Dynamic losses result from disturbances caused by changes in the airflow direction or area. These losses occur at fittings such as elbows, tees, transitions, entrances, and exits. For ease of calculation, the loss

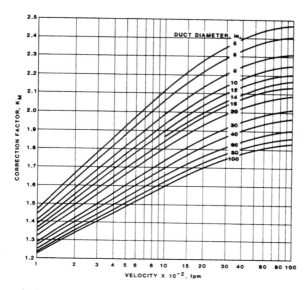

A. Roughness Category—Rough (ϵ = 0.01 ft)

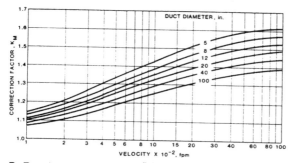

B. Roughness Category—Medium Rough (ϵ = 0.003 ft)

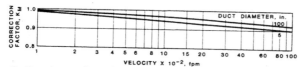

C. Roughness Category—Medium Smooth (ϵ = 0.0003 ft)

(ϵ = 0.0003 ft)

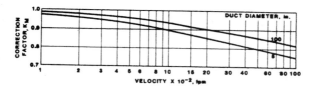

D. Roughness Category—Smooth (ϵ = 0.0001 ft)

Figure 5.25 Correction factors for duct roughness. (*Reprinted by permission from ASH-RAE Handbook*, 1993 Fundamentals, *Chap. 33.*)

TABLE 5.2 Equivalent Rectangular Duct Dimensions

Duct diameter, in	Rectangular size, in	Aspect ratio														
		1.00	1.25	1.50	1.75	2.00	2.25	2.50	2.75	3.00	3.50	4.00	5.00	6.00	7.00	8.00
6	Width	—	6													
	Height	—	5													
7	Width	6	8													
	Height	6	6													
8	Width	7	9	9	11											
	Height	7	7	6	6											
9	Width	8	9	11	11	12	14									
	Height	8	7	7	6	6	6									
10	Width	9	10	12	12	14	14	15								
	Height	9	8	8	7	7	6	6								
11	Width	10	11	12	14	14	16	17	17	18	21					
	Height	10	9	8	8	7	7	7	6	6	6					
12	Width	11	13	14	14	16	16	18	19	21	21	24				
	Height	11	10	9	8	8	7	7	7	7	6	6				
13	Width	12	14	15	16	18	18	20	19	21	25	24	30			
	Height	12	11	10	9	9	8	8	7	7	7	6	6			
14	Width	13	14	17	18	18	20	20	22	24	25	28	30	36		
	Height	13	11	11	10	9	9	8	8	8	7	7	6	6		
15	Width	14	15	17	18	20	20	23	25	24	28	28	35	36	42	
	Height	14	12	11	10	10	9	9	9	8	8	7	7	6	6	
16	Width	15	16	18	19	20	23	23	25	27	28	32	35	42	42	48
	Height	15	13	12	11	10	10	9	9	9	8	8	7	7	6	6
17	Width	16	18	20	21	22	25	25	28	27	32	32	35	42	49	48
	Height	16	14	13	12	11	11	10	10	9	9	8	7	7	7	6
18	Width	16	19	21	23	24	25	28	28	30	32	36	40	42	49	56
	Height	16	15	14	13	12	11	11	10	10	9	9	8	7	7	7
19	Width	17	20	21	23	24	27	28	30	30	35	36	40	48	49	56
	Height	17	16	14	13	12	12	11	11	10	10	9	8	8	7	7
20	Width	18	20	23	25	26	27	30	30	33	35	40	45	48	56	56
	Height	18	16	15	14	13	12	12	11	11	10	10	9	8	8	7
21	Width	19	21	24	26	28	29	30	33	33	39	40	45	54	56	64
	Height	19	17	16	15	14	13	12	12	11	11	10	9	9	8	8

22	Width	20	23	26	26	28	32	33	36	36	36	39	44	50	54	56	64
	Height	20	18	17	15	14	14	13	13	13	12	11	11	10	9	8	8
23	Width	21	24	28	26	30	32	35	36	36	39	42	44	50	54	63	64
	Height	21	19	16	17	15	14	14	13	13	12	12	11	10	9	9	8
24	Width	22	25	30	27	32	34	35	39	39	42	42	48	55	60	63	72
	Height	22	20	17	18	16	15	14	14	13	13	13	12	11	10	9	9
25	Width	23	25	30	29	32	36	38	39	42	46	46	48	55	60	70	72
	Height	23	20	17	19	16	15	15	14	14	13	13	12	11	10	10	9
26	Width	24	26	32	30	34	36	38	41	42	42	52	52	55	66	70	72
	Height	24	21	18	20	17	16	15	15	14	14	13	13	11	11	10	9
27	Width	25	28	33	30	36	38	40	41	45	45	49	52	60	66	70	80
	Height	25	22	19	20	18	17	16	15	15	14	14	13	12	11	10	10
28	Width	26	29	35	32	36	38	43	44	44	49	49	56	60	66	77	80
	Height	26	23	20	21	18	17	17	16	15	15	14	14	12	11	11	10
29	Width	27	30	35	33	38	41	43	44	48	48	53	56	65	72	77	88
	Height	27	24	20	22	19	18	17	16	16	15	15	14	13	12	11	11
30	Width	27	31	37	35	40	43	45	47	48	53	53	60	65	72	77	88
	Height	27	25	21	23	20	19	18	17	17	16	15	15	13	12	11	11
31	Width	28	31	39	35	40	43	45	50	51	56	56	60	70	78	84	88
	Height	28	25	22	23	20	19	18	17	17	16	16	15	14	13	12	11
32	Width	29	33	39	36	42	45	48	50	54	54	56	60	70	78	84	96
	Height	29	26	22	24	21	20	18	18	18	17	16	15	14	13	12	12
33	Width	30	34	40	38	44	47	50	52	54	60	60	64	75	78	91	96
	Height	30	27	23	25	22	21	19	19	18	18	17	16	15	13	13	12
34	Width	31	35	42	39	44	47	50	52	57	60	63	64	75	84	91	96
	Height	31	27	24	25	22	21	20	19	19	18	17	16	15	14	13	12
35	Width	32	36	42	39	46	50	53	55	55	63	63	68	75	84	91	104
	Height	32	28	24	26	23	22	20	20	19	18	18	17	15	14	13	13
36	Width	33	36	44	41	48	50	53	55	60	63	63	68	80	90	98	104
	Height	33	29	25	27	24	22	21	20	20	19	18	17	16	15	14	13
38	Width	35	39	47	44	50	54	58	61	63	67	72	72	85	96	105	112
	Height	35	31	27	29	25	24	23	22	22	21	20	19	18	17	16	14

SOURCE: Abstracted, with permission, from the ASHRAE Handbook, *1993 Fundamentals*, p. 32.8, Table 2.

is assumed to occur entirely at the fitting, and friction is excluded. Friction in the straight duct run is calculated from the centerlines of fittings. Where one fitting closely follows another, the data may be incorrect because such situations are unpredictable.

The dynamic pressure loss is calculated from the velocity pressure and a coefficient. Most coefficients have been determined experimentally—and a great deal of research has gone into this subject. The ASHRAE Handbook, *1993 Fundamentals,* Chap. 32,[3] contains coefficients for more than 100 fitting types. Dynamic losses will usually be a significant fraction of total duct pressure losses.

5.3.5 Fire and smoke control

Because a duct system can provide a path for the spread of smoke, hot gases, and fire within a building, fire protection must be designed into the duct system. In most buildings, this means providing fire and/or smoke dampers at all points where a duct penetrates a fire wall or smoke barrier, plus the installation of fire or smoke detectors at air-handling units. Most fire safety codes require compliance with National Fire Protection Association (NFPA) Standard 90A.[5] When smoke or fire is detected, the affected dampers close and the supply fans stop.

However, an "ordinary" HVAC system is not an engineered smoke management system, although such a system can be used for HVAC. An engineered smoke management system usually requires a higher airflow rate than a normal HVAC system and is provided with controls and dampers to evacuate smoke from the zones where it is being produced while providing a positive pressure in adjacent zones. For details of the design of these systems, see Ref. 6.

A fire damper is designed with either a blade or a drop-down curtain held open by a fusible link (Fig. 5.26). When the link melts, the damper closes automatically. The damper must be reopened manually by replacing the link. A smoke damper is held open by a motor operator which fails to the damper-closed position. A smoke detector is connected to close one or more dampers when smoke is present. When the detector is reset, the dampers will open. Combination fire and smoke dampers are also made (Fig. 5.27). Most building codes require that fire and smoke dampers be tested and rated by a nationally recognized laboratory such as Factory Mutual (FM) or Underwriters Laboratories (UL). Fire dampers are rated for 1-, 2-, or 3-h resistance to match the barrier in which they are installed. Smoke dampers are rated for temperature degradation and leakage.

Access doors in the structure and duct must be provided at all fire and smoke dampers to allow replacement of failed linkages.

Figure 5.26 Fire damper. (*Courtesy of Ruskin Mfg. Div. of Phillips Industries, Inc.*)

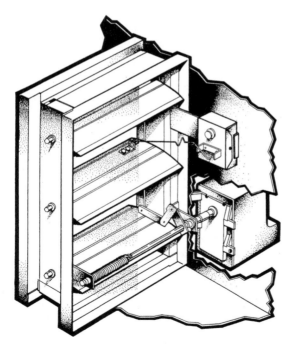

Figure 5.27 Combination fire and smoke damper. (*Courtesy of Ruskin Mfg. Div. of Phillips Industries, Inc.*)

5.3.6 Duct insulation

Most energy codes, many of which are derived from ASHRAE Standard 90A,[7] require thermal insulation of duct systems, plenums, and enclosures, with some exceptions. With or without insulation, there will be heat loss or gain as the air flows through the system, and this may make significant changes in the air quantities required for zones located some distance down the duct. The loss or gain through the wall of a duct section may be calculated by

$$Q = \left[\frac{UPL}{12} \left(\frac{t_e + t_l}{2} - t_a \right) \right] \tag{5.10}$$

and the leaving temperature may be calculated from

$$t_l = \frac{t_e(y - 1) + 2t_a}{y + 1} \tag{5.11}$$

where $y = \begin{cases} \dfrac{2.4AVd}{UPL} & \text{rectangular ducts} \\ \dfrac{0.6DVd}{UL} & \text{round ducts} \end{cases}$

A = cross-sectional area of duct, in^2
V = average velocity, ft/mm
D = diameter of duct, in
L = duct length, ft
Q = heat gain or loss, Btu/h
U = overall heat transfer coefficient of duct wall, Btu/(h · ft^2 · °F) (see Fig. 5.28)
P = perimeter of duct, in
d = density, lb/ft^3
t_e = temperature of air entering duct section
t_i = temperature of air leaving duct section
t_a = temperature of air surrounding duct section

Figure 5.28 shows heat transfer coefficients for insulated and uninsulated ducts. Note that when internal insulation is used, the external duct dimensions must be increased to compensate. Also, the roughness factor for internal duct lining may differ significantly from the "standard" roughness of 0.0003. In calling out duct sizes, it is common practice to present the net free duct size and add the internal insulation thickness to it, i.e., 24 × 12 in + 1 in AL. In some applications code requirements for indoor air quality may prohibit the use of internal duct insulation.

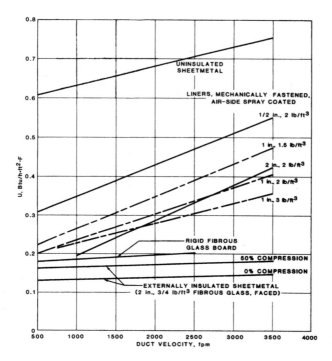

A. Rigid ducts

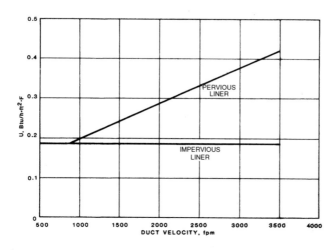

B. Flexible ducts

Figure 5.28 Duct heat transfer coefficients. (*Reprinted by permission from the ASHRAE Handbook,* 1997 Fundamentals, *Chap. 32, Fig. 13, p. 32.16.*)

5.3.7 Duct system leakage

Air leakage from the duct system can degrade system heating or cooling performance and can waste energy. The SMACNA manuals[4] describe jointing and sealing methods which will minimize leakage. If these are adhered to, the duct system will not leak significantly. However, unless properly provided for in specifications and installation, plenum walls and doors, piping, damper, and control device penetrations, and many equipment elements may contribute a large amount of leakage. Any air leak means a loss of energy—not only the thermal energy required to heat or cool the air but also the fan work required to move the air.

5.3.8 Duct design velocities

A very simple example of duct layout and sizing is shown in Figs. 5.29 and 5.30 and Table 5.4. Figure 5.29 is a plan of a partial duct system. The circled numbers identify nodal points. Numbers enclosed in squares identify fittings, each of which must be calculated separately by using the tables in the ASHRAE Handbook.[3] Data for diffusers and grilles are obtained from manufacturer's catalogs. Duct velocities and unit losses are obtained from the friction chart in Fig. 5.24 (no roughness corrections are used in this example). Table 5.4 is a tabulation of the static pressure losses for all components of the system. The velocity pressure is calculated from Eq. (5.9). When pressure losses have been calculated, a profile of the static pressure gradients can be drawn, as in Fig. 5.30. Where an unbalance exists at a junction, it will be necessary to adjust a balancing damper to compensate. Ideally, however, the system should be designed to make the use of balancing dampers minimal or unnecessary. This is especially true for VAV systems, since the effect of the balancing damper varies with the velocity changes.

5.3.9 Duct sizing

The three methods commonly used for sizing HVAC duct systems are equal-friction, velocity reduction, and static regain. A fourth method—constant velocity—is used for the design of systems which convey particulates.

With the equal-friction method, the ducts are sized for a constant pressure loss per unit length of duct.

The velocity reduction method is implemented by selecting an initial velocity at the fan discharge and then arbitrarily reducing the velocity at each junction (branch). The return duct is sized in a similar manner, with increasing velocities as the duct approaches the air-handling

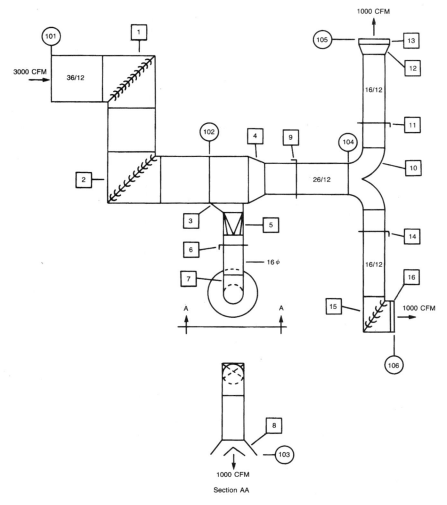

Figure 5.29 Duct sizing example.

unit. In practice, this can yield results similar to those obtained by the equal-friction method.

The static regain method is based on duct sizing such that the static pressure increase (static regain) at each takeoff offsets the pressure loss in the succeeding section of duct. Because static regain is achieved by decreasing the velocity, this method will result in larger ducts than the other methods. Properly designed, this method will result in approximately equal static pressures at the entrances to all branches.

In all three methods, the system should be designed and calculated as described below. Resize ducts as necessary to achieve an approxi-

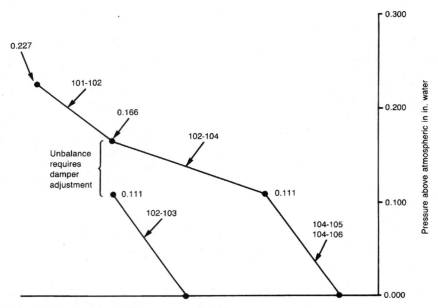

Figure 5.30 Static pressure gradient for Fig. 5.29.

mate balance at each junction. While balancing dampers can be used and are often needed, noise generated by the damper increases with increasing pressure reduction.

As can be seen, even a simple system takes considerable time to analyze and calculate. Computer programs are available.

Every duct system should be laid out and analyzed carefully to avoid balancing problems at installation. Under the worst conditions, it may be impossible to balance the system, or proper balancing may create unacceptable noise. At that point a retrofit will require much greater expense than the cost of proper design.

5.3.10 Duct pressure gradient analysis

The analysis and graphical presentation of the pressure gradients in a duct system can be a very useful tool for design and troubleshooting. Calculations are based on the equations for incompressible fluid flow, since air at normal HVAC pressures is essentially incompressible. The basic rules are:

1. All system losses are total pressure losses.

2. *Total pressure* equals *static pressure* plus *velocity pressure*—TP = SP + VP.

TABLE 5.3 Equivalent Spiral Flat Oval Duct Dimensions

Duct diameter, in	Major axis (a), in / Minor axis (b), in															
	3	4	5	6	7	8	9	10	11	12	14	16	18	20	22	24
5	8															
5.5	9	7														
6	11	9														
6.5	12	10	8													
7	15	12	10	8												
7.5	19	13	—	9												
8	22	15	11	—												
8.5		18	13	11	10											
9		20	14	12	—	10										
9.5		21	18	14	12	—										
10			19	15	13	11										
10.5			21	17	15	13	12									
11				19	16	14	—	12								
11.5				20	18	16	14	—								
12				23	20	17	15	13								
12.5				25	21	—	—	15	14							
13				28	23	19	17	16	—	14						
13.5				30	—	21	18	—	16	—						
14				33	—	22	20	18	17	15						
14.5				36	—	24	22	19	—	17						
15				39	—	27	23	21	19	18						
16				45	—	30	—	24	22	20	17					
17				52	—	35	—	27	24	21	19					
18				59	—	39	—	30	—	25	22	19				

SOURCE: Abstracted, with permission, from the ASHRAE Handbook, *1993 Fundamentals*, p. 32.10.

3. Total pressure loss means energy loss. Fan work energy must be used to make up this loss.

4. Total pressure must be used in duct calculations.

5. Velocity and velocity pressure are related by Eq. (5.9).

6. VP is always positive in the direction of flow.

7. TP decreases in the direction of flow of turbulence and friction.

8. Methods of calculating pressure losses are described in Sec. 5.3.9.

With reference to the fan characteristics we may use the following terms:

FTP = fan total pressure

FSP = fan static pressure

FVP = fan velocity pressure at the fan outlet

TABLE 5.4   Calculations for Fig. 5.29

Section from/to node	Fitting number	Fitting type	Airflow, ft³/min	Duct size	Velocity, ft/min	Velocity pressure, in H₂O	C_0	Duct length, ft	Duct DP/100 ft in H₂O	Fitting/ duct DP in H₂O	Section DP in H₂O
101 102	1	Elbow	3000	36/12	1180	0.087	0.15			0.013	
	2	Elbow	3000	36/12	1180	0.087	0.15			0.013	
		Duct	3000	36/12	1180	0.087		40	0.087	0.035	0.061
102 103	3	Branch	1000	16/12	830	0.043	0.32			0.014	
	5	Trans.	1000	16/12-16	830	0.043	0.25			0.011	
	6	Damper	1000	16	740	0.034	0.20			0.007	
	7	Elbow	1000	16	740	0.034	0.22			0.061	
	8	Diffuser	1000	16 neck							
		Duct	1000	16	740	0.034		20	0.053	0.011	0.111
102 104	3	Main	3000/2000	36/12	1180	0.087	0.28			0.024	
	4	Trans.	2000	36/26/12	1180/1050	0.069	0.05			0.003	
	9	Damper	2000	26/12	1050	0.069	0.04			0.003	
		Duct	2000	26/12	1050	0.069		30	0.083	0.025	0.055
104 105	10	Branch	2000/1000	26/16/12	1050/830	0.069	0.52			0.036	
		Duct	1000	16/12	830	0.043		20	0.070	0.014	
	11	Damper	1000	16/12	830	0.043	0.04			0.002	
	12	Trans.	1000	16/24/12	830/560	0.043	0.15			0.006	
	13	Grille	1000	24/12						0.053	0.111
104 106	10	Branch	2000/1000	26/16/12	1050/830	0.069	0.52			0.036	
		Duct	1000	16/12	830	0.043		20	0.070	0.014	
	14	Damper	1000	16/12	830	0.043	0.04			0.002	
	15	Elbow	1000	16/24/12	830/560	0.043	0.15			0.006	
	16	Grille	1000	24/12						0.053	0.111

$$FTP = FTP_{outlet} - FTP_{inlet}$$

$$FSP = FTP - FVP$$

The basic plotting grid for duct pressure gradients is shown in Fig. 5.31. The zero reference is atmospheric pressure, usually within the conditioned space. The scale for positive and negative pressure values should be adjusted for the actual values encountered and is typically in inches of water.

Figure 5.32 is a pressure gradient diagram for a very simple system consisting of an inlet duct, a fan, and an outlet duct. Notice the inlet and outlet losses when the velocity goes to zero.

Very few systems are as simple as that shown in Fig. 5.32. Figure 5.33 shows a more typical HVAC system with 100 percent outside air, filter, cooling and heating coils, and a simple supply duct. Notice, again, that VP is always positive while TP and SP will usually be negative on the fan inlet side and positive on the fan outlet side. For velocities up to 2000 ft/min the pressure losses in ducts and plenums are usually small compared to losses due to coils and filters. Poorly designed duct and fittings may have high losses due to turbulence.

Figure 5.33 uses a simple supply duct with a single diffuser. Most duct systems are more complex, with several branch ducts and diffusers. In a carefully designed system each branch must be calculated from the outlet back to its junction with the main. Then the main duct must be adjusted to provide the required static pressure at the branch.

Figure 5.31 Basic plotting grid for duct pressure gradient diagrams — values typically in inches of water.

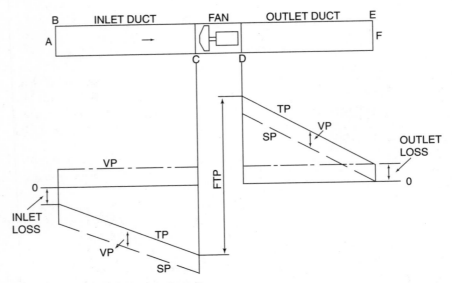

Figure 5.32 Basic pressure gradient diagram.

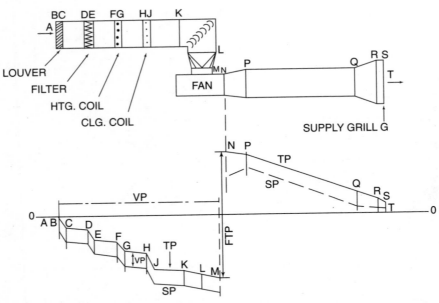

Figure 5.33 Pressure gradients for simple system with fan, coils, filters, etc.

Balancing dampers can compensate only for small design errors. If the design error is too great, the increase in static required to compensate may make it impossible to use the specified fan. There are also "system effects," discussed below, which affect fan selection and performance.

Fan selection for a given system is determined by reference to two "curves." The first is the manufacturer's curve. A generalized curve for a backward-inclined centrifugal fan is shown in Fig. 5.34. The best point of selection is on the pressure curve to the right of its peak, over the range of maximum efficiencies—for this fan in the range of 50 to 70 percent of wide open volume. For a specific fan scales will be volume in cfm and pressures in inches of water. If we plot a duct system curve on the same scale, as shown in Fig. 5.35, that curve will be a second-degree arc in accordance with the fluid flow laws which state that friction losses vary as the square of the velocity. If we overlay the two curves, we can determine the proper fit of fan and duct system. This is essentially what is done when selecting a fan from the catalog for a specific pressure and cfm, with due regard for efficiency.

System effects have been mentioned and are described in some detail in AMCA 201. These relate to the fan inlet and outlet conditions imposed by the system. As compared to the test conditions used in rating the fans, the field conditions are usually less than ideal and the fan performance is affected. The result is shown in Fig. 5.36. When the system is installed and operated the system effects may increase the effective pressure losses, with the result that the fan operating point moves back and up along the pressure curve, resulting in a decrease in supply air volume. The typical field adjustment is to increase the

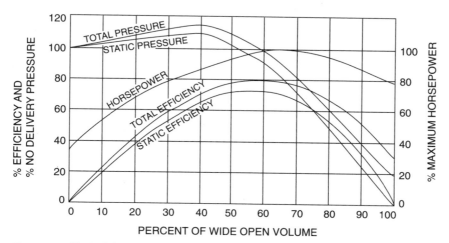

Figure 5.34 Typical fan curves, backward-inclined fan.

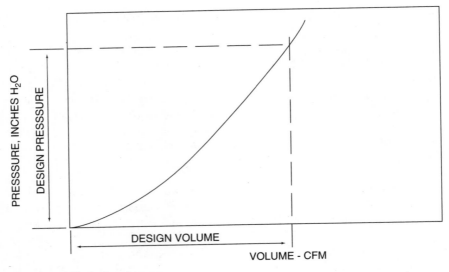

Figure 5.35 Typical system curve.

fan speed, which moves the pressure curve to the right, to get back to the design volume. This may result in overloading the motor or requiring a motor replacement. A careful pressure gradient analysis of the system will often indicate some simple duct changes which will allow the system to operate at or near the original design point.

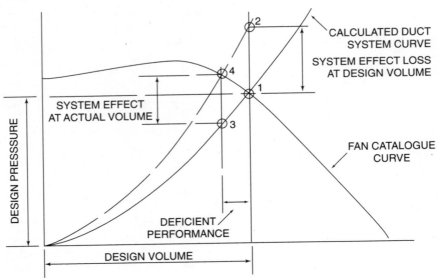

Figure 5.36 Deficient system performance. (*Reprinted by permission from AMCA Publication 201, Fans and Systems.*)

5.4 Diffusers, Grilles, and Registers

A *diffuser* is an outlet through which the air supply passes from the duct system to the space being conditioned. A *grille* is a covering for an opening through which air passes. A *register* is a grille with a damper for volume control. While these are the technically correct definitions, supply outlets in walls and floors commonly are referred to as grilles or registers and ceiling outlets as diffusers (although manufacturers also catalog sidewall diffusers). A diffuser may have a volume damper but is still called a diffuser. Return-air inlets are always called grilles or registers.

5.4.1 Air diffusion

The airstream leaves a supply outlet as a jet. This air jet creates motion in the surrounding air because of friction, which induces some of the surrounding air to become part of the jet. As a result of this induction and friction, the jet becomes larger and more diffuse and slows down, until finally it loses all noticeable form and velocity. In a free jet—one not influenced by cross drafts, adjacent surfaces, or obstructions—this effect is predictable with a fair degree of accuracy and is dependent on the initial velocity, size, and temperature.

The *throw* is the distance that the stream travels before velocity is reduced to some defined speed, usually 50, 75, or 100 ft/min. The threshold at which people sense air movement is about 50 ft/min. The *drop* is the distance that the lower edge of a horizontal jet falls below the outlet at the end of its throw. The *spread* is the divergence of the airstream after it leaves the outlet. Figure 5.37 shows these effects for an ideal case. Most HVAC applications are not ideal.

The throw, spread, and drop are affected by the outlet design. All grilles and diffusers have a fixed or adjustable pattern which controls the initial direction and spread of the airstream. Ceiling diffusers spread the air out along the ceiling or throw it downward in a tight jet, or do something in between. Grilles may have two-way spread (one plane only) or four-way spread (both planes). Adjusting for wide spread decreases the throw. The *aspect ratio* (ratio of length to width) of an outlet also affects the jet properties. The smaller dimension tends to govern the throw.

When the outlet is located in the wall near a floor or ceiling (or in the ceiling or floor near a wall), the jet tends to hug the adjacent surface. Cold air supply streams tend to fall because of higher air density, while hot airstreams are buoyant and tend to rise. Thus, ideally, warm air should be supplied horizontally from low sidewall outlets, and cold air from high sidewall outlets or ceiling diffusers ad-

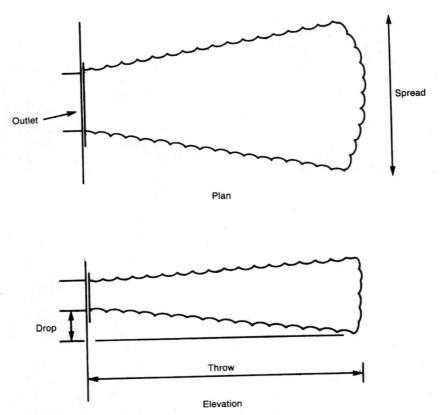

Figure 5.37 Throw, drop, and spread in a free air jet.

justed to discharge the air horizontally. One of the most satisfactory compromises, where the air supply temperature varies seasonally, is a floor outlet near the outside wall. Interferences, such as beams, columns, and light fixtures, can degrade the air distribution pattern.

5.4.2 Selecting diffusers

The most important considerations in diffuser and/or grille selection are (1) the air motion in the occupied level, (2) air temperature gradient in the space (stratification), (3) noise, and (4) pressure drop. Manufacturers' catalogs provide information on the throw, drop, noise level, and pressure drop for a specified airflow rate and a variety of sizes and types of outlets. The most common error in diffuser installation involves location and throw such that the airstream is forced down a wall into the occupied zone while still at a high velocity, resulting in drafty and uncomfortable conditions. Also avoid locating return grilles so that the supply air "short-circuits" to the return instead of diffusing through the space. Noise levels should always be suitable for the occupancy and use of the space (see Chap. 20). Many manufacturers and the ASHRAE Handbook[8] provide detailed discussions of air distribution theory.

5.5   Louvers

A *louver* is a type of grille which covers an opening used for air intake or air transfer. A major application is the storm louver, used in the outside wall of a building for air intake or exhaust. This device is designed to keep out rain, etc., while allowing air to pass through (Fig. 5.38).

Within a building the storm resistance is not needed; but where the louver is used for air transfer through doors or walls, the louver blades should be arranged to be sight-proof. Exterior louvers are usually selected for a face velocity of 500 to 700 ft/min, while transfer louvers and grilles are selected for a low pressure drop, usually resulting in a face velocity of 200 to 300 ft/min.

5.6   Dampers

A *damper* is an adjustable obstruction in a duct used to control or balance airflow. Types of dampers used in HVAC work include butterfly, multiblade opposed-blade and parallel-blade, splitter, gate or slide, and shutter. Criteria for damper selection include leakage when

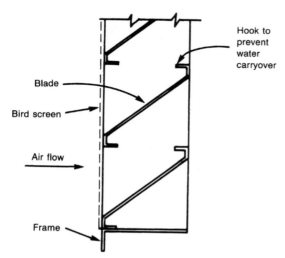

Figure 5.38 Cross section of a storm louver.

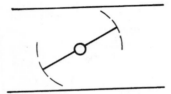

Figure 5.39 Butterfly damper.

closed, control characteristics, and cost. Dampers may be operated manually or by means of motor operators and can be pneumatic, hydraulic, electric, or electronic.

Butterfly dampers (Fig. 5.39) are single-blade devices, pivoting around a central axle, for a round or rectangular duct. In quality and cost, they range from shop-fabricated from a piece of sheet metal—with a high leakage rate when closed—to a manufactured damper and frame with essentially no leakage when closed. The former class predominates because this type of damper is used mostly for balancing where tight shutoff is not required. Splitter dampers (Fig. 5.40) are used for balancing at Y junctions. They are not recommended; butterfly dampers in each branch are preferred. Splitter dampers are fabricated as part of the fitting.

Gate or slide dampers (Fig. 5.41) are used primarily for shutoff in exhaust systems handling particulates. When open, the damper is completely out of the airstream; when the damper is closed, there is very little leakage.

Shutter dampers (Fig. 5.42) are used mostly for fire and smoke control. The advantage of this design is that there is no obstruction to airflow in the open position.

A multiblade damper consists of several butterfly dampers arranged side by side, to work in concert. Multiblade dampers are sometimes used for manual balancing, but most often are used for automatic control of airflow in two-position or modulating modes. Parallel-blade dampers (Fig. 5.43) are arranged so that all the blades turn simultaneously in the same direction. Opposed-blade dampers (Fig. 5.44) are arranged so that adjacent blades turn in opposite directions. The different arrangements lead to different control characteristics as the dampers modulate, as can be seen in the figures. In the half-closed

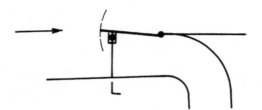

Figure 5.40 Splitter damper.

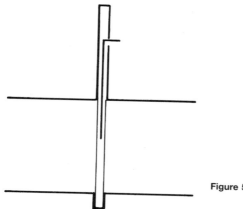

Figure 5.41 Gate (slide) damper.

position, the parallel-blade damper still has a fairly large opening for flow, although all the air is directed toward one side. The opposed-blade damper has a much better throttling characteristic, and the air is uniformly distributed. For a discussion of the control characteristics of the two damper types, see Sec. 8.3.3.2. The opposed-blade damper is preferred for modulating control, while the parallel-blade damper is used for two-position control because it is less expensive. These dampers are available with standard (relatively high) or low leakage ratings.

Special dampers are also made but generally fall into one of the above classes.

Manual damper operators are usually simple lever arms with some means of indicating damper position. On very large dampers, gear operators may be used. Automatic operators include (1) pneumatic "motors," piston-and-spring systems using low-pressure air (18 to 25 lb/in²); (2) hydraulic pistons using high-pressure air (100 lb/in² or more) or some other hydraulic fluid; (3) electric motors using two-position, floating, or modulating action; and (4) electronic motors. The most commonly used are pneumatic, valued for simplicity and low cost, and electric.

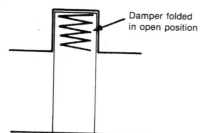

Damper folded
in open position

Figure 5.42 Shutter damper.

Figure 5.43 Parallel-blade damper.

5.7 Filters

A *filter* is a device for removing contaminants from a fluid stream — in HVAC, the air system. While most filters are designed to remove dust to varying degrees, some filters remove gaseous contaminants. Filters are tested and rated under ASHRAE standards.[9]

Particulate removal filters are classified under three standards: weight arrestance, dust spot, and di-octyl-phthalate (DOP) smoke. Weight-arrestance filters are rated on the basis of the removal of a portion of a standard sample of dust measured by weight. This, of

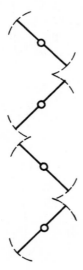

Figure 5.44 Opposed-blade damper cross-sectional view.

course, puts the emphasis on removing the large particles and considerable amounts of fine dust that get by this type of filter, which is used primarily in residential and light commercial applications. Most types of "throwaway" and "cleanable" filters are in this class. These are also called *roughing* filters when used ahead of other types.

Dust spot filters (Fig. 5.45) are tested and rated by introducing a standard sample of "atmospheric" dust upstream of the filter. Downstream of the filter, a fine paper filter is installed, and after a known amount of dust has been fed into the system, the paper filter is tested for a decrease in light transmission due to dust which was not removed by the test filter. Currently, many specifications require filters with 40 to 50 percent dust spot efficiencies, which will provide a high degree of cleanliness. Electrostatic filters (Fig. 5.46) include fine wires with a high-voltage charge installed in the airstream. This puts a static charge on the particles of dust, which are then attracted to oppositely charged plates. Downstream roughing filters may be added to pick up coagulates. Dust spot efficiencies of 70 to 80 percent or higher can be obtained. The filters must be washed with detergent periodically.

Filters rated under the DOP test (Fig. 5.47) are used for "absolute" filtration, where over 99 percent of dust particles over 0.3 μm are removed; they are used in hospitals, laboratories, and clean rooms.

Figure 5.45 Dust spot filter. (*Courtesy of American Air Filter.*)

Figure 5.46 Electrostatic filter. (*Courtesy of American Air Filter.*)

Tables 5.5, 5.6, and 5.7 show how the various types of filter materials compare. To obtain rated efficiencies, the filters must be properly installed in frames designed to minimize leakage. For very high efficiencies, all joints are gasketed.

The pressure drop through a filter is closely related to its efficiency—higher efficiencies require higher pressure drops, with a consequent penalty in fan horsepower. As the filter loads up with dirt, its efficiency and pressure drop increase. (As the filter gets dirtier the

Figure 5.47 DOP filter. (*Courtesy of Cambridge Filter Corp.*)

TABLE 5.5 Comparative Performance of Viscous Impingement and Dry Media Filters

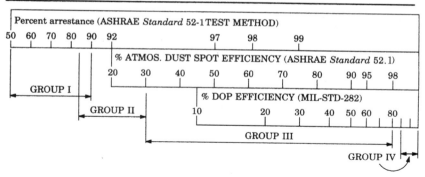

Group I
Panel-type filters of spun glass, open cell foams, expanded metal and screens, synthetics, textile denier woven and nonwoven, or animal hair.

Group II
Pleated panel-type filters of fine denier non-woven synthetic and synthetic-natural fiber blends, or all natural fiber.

Group III
Extended surface supported and non-supported filters of fine glass fibers, fine electret synthetic fibers, or wet-laid paper of cellulose-glass or all-glass fibers.

Group IV
Extended-area pleated HEPA-type filters of wet-laid ultra fine glass fiber paper. Biological grade air filters are generally 95% DOP efficiency; HEPA filters are 99.97 and 99.99%; and ULPA filters are 99.999%.

Notes:
1. Group numbers have no significance other than their use in this table.
2. Correlations between the test methods shown are approximations for general guidance only.

SOURCE: Reprinted by permission from ASHRAE Handbook, *1996 HVAC Systems and Equipment,* p. 24.7.

added dirt helps to increase the filter efficiency while at the same time it increases the resistance to air flow.) Every filter bank should be provided with a good pressure-drop indicator to remind operating personnel to clean or change the filters. Typical pressure ranges (clean to dirty) are as follows: weight arrestance, 0.15 to 0.40 in at 500 ft/min; dust spot, 0.25 to 1.0 in at 500 ft/min; and DOP, 1.0 to 2.0 in at 250 ft/min.

For removal of gaseous contaminants and odors, filters containing various absorbents are used. The most common is activated carbon. Another effective material is activated alumina impregnated with potassium permanganate. Wash-down "scrubbers" are also used. For grease removal (in kitchen hoods, etc.), cleanable metallic filter media are used. Contaminant filters must be cleaned or disposed of properly in accordance with toxic waste regulations.

TABLE 5.6 Performance of Dry Media Filters

Filter media	ASHRAE weight arrestance, %	ASHRAE atmospheric dust spot efficiency, %	MIL-STD 282 DOP efficiency, %	ASHRAE dust-holding capacity, g/(1000 ft³/min) per cell
Finer open cell foams and textile denier nonwovens	70–80	15–30	0	180–425
Thin, paperlike mats of glass fibers, cellulose	80–90	20–35	0	90–180
Mats of glass fiber multiply cellulose, wool felt	85–90	25–40	5–10	90–180
Mats of 5- to 10-μm fibers, ¼ to ½ in (6 to 12 mm) thickness	90–95	40–60	15–25	270–540
Mats of 3- to 5-μm fibers, ¼ to ½ in (6 to 20 mm) thickness	>95	60–80	35–40	180–450
Mats of 1 to 4-μm fibers, mixtures of various fibers and asbestos	>95	80–90	50–55	180–360
Mats of 0.5- to 2-μm fibers (usually glass fibers)	NA	90–98	75–90	90–270
Wet-laid papers of mostly glass and asbestos fibers <1-μm diameter (HEPA filters)	NA	NA	95–99.999	500–1000
Membrane filters (membranes of cellulose acetate, nylon, and the like, having holes 1-μm diameter or less)	NA	NA	~100	NA

NA: Indicates that test method cannot be applied to this level of filter.
SOURCE: Reprinted by permission from ASHRAE Handbook, *1983 Equipment*, Chap. 10, Table 2, p. 10.7.

TABLE 5.7 Performance of Renewable Media Filters (Steady-State Values)

Description	Type of media	ASHRAE weight arrestance, %	ASHRAE atmospheric dust-spot efficiency, %	ASHRAE dust-holding capacity, g/ft^2	Approach velocity, ft/min
20 to 40 μm glass and synthetic fibers, 2 to 2½ in thick	Viscous impingement	70–82	<20	60–180	500
Permanent metal media cells or overlapping elements	Viscous impingement	70–80	<20	NA (permanent media)	500
Coarse textile denier nonwoven mat, ½ to 1 in thick	Dry	60–80	<20	15–70	500
Fine textile denier nonwoven mat, ½ to 1 in thick	Dry	80–90	<20	10–50	200

SOURCE: Reprinted by permission from *ASHRAE Handbook 1996, HVAC Systems and Equipment,* Chap. 24, Table 1, p. 24.8.

5.8 Air Distribution with High Flow Rates

It has been noted that airflow rates in excess of 3 $(ft^3/min)/ft^2$ require special care in the design of the air distribution system. Clean rooms and mainframe computer rooms typically require 8 to 10 $(ft^3/min)/ft^2$, and these airflow rates cannot be handled by conventional means. A typical computer room installation is described in Sec. 11.3.3.

In a clean room, the high airflow rate is caused by the need to maintain the cleanliness level and not by high heat gains; a ΔT of 5 to 10°F is typical. The laminar-flow method of distribution is commonly used, either horizontal or vertical. In the vertical arrangement, most of the ceiling, except for lights, is used for airflow, with DOP filters as the diffusers (Fig. 5.48). Return air is removed at the floor line. Then airflow velocity is quite uniform and low across the entire room. Diffuser outlet velocities will be 30 to 40 ft/min.

For horizontal laminar flow, one entire sidewall is used for supply and the opposite wall for return. Again, outlet velocities will be low, usually less than 50 ft/min, which is the threshold level for human sensing. Because a low ΔT is needed in the room, a separate air-handling unit for temperature and humidity control is sometimes provided (Fig. 5.49). Separate fan and filter "modules" are used for room air circulation.

5.9 Stratification

Stratification is the separation of air into layers having different temperatures. The HVAC designer is concerned with two kinds of stratification: that which occurs in the conditioned space as a result of incorrect air distribution, and that which occurs in AHUs and ducts as a result of inadequate mixing.

The first affects comfort, and is discussed elsewhere in this chapter. The second is a result of system geometry. Both can be corrected or minimized by proper design.

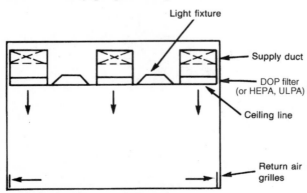

Figure 5.48 Vertical laminar-flow air distribution.

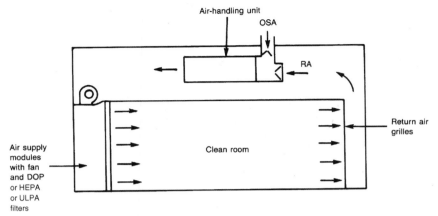

Figure 5.49 Horizontal laminar-flow air distribution with separate AHU for temperature and humidity control.

The primary cause of stratification in an AHU is inadequate mixing of outside and return airstreams. The typical "mixing box" does not typically provide good mixing. Several situations are shown in Fig. 5.50. Figure 5.50a illustrates the worst case. The two airstreams enter side by side, and essentially no mixing occurs. This lack of mixing will

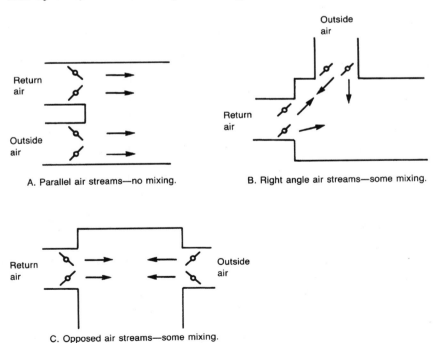

Figure 5.50 Typical mixing-box configurations.

carry through filters, coils, and even fans, creating problems with sensing and control as well as causing nuisance trips of freeze protection sensors — which are not really nuisance trips, because otherwise a portion of the coil might freeze.

Figure 5.50*b* shows a somewhat better condition but only if parallel-blade dampers are used, resulting in poor control of volume flows. Figure 5.50*c* gives still better results and allows the use of opposed-blade dampers. However, if the distance between the inlets is more than 4 or 5 ft, there will still be a great deal of stratification.

A sure way to eliminate stratification is by means of a mechanical "mixer" (Fig. 5.51), which imparts rotation and turbulence to the air-stream, or by means of propeller fans, which are mounted to blow across the duct or plenum, as in Fig. 5.52. Note that good mixing is the result of turbulence which creates some energy loss in the air-stream and increases the pressure requirement at the supply fan.

5.10   Noise Control

Noise is generated in HVAC systems by motors and mechanical devices, principally fans and pumps, as well as by the airstream itself. The subject of acoustics and noise control is discussed in Chap. 20.

Figure 5.51 Mechanical air mixer, the Air Blender. (*Manufactured by Blender Products, Inc.*)

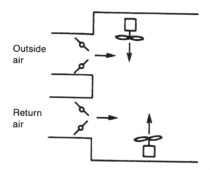

Figure 5.52   Using propeller fans for air mixing.

5.11   Indoor Air Quality

The issue of indoor air quality[10] has taken on much significance in recent years as the effects of air quality on human health and even life safety have become better known. There are many factors associated with air quality. Concern is for airborne particulates and for gaseous contaminants; some may be introduced from the outside, but many are generated inside by habitation, by process, or by off-gassing of synthetic construction or furnishing materials. Many office machines give off vapors or particulates which may offend the occupants. In recent years, smoking has become a less common problem than it once was.

An extreme case of poor indoor air quality may lead to a facility being diagnosed as having "sick building syndrome." In such a case, occupants may object to odors or moisture levels, or may complain of irritated air passageways, irritated eyes, dry, itching skin, workplace induced illness, etc. Poor lighting sometimes exacerbates the perception of a poor air quality condition.

Traditionally, internally generated air contaminants in buildings have been diluted to within acceptable levels by the introduction of at least a percentage of outside air into a generally recirculating, ventilating air stream. In the past when buildings were not so air tight, infiltration also helped to maintain an acceptable indoor environment.

In current practice, designers should regularly address the issues of indoor air quality in the building HVAC system design. All HVAC systems should include capability for the introduction of some outdoor air, either in minimum quality based on occupancy (15 cfm/person) or based on direct measurement of a pollutant of concern, such as carbon dioxide (800–1000 ppm maximum). If ambient outdoor air poses its own problems, special filtration may be required. Designers should become familiar with local conditions, expectations, and reasonable solutions. The ASHRAE publication "Ventilation for Acceptable Indoor Air Quality" treats the subject in some detail, and offers different approaches for determining required ventilation rates for air quality. The topic is not without controversy.

5.12   Summary

In this chapter we discussed air transport systems and air distribution, including fans, ductwork, grilles, registers, diffusers, dampers, and filters. Air-handling systems are discussed in Chap. 11.

References

1. Air Moving and Conditioning Association (AMCA), Inc., Standard 210-85, *Laboratory Method* of *Testing Fans for Rating,* 1985, Arlington Heights, IL.
2. AMCA, No. 201-AMCA Fan Application Manual, *Fans and Systems,* 1990.
3. ASHRAE Handbook, *1997 Fundamentals,* Chap. 32, "Duct Design."
4. Sheet Metal and Air Conditioning Contractors' National Association (SMACNA), Standards for duct construction, several volumes, 1971 to 1996, Vienna, VA.
5. National Fire Protection Association, national fire codes, several volumes, issued annually, Quincy, MA.
6. J. H. Klote and James A. Milke, Jr., *Design of Smoke Management Systems,* 1992 [available from ASHRAE/SFPE (Society of Fire Protection Engineers)], Atlanta, GA.
7. ASHRAE, Standard 90.1-1989, *Energy Conservation in New Building Design.*
8. ASHRAE Handbook, *1997 Fundamentals,* Chap. 31, "Space Air Distribution."
9. ASHRAE, Standard 52.1-1992, *Method of Testing Air-Cleaning Devices Used in General Ventilation for Removing Particulate Matter.*
10. ASHRAE Proceedings, IAQ (Indoor Air Quality), "Managing Indoor Air Quality for Health and Energy Conservation," 1986.

6

Design Procedures: Part 4

Fluid-Handling Systems

6.1 Introduction

All air-handling units (AHUs) and many terminal units, if they are
not self-contained, require a source of heating and/or cooling energy.
This source is called a *central plant,* and the means by which thermal
energy is transferred between the central plant and the AHU is usually a fluid conveyed through a piping system. The fluids used in
HVAC practice are steam, hot or cold water, brine, refrigerant, or a
combination of these. The equipment used to generate the thermal
energy is described in Chap. 7. In this chapter we discuss the transport systems.

6.2 Steam

Steam is water in vapor form. Because it expands to fill the piping
system, steam requires no pumping except for condensate return and
boiler feed. The specific heat of water vapor is quite low, but the latent
heat of vaporization is high. As a result, steam conveys heat very
efficiently.

Steam may be used directly at the AHU (in steam-to-air, finned-
tube coils), or a steam-to-water heat exchanger may be used to provide
the hot water used in AHU coils or in radiation. Steam radiation is
also employed. When used directly, steam pressures are usually 15
lb/in^2 gauge or less. When used with a heat exchanger, steam pressures up to 100 lb/in^2 gauge are common. Higher pressures allow

smaller piping but create expansion and support problems. Industrial plants often use high-pressure steam.

6.2.1 Steam properties

Table 6.1 shows thermodynamic properties of water at saturation temperatures and corresponding pressures from 0 to 250°F. Complete tables in the American Society of Mechanical Engineers (ASME) steam tables[1] cover a range from 32 to 700°F. Other tables cover superheated steam. The ASHRAE Handbook *Fundamentals*[2] extends the "at saturation" table down to −80°F.

The table indicates that there is a correspondence between saturation temperature and absolute pressure. Thus, the normal (sea-level) boiling point of 212°F corresponds to the standard sea-level pressure of 14.71 lb/in². At higher altitudes (and lower atmosphere pressures), the boiling temperature decreases until in Albuquerque, New Mexico, or Denver, Colorado, 1 mi above sea level, it takes 4 or 5 min to boil a 3-min egg.

The steam property of greatest interest to the HVAC designer is enthalpy, particularly the enthalpy of evaporation, or the latent heat of vaporization h_{fg}. This is the amount of heat, in Btu per pound, which

TABLE 6.1 Thermodynamic Properties of Water at Saturation

Temp., °F	Absolute pressure, lb/in²	Specific volume, ft³/lb			Enthalpy, Btu/lb		
		Sat. liquid v_f	Evap. v_{fg}	Sat. vapor v_g	h_f	Evap. h_{fg}	Sat. vapor h_g
0	0.018502	0.01743	14,797	14,797	−158.94	1,220.00	1,061.06
15	0.039598	0.01745	7,139	7,139	−151.76	1,219.43	1,067.67
30	0.080859	0.01747	3,606	3,606	−144.35	1,218.61	1,074.26
32	0.088643	0.01747	3,302	3,302	−143.35	1,218.49	1,075.14
32	0.08865	0.01602	3,302.07	3,302.09	−0.02	1,075.15	1,075.14
45	0.14755	0.01602	2,035.91	2,035.92	13.05	1,067.79	1,080.84
60	0.25635	0.01604	1,206.30	1,206.32	28.07	1,059.32	1,087.39
75	0.43009	0.01606	739.42	739.44	43.06	1,050.85	1,093.92
90	0.69890	0.01610	467.52	467.53	58.04	1,042.36	1,100.40
105	1.10304	0.01615	304.08	304.10	73.02	1,033.81	1,106.83
120	1.69473	0.01620	202.98	202.99	88.00	1,025.20	1,113.20
135	2.54053	0.01627	138.74	138.76	102.99	1,016.52	1,119.50
150	3.77283	0.01634	96.93	96.94	117.98	1,007.73	1,125.72
165	5.3423	0.01642	69.084	69.101	133.00	998.84	1,131.84
180	7.5194	0.01651	50.155	50.171	148.04	989.81	1,137.85
195	10.3958	0.01660	37.035	37.052	163.10	980.63	1,143.73
212	14.7097	0.01671	27.763	27.780	180.20	970.03	1,150.23
220	17.2013	0.01677	23.118	23.134	188.27	964.95	1,153.22
230	20.7961	0.01684	19.355	19.372	198.37	958.52	1,156.89
240	24.9869	0.01692	16.298	16.314	208.49	952.00	1,160.48
250	29.8457	0.01700	13.798	13.815	218.63	945.36	1,163.99

SOURCE: Abstracted from ASHRAE Handbook, *1997 Fundamentals*, Chap. 6, Table 3. Used by permission.

must be added to change the state of the water from liquid to vapor with no change in temperature. This same amount is removed and used, in a heat exchanger, when steam is condensed. Note that while liquid water has an enthalpy change of about 1 Btu/lb per degree of temperature change and steam has much less than that, the change-of-state enthalpy is 970 Btu/lb at 212°F. This is what makes steam so efficient as a conveyor of heat. In calculating the steam quantity (pounds per hour) required for a specific application, use the latent heat h_{fg}.

Steam *quality* refers to the degree of saturation in a mixture of steam and free water. As indicated in Table 6.1, there is a saturation pressure (or "vapor" pressure) corresponding to each absolute temperature. When the pressure and temperature match, the steam is said to be *saturated*, with a quality of 100 percent. When steam flows in a piping system, there is always some heat loss through the pipe wall, with a consequent reduction in temperature. If the steam was initially saturated, some will condense into waterdroplets that will be carried along with the flow. Then the steam quality will be less than 100 percent. Steam containing free water is *wet steam*. The free moisture can cause problems in some types of equipment, such as turbines. Steam lines must be sloped downward in the direction of flow, so that condensed water can be carried along to a point where it can be extracted. When the steam temperature exceeds the saturation temperature, the steam is *superheated*. Superheated steam is useful where free moisture is to be avoided, such as in some turbines.

6.2.2 Pressure reduction

When steam is distributed from a central plant, it may be desirable to use higher pressures for distribution, resulting in smaller piping. Then it is usually necessary to use *pressure-reducing valves* (PRVs) to provide a suitable point-of-use pressure. A typical pressure-reducing station is shown in Fig. 6.1. To provide better control, it is common practice to use two PRVs in parallel, one sized for one-third and the other sized for two-thirds of the load, respectively, and sequenced so that the smaller valve opens first. This allows the larger valve to work against smaller pressure differentials, which helps to avoid wire drawing of the valve seat at low loads. A manual bypass with a globe valve is provided for emergency use. The PRV should have an internal or external pilot, for accurate control of downstream pressure regardless of upstream changes.

The maximum pressure drop through any steam PRV at the design flow rate is about one-half of the entering pressure; more exactly, the ratio of downstream to upstream pressure cannot be less than 0.53.

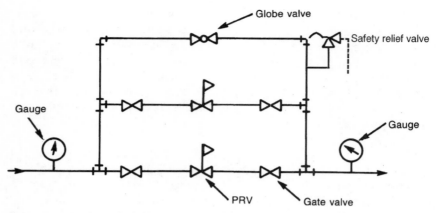

Figure 6.1 Pressure-reducing station.

This is due to the physical laws governing flow of compressible fluids through orifices. If greater pressure reductions are required, it is necessary to use two or more stages, as shown in Fig. 6.2.

6.2.3 Steam condensate

Condensate is usually returned to the boiler for reuse. In small systems, this can sometimes be done by gravity. In most systems, pumping is required. The condensate flows by gravity to a collecting tank from which it is pumped directly to the boiler or to a boiler feed system, as described in Chap. 7.

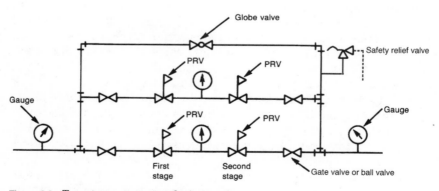

Figure 6.2 Two-stage pressure-reducing station.

6.3 Water

Water is used extensively in modern cooling and heating practice because it is an effective heat transport medium and because it is considered simple to deal with. Because the water system can be essentially closed, there are fewer corrosion and water treatment problems than with steam. Except in high-rise buildings, system static pressures are low and temperature changes are not severe, allowing the use of low-cost materials and simple piping support systems. An exception is high-temperature hot water, discussed later in this chapter.

6.3.1 Water properties

Refer to Tables 6.1 and 6.2 for water properties over a wide range of temperatures and corresponding pressures. The enthalpy of water over this range changes at a rate of about 1 Btu/(lb · °F). For design purposes, this value can be used without significant error. The density of water varies from 62.3 lb/ft^3 at 70°F to 60.1 lb/ft^3 at 200°F. For HVAC design purposes, the value of 62.3 lb/ft^3 is commonly used; it is sufficiently accurate over a range from 32 to 100°F but should be compensated for at higher temperatures. Based on 7.5 gal/ft^3, 1 gal weighs about 8.3 lb. Then

1 Btu/(lb · °F) × 8.3 lb/gal × 60 min/h

$$= 500 \text{ Btu/[h} \cdot (\text{gal/min}) \cdot \text{°F]} \quad (6.1)$$

which is a constant commonly used in calculating water flow quantities.

To determine the water quantity required to serve a given load, divide the load, in Btu per hour, by 500 and by the desired water temperature drop or rise in degrees Fahrenheit. Typical numbers are 8, 10, and 20°F for cooling (resulting in a divisor of 4000, 5000, and 10,000, respectively) and 20 to 40°F for heating (a divisor of 10,000 to 20,000).

Another measure of water quantity is gallons per minute per ton-hour of refrigeration. Because 1 ton · h equals 12,000 Btu/h, a 10°F rise in the chilled water temperature works out to 2.4 gal/(min − ton). An 8°F rise requires 3 gal/(min − ton), and a 20°F rise is 1.2 gal/(min − ton). On the condensing water side, it is assumed that heat rejection in a vapor compression machine is approximately 15,000 Btu/(ton − h) and a 10°F rise requires 3 gal/(min − ton − h). The actual heat rejection will vary with the refrigeration system efficiency and will usually be somewhat less than 15,000 Btu/(ton − h), except

TABLE 6.2 Properties of Water, 212 to 400°F

Temperature		Pressure		Density		Specific heat		Total heat above 32°F (0°C)				Dynamic (absolute) viscosity
°F	°C	lb/in² absolute	kPa	lb/ft³	kg/m³	Btu/(lb·°F)	kJ/(kg·°C)	Btu/lb³	kJ/kg	Btu/ft³	kJ/m³	CP (mPa·s)
212	100.0	14.70	101.3	59.81	957.0	1.007	4.219	180.07	419.6	10,770	401,721	0.2838
220	104.4	17.19	118.4	59.63	954.1	1.009	4.228	188.13	438.3	11,216	418,357	0.2712
230	110.0	20.78	143.2	59.38	950.1	1.010	4.232	198.23	461.9	11,770	439,021	0.2567
240	115.6	24.97	172.0	59.10	945.6	1.012	4.240	208.34	485.4	12,313	459,275	0.2436
250	121.1	29.83	205.5	58.82	941.1	1.015	4.253	218.48	509.1	12,851	479,342	0.2317
260	126.7	35.43	244.1	58.51	936.2	1.017	4.261	228.64	532.7	13,378	498,999	0.2207
270	132.2	41.86	288.4	58.24	931.8	1.020	4.274	238.84	556.5	13,910	518,843	0.2107
280	137.8	49.20	339.0	57.94	927.0	1.022	4.282	249.06	580.3	14,430	538,239	0.2015
290	143.3	57.56	396.6	57.64	922.2	1.025	4.295	259.31	604.2	14,947	557,523	0.1930
300	148.8	67.01	461.7	57.31	917.0	1.032	4.324	269.59	628.1	15,450	576,285	0.1852
310	154.4	77.68	535.2	56.98	911.7	1.035	4.337	279.92	652.2	15,950	594,935	0.1779
320	160.0	89.66	617.8	56.66	906.6	1.040	4.358	290.28	676.4	16,437	613,100	0.1712
330	165.6	103.06	710.1	56.31	901.0	1.042	4.366	300.68	700.6	16,931	631,526	0.1649
340	171.1	118.01	813.1	55.96	895.4	1.047	4.387	311.13	724.9	17,409	649,356	0.1591
350	176.7	134.63	927.6	55.59	889.4	1.052	4.408	321.63	749.4	17,879	666,887	0.1536
360	182.2	153.04	1054.5	55.22	883.5	1.057	4.429	332.18	774.0	18,343	684,194	0.1484
370	187.8	173.37	1194.5	54.85	877.6	1.062	4.450	342.79	798.7	18,802	701,315	0.1436
380	193.3	195.77	1348.9	54.47	871.5	1.070	4.483	353.45	823.5	19,252	718,100	0.1391
390	198.9	220.37	1518.4	54.05	864.8	1.077	4.513	364.17	848.5	19,681	734,101	0.1349
400	204.4	247.31	1704.0	53.65	858.4	1.085	4.546	374.97	873.7	20,117	750,364	0.1308

SOURCE: Reprinted by permission from *Thermodynamic Properties of Steam*, J. H. Keenan and F. G. Keyes, published by John Wiley and Sons, Inc., 1936 edition. Subsequent editions have equivalent data.

that for absorption refrigeration, rejection will be 20,000 to 30,000 Btu/(ton − h).

6.4   High-Temperature Water

High-temperature water (HTW) systems operate with supply water temperatures over 350°F and with a pressure rating of 300 to 350 lb/in^2 gauge. Maximum temperatures are about 400°F in order to stay within the 300 lb/in^2 gauge limit on pipe and fittings. Medium-temperature systems operate with supply water temperatures between 250 and 350°F, which allows the use of 150 lb/in^2 gauge rating on piping systems. Table 6.2 lists properties of water at temperatures up to 400°F.

Systems must be kept tight because the water at this temperature will flash instantly to steam at any leak. Large temperature drops are typical—150 to 200°F is normal. The system must be carefully pressurized to above the saturation pressure corresponding to the water temperature, to prevent the water from flashing into steam.

Heat exchangers are used to provide lower-temperature hot water or steam for HVAC use. HTW may be used directly for generation of domestic hot water. Most jurisdictions require double-wall heat exchangers to guarantee protection from tube failure and cross-contamination. It is common to place user equipment in series, taking part of the HTW temperature drop through each device (Fig. 6.3).

6.5   Secondary Coolants
(Brines and Glycols)

Brine is a mixture of water and any salt, with the purpose of lowering the freezing point of the mixture. In HVAC practice, the term is also applied to mixtures of water and one of the glycols. Brines are used as heat transfer fluids when near- or subfreezing temperatures are encountered. Brines may be used directly in cooling coils of air-handling units or, through heat exchangers, may be used to provide chilled water. Brines are also commonly used in runaround heat reclaim systems (see Chap. 7). Heating systems exposed to subfreezing air may use a glycol solution as a circulating medium.

6.5.1   Properties of secondary coolants

Calcium and sodium chloride solutions in water have been the most common brines. Properties of pure brines are shown in Tables 6.3 and 6.4. For commercial-grade brines, use the formulas in the footnotes to the tables. Note particularly that the specific heat decreases as the

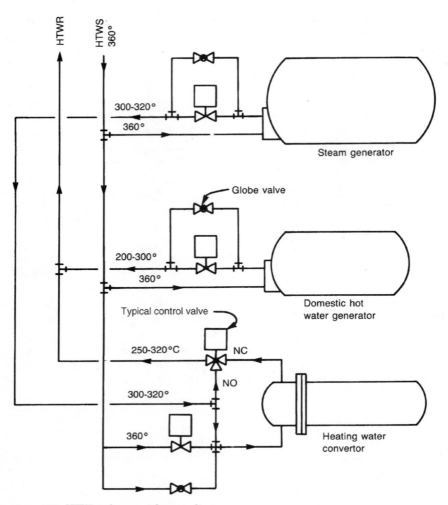

Figure 6.3 HTW end use, with cascading.

percentage of the salt increases. Thus, a 25% solution of calcium chloride will lower the freezing point of the mixture to −21°F and will decrease the specific heat to 0.689 Btu/(lb · °F). This means that the solution will transfer only about two-thirds of the heat transferred by pure water at the same mass flow rate and temperature difference. This is partially offset by the increased mass of the mixture in pounds per gallon.

However, the viscosity of the brine increases (Fig. 6.4) while the thermal conductivity decreases (Fig. 6.5) as the percentage of salt

TABLE 6.6 Properties of Pure Calcium Chloride Brine

Pure $CaCl_2$, % by mass	Specific gravity‡	Baume density at 60°F	Specific heat 60°F Btu/(lb·°F)	Crystallization starts, °F	Mass per unit volume† at 60°F — lb/gal $CaCl_2$	lb/gal Brine	lb/ft³ $CaCl_2$	lb/ft³ Brine	Specific gravity at various temperatures (refers to water at 60°F) −4°F	14°F	32°F	50°F
0	1.000	0.0	1.000	32.0	0.000	8.34	0.00	62.40				
5	1.044	6.1	0.924	27.7	0.436	8.717	3.26	65.15			1.043	1.042
6	1.050	7.0	0.914	26.8	0.526	8.760	3.93	65.52			1.052	1.051
7	1.060	8.2	0.898	25.9	0.620	8.851	4.63	66.14			1.061	1.060
8	1.069	9.3	0.884	24.6	0.714	8.926	5.34	66.70			1.071	1.069
9	1.078	10.4	0.869	23.5	0.810	9.001	6.05	67.27			1.080	1.078
10	1.087	11.6	0.855	22.3	0.908	9.076	6.78	67.83			1.089	1.087
11	1.096	12.6	0.842	20.8	1.006	9.143	7.52	68.33			1.098	1.096
12	1.105	13.8	0.828	19.3	1.107	9.227	8.27	68.95			1.108	1.105
13	1.114	14.8	0.816	17.6	1.209	9.302	9.04	69.51			1.117	1.115
14	1.124	15.9	0.804	15.5	1.313	9.377	9.81	70.08			1.127	1.124
15	1.133	16.9	0.793	13.5	1.418	9.452	10.60	70.64		1.139	1.137	1.134
16	1.143	18.0	0.779	11.2	1.526	9.536	11.40	71.26		1.149	1.146	1.143
17	1.152	19.1	0.767	8.6	1.635	9.619	12.22	71.89		1.159	1.156	1.153
18	1.162	20.2	0.756	5.9	1.747	9.703	13.05	72.51		1.169	1.166	1.163
19	1.172	21.3	0.746	2.8	1.859	9.786	13.90	73.13		1.180	1.176	1.173
20	1.182	22.1	0.737	−0.4	1.970	9.853	14.73	73.63		1.190	1.186	1.183
21	1.192	23.0	0.729	−3.9	2.085	9.928	15.58	74.19				
22	1.202	24.4	0.716	−7.8	2.208	10.037	16.50	75.00	1.215	1.211	1.207	1.203
23	1.212	25.5	0.707	−11.9	2.328	10.120	17.40	75.63				
24	1.223	26.4	0.697	−16.2	2.451	10.212	18.32	76.32	1.236	1.232	1.228	1.224
25	1.233	27.4	0.689	−21.0	2.574	10.295	19.24	76.94				
26	1.244	28.3	0.682	−25.8	2.699	10.379	20.17	77.56				
27	1.254	29.3	0.673	−31.2	2.827	10.471	21.15	78.25				
28	1.265	30.4	0.665	−37.8	2.958	10.563	22.10	78.94				
29	1.276	31.4	0.658	−49.4	3.090	10.655	23.09	79.62				
29.87	1.290	32.6	0.655	−67.0	3.16	10.75	23.65	80.45				
30	1.295	33.0	0.653	−50.8	3.22	10.80	24.06	80.76				
32	1.317	34.9	0.640	−19.5	3.49	10.98	26.10	82.14				
34	1.340	36.8	0.630	+4.3	3.77	11.17	28.22	83.57				

*Mass of Type I (77% min) $CaCl_2$ = (mass of pure $CaCl_2$)/(0.77). Mass of Type 2 (94% min) $CaCl_2$ = (mass of pure $CaCl_2$)/(0.94).

†Mass of water per unit volume = Brine mass minus $CaCl_2$ mass.

‡Specific gravity is solution at 60°F referred to water at 60°F.

SOURCE: Reprinted by permission from ASHRAE Handbook, 1997 Fundamentals, Chap. 20, Table 1, p. 20.2.

TABLE 6.4 Properties of Pure Sodium Chloride Brine

Pure NaCl, % by mass	Specific gravity‡	Baume density at 60°F	Specific heat 59°F Btu/(lb·°F)	Crystallization starts, °F	Mass per unit volume† at 60°F lb/gal NaCl	Brine	lb/ft³ NaCl	Brine	Specific gravity at other temperatures (refers to water at 60°F) 14°F	32°F	50°F	68°F
0	1.000	0.0	1.000	32.0	0.000	8.34	0.000	62.4				
5	1.035	5.1	0.938	26.7	0.432	8.65	3.230	64.6		1.0382	1.0366	1.0341
6	1.043	6.1	0.927	25.5	0.523	8.71	3.906	65.1		1.0459	1.0440	1.0413
7	1.050	7.0	0.917	24.3	0.613	8.76	4.585	65.5		1.0536	1.0515	1.0486
8	1.057	8.0	0.907	23.0	0.706	8.82	5.280	66.0		1.0613	1.0590	1.0559
9	1.065	9.0	0.897	21.6	0.800	8.89	5.985	66.5		1.0691	1.0665	1.0633
10	1.072	10.1	0.888	20.2	0.895	8.95	6.690	66.9		1.0769	1.0741	1.0707
11	1.080	10.8	0.879	18.8	0.992	9.02	7.414	67.4		1.0849	1.0817	1.0782
12	1.087	11.8	0.870	17.3	1.090	9.08	8.136	67.8		1.0925	1.0897	1.0857
13	1.095	12.7	0.862	15.7	1.188	9.14	8.879	68.3		1.1004	1.0933	1.0971
14	1.103	13.6	0.854	14.0	1.291	9.22	9.632	68.8		1.1083	1.1048	1.1009
15	1.111	14.5	0.847	12.3	1.392	9.28	10.395	69.3	1.1195	1.1163	1.1126	1.1086
16	1.118	15.4	0.840	10.5	1.493	9.33	11.168	69.8	1.1277	1.1243	1.1205	1.1163
17	1.126	16.3	0.833	8.6	1.598	9.40	11.951	70.3	1.1359	1.1323	1.1284	1.1241
18	1.134	17.2	0.826	6.6	1.705	9.47	12.744	70.8	1.1442	1.1404	1.1363	1.1319
19	1.142	18.1	0.819	4.5	1.813	9.54	13.547	71.3	1.1535	1.1486	1.1444	1.1398
20	1.150	19.0	0.813	+2.3	1.920	9.60	14.360	71.8	1.1608	1.1568	1.1542	1.1478
21	1.158	19.9	0.807	-0.0	2.031	9.67	15.183	72.3	1.1692	1.1651	1.1606	1.1559
22	1.166	20.8	0.802	-2.3	2.143	9.74	16.016	72.8	1.1777	1.1734	1.1688	1.1640
23	1.175	21.7	0.796	-5.1	2.256	9.81	16.854	73.3	1.1862	1.1818	1.1771	1.1721
24	1.183	22.5	0.791	+3.8	2.371	9.88	17.712	73.8	1.1948	1.1902	1.1854	1.1804
25	1.191	23.4	0.786	+16.1	2.488	9.95	18.575	74.3				
25.2	1.200			+32.0								

*Mass of commercial NaCl required = (mass of pure NaCl required)/(% purity).

†Mass of water per unit volume = brine mass minus NaCl mass.

‡Specific gravity is solution at 59°F referred to water at 39°F.

SOURCE: Reprinted by permission from ASHRAE Handbook, *1997 Fundamentals*, Chap. 20, Table 1, p. 20.3.

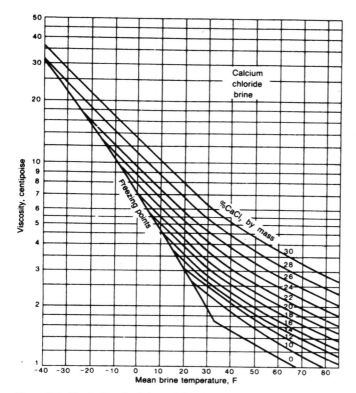

Figure 6.4 Viscosity of calcium chloride brine. (*Reprinted by permission from ASHRAE Handbook*, 1997 Fundamentals, *Chap. 20, Fig. 7, p. 20.4.*)

increases. Compared to pure water, this results in a higher pumping head and lower heat transfer rate. Such a brine is less efficient than water as a heat-conveying medium. The tables indicate a percentage solution at which a minimum freezing temperature is obtained. This is the *eutectic point*. Brine solutions are corrosive, particularly when exposed to air or carbon dioxide. Inhibitors are recommended. Chromate solutions are typically prohibited. Other chemicals, such as sodium nitrite or sodium borate, may be used. A qualified water treatment expert should be consulted.

Solutions of ethylene glycol or propylene glycol in water are used extensively. With proper inhibitors to prevent corrosion, these solutions can lower the mixture's freezing point to well below 0°F (Fig. 6.6). As with the salt solutions, the thermal conductivity and specific heat of the mixture decrease and the viscosity increases with an increase in the percentage of glycol. These are similar to the antifreeze

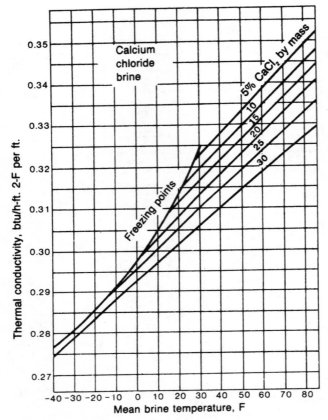

Figure 6.5 Thermal conductivity of calcium chloride brine. (*Reprinted by permission from ASHRAE Handbook,* 1997 Fundamentals, *Chap. 20, Fig. 4, p. 20.3.*)

solutions used in automobile cooling systems. Inhibitors must be checked and maintained periodically.

Common refrigerants may also be used as a secondary coolant. That is, liquid refrigerant may be pumped through distribution piping. Refrigerants have the advantages of low freezing points and low viscosity, but also have low specific heats.

6.6 Piping Systems

Piping systems are the means by which the thermal energy of fluids is transported from one place to another. The type of fluid and its temperature and pressure influence and limit the choice of piping materials. Most systems are closed; i.e., the fluid is continually recircu-

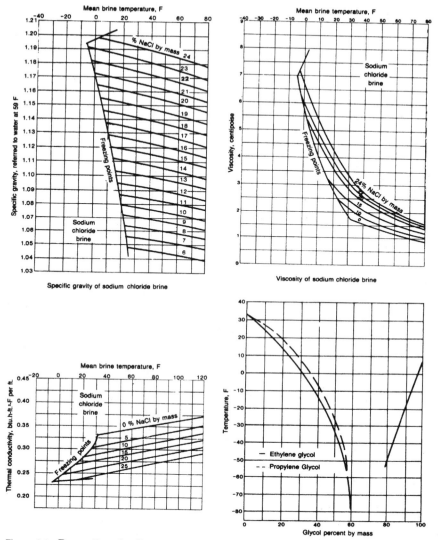

Figure 6.6 Properties of sodium chloride brine solutions, and freezing points of aqueous solutions of ethylene glycol and propylene glycol. (*Reprinted by permission from ASH-RAE Handbook, 1997 Fundamentals, Chap. 20.*)

lated and no makeup water is required except to replace that lost due to leaks. Steam systems are partly to completely open—as when the steam is used for a process or humidification—and require continuous makeup water. Cooling-tower water systems are open and need makeup water to replace the water evaporated in the tower.

Closed systems require some means of compensating for the changes in volume of the fluid due to temperature changes. Expansion (compression) tanks are used.

Piping must be properly supported, with compensation for expansion due to temperature changes and anchors to prevent undesired movement.

6.6.1 Piping materials

By far the most common material used in HVAC piping systems is black steel (low-carbon steel). Table 6.5 covers dimensional data for steel pipe. Pressure ratings vary with the pipe size (greater for smaller pipes), but in general, standard-weight pipe can be used for working pressures up to 300 lb/in^2 gauge, extra-strong pipe to 450 lb/in^2 gauge, and double-extra-strong pipe to 1000 lb/in^2 gauge or more. Pipes of 14-in and larger outside diameter (OD) are made with thinner walls for the lower pressures which are often acceptable, as well as with thicker walls for higher pressures.

Another standard defines pipe sized by *schedule number*. In this system, schedule 40 is the same as standard weight, and schedule 80 is the same as extra-strong, up to 6 in in size. Sizes of 8, 10, and 12 in standard weight are the same as schedule 30.

Black steel is often preferred because it is strong, is readily available, can be used over a wide range of temperatures and pressures, and is easy to assemble and join by several common methods. If proper inhibitors are used in the steam, water, and brine, black steel corrodes very little; and in closed systems it will tend to stabilize in a neutral, noncorrosive state. Unfortunately, very few systems remain completely closed for very long, so at least some water treatment is necessary.

Another popular piping material for HVAC systems is copper, usually in tubing form. Copper pipe has the same outside diameter as steel pipe, with slightly thinner walls. Dimensions of copper tubing are shown in Table 6.6. Type L is most commonly used and is suitable for pressures of up to about 250 to 300 lb/in^2 gauge.

Other materials include fiberglass-reinforced plastic (FRP), ultra-high molecular weight polyethylene (UHPE), polypropylene (PP) polybutylene, (PB), polyvinyl chloride (PVC), and chlorinated polyvinyl chloride (CPVC). These have excellent corrosion resistance and low flow resistance, but have lower pressure and temperature ratings than steel or copper. Complete data on these materials are available from the manufacturers. PVC is often used for equipment drain lines.

Galvanized-steel piping is used occasionally. The dimensions are the same as for black-steel pipe. Cast-iron piping is seldom used in sizes

TABLE 6.5 Steel Pipe Dimensions and Weights

Nominal size, in	Outside diam., in	Standard-weight pipe		Extra-strong pipe		Double-extra-strong pipe		Volume*
		Thickness, in	Weight, lb/ft	Thickness, in	Weight, lb/ft	Thickness, in	Weight, lb/ft	
¼	0.540	0.088	0.43	0.119	0.54			0.0007
½	0.840	0.109	0.85	0.147	1.09	0.294	1.71	0.0021
¾	1.050	0.113	1.13	0.154	1.47	0.308	2.44	0.0037
1	1.315	0.133	1.68	0.179	2.17	0.358	3.66	0.0060
1¼	1.660	0.140	2.28	0.191	3.00	0.382	5.21	0.0104
1½	1.900	0.145	2.73	0.200	3.63	0.400	6.41	0.0141
2	2.375	0.154	3.68	0.218	5.02	0.436	9.03	0.0233
2½	2.875	0.203	5.82	0.276	7.66	0.552	13.70	0.0332
3	3.500	0.216	7.62	0.300	10.25	0.600	18.58	0.0513
4	4.500	0.237	10.89	0.337	14.98	0.674	27.54	0.0884
5	5.563	0.258	14.81	0.375	20.78	0.750	38.55	0.1389
6	6.625	0.280	19.19	0.432	28.57	0.864	53.16	0.2006
8	8.625	0.322	28.81	0.500	43.89	0.875	72.42	0.3475
10	10.750	0.365	41.13	0.500	54.74			0.5476
12	12.750	0.375	50.71	0.500	65.42			0.7855

*Volume in cubic feet of water per foot of pipe length, standard weight. Also 8-, 10-, and 12-in pipe is made with thinner walls, but these are nonstandard. Intermediate sizes such as 3½ in are also made, but seldom used. And ⅛ and ⅜ in are also made. Larger sizes, 14 to 30 in, have nominal size equal to outside diameter but are not part of this standard.

TABLE 6.6   Copper Tubing Dimensions (in inches)

Nominal size	Type K		Type L		Type M	
	Inside diameter	Wall thickness	Inside diameter	Wall thickness	Inside diameter	Wall thickness
⅜	0.402	0.049	0.430	0.035	0.450	0.025
½	0.527	0.049	0.545	0.040	0.569	0.028
¾	0.745	0.065	0.785	0.045	0.811	0.032
1	0.995	0.065	1.025	0.050	1.055	0.035
1¼	1.245	0.065	1.265	0.055	1.291	0.042
1½	1.481	0.072	1.505	0.060	1.527	0.049
2	1.959	0.083	1.985	0.070	2.009	0.058
2½	2.435	0.095	2.465	0.080	2.495	0.065
3	2.907	0.109	2.945	0.090	2.981	0.072
3½	3.385	0.120	3.425	0.100	3.459	0.083
4	3.857	0.134	3.905	0.110	3.935	0.095

Note: Tubing is available up to 8 in.

less than 4 in, although cast-iron fittings are available down to 1-in size. Wrought-iron piping has been used extensively in the past for steam condensate, but it is seldom used anymore because of the extra cost.

6.6.2   Pipe fittings

Pipe fittings include elbows, tees, wyes, couplings, unions, reducers, plugs, caps, and bushings. Elbows may be 45°, 90°, or even 180°, reducing or nonreducing, with short or long radius. Tees and wyes may also be reducing types. Special fittings are available to prevent electrolytic corrosion when dissimilar piping materials are joined. A standard manufacturer's catalog can be consulted for dimensions and types of fittings.

6.6.3   Joining methods

Steel pipe joints may be welded, threaded, grooved, or flanged. Welding is typical on piping 3 in in diameter and larger and should be done in accordance with ASME power piping standards,[3] using certified welders. The grooved joint is made by using a gasketed clamp which locks into grooves cut or rolled near the end of the pipe section or fitting. Gasket materials must be suitable for the temperature, pressure, and fluid.

Copper pipe joints may be brazed, threaded, grooved, or flanged. Copper tubing is joined by soldering or brazing or by the use of compression or grooved fittings.

FRP and PVC piping are usually joined by use of solvent cement. Flanged joints are also employed. Some other plastics are joined by heat fusion.

All piping systems must be provided with unions (screwed or flanged) at connections to equipment and valves.

6.6.4 Supports, anchors, guides, and expansion

Spacing of pipe supports is a function of pipe size and material. The principal objectives are to avoid sagging and to maintain a uniform slope to allow good drainage. Steam lines must be trapped at low points to provide for removal of condensate. There are many different support systems available; a complete discussion is beyond the scope of this book.

The length of all piping will change as the temperature of the fluid changes. With steam or high-temperature water, the changes can be great. For example, steel has a linear coefficient of expansion of 0.00000633 per degree Fahrenheit. If a steel pipeline 100 ft long is installed in an ambient of 50°F and is later filled with saturated steam at 15 lb/in^2 gauge (250°F), the pipe length will increase about 1.52 in. If left unrestrained, the pipe may move in unacceptable ways. If the pipe is restrained, large forces will be developed and either the pipe or the restraints may break. The expansion must be compensated by means of expansion joints, loops, or elbows.

Expansion joints may be of the bellows (Fig. 6.7), slide, or ball-joint type (Fig. 6.8). Joints are simpler than loops, but slide joints may develop leaks over time unless packing is maintained or replaced. Bellows joints need no packing but may eventually fail due to fatigue.

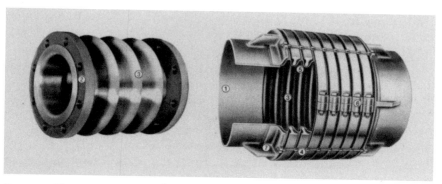

Figure 6.7 Expansion joint, bellows type. Left: plain; right: with equalizing rings. (*Courtesy of Adsco Manufacturing Corp.*)

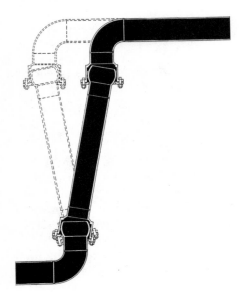

Figure 6.8 Expansion joint, ball-joint offset method. (*Courtesy of Aeroquip Corp., Jackson, Mich.*)

Expansion may also be controlled by means of loops or elbows. Figure 6.9 shows a simple piping system with an expansion loop and expansion elbow. The design provides for flexibility so that the pipe can bend without exceeding the allowable stress of the pipe material. Information on the design of expansion loops and elbows can be found in many references (see Ref. 4) as well as from some pipe fitting manufacturers.

For the expansion to be properly controlled, it is necessary to provide a point of reference, with no movement. This is called an *anchor*, and the pipe must be fastened at this point strongly enough to resist the forces generated by expansion. Failure of an anchor can be disastrous. In addition, guides must be provided to prevent unwanted lateral movement in the pipeline. A guide restrains the pipe laterally while allowing it to move lengthwise. The pipe must be free to move on other, intermediate supports.

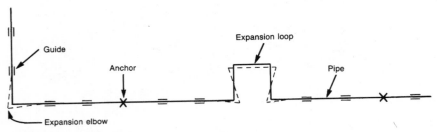

Figure 6.9 Expansion loop and expansion elbow.

6.6.5 Valves

A *valve* is a device for controlling the flow of fluid in a pipeline. Control may mean limiting or throttling flow, preventing backflow, or completely stopping flow. Automatic control valves are discussed in Chap. 8. Manually operated valves are discussed here.

There are a great many types and configurations of manual valves. They can be grouped into a few general classes. Stop valves are used for shutoff of flow. The primary reason is to allow isolation of equipment or sections of piping for repair or replacement. Throttling valves can be adjusted to control flow quantities within limits which depend on the system pressure variations. Backflow prevention valves, including check valves, are used to prevent flow in the wrong direction. Reverse flow may occur as a result of pressure changes and may degrade system performance or may even be dangerous. Pressure-reducing valves provide control of downstream pressure regardless of upstream pressure variations, as long as upstream pressure exceeds downstream pressure. Pressure relief valves are safety devices which open to relieve excessive pressures which might damage the system.

Traditionally, the most common stop valve has been the gate valve. In the full-open position, the gate is out of the way and resistance to flow is minimal. In the fully closed position, the gate seats tightly and flow is effectively stopped. The gate valve is not a good throttling device. Gate valves are made in many sizes, configurations, and materials to handle almost any fluid or pressure.

In larger piping over 3 or 4 in, it may be less expensive to use a butterfly valve. Butterfly valves are made in flange, wafer, or tapped lug configuration. Do not use a wafer valve for dead-end service because it is held in place by clamping between the adjoining pipes. The tapped lug body works as a flange union joint and can be used for dead-end service. Butterfly valves are available in a more limited range of pressure ratings and materials compared to gate valves.

For throttling control, the globe-type valve is often recommended. Globe valves are made in many configurations, but all have a shaped plug, such that gradual throttling can be accurately accomplished. Many different sizes, materials, and pressure ratings are available.

A needle valve is similar in principle to a globe valve, but with a needlelike plug. Needle valves are used mostly in small sizes for fine-tuning very small flows.

A plug valve has a cylindrical slotted plug, frequently tapered, which fits into the valve body tightly enough to prevent leakage. When it is rotated so that the slot is aligned with the body ports, flow is unimpeded. At right angles, flow is stopped. By rotating the plug to an intermediate position, flow can be modulated. Plug valves are commonly used for "balancing" system flows, and some models have a memory marker so that the valve can be used for shutoff and later returned to the proper balance point.

The ball valve has gained great popularity in recent years. A ball valve is similar to a plug valve but has a spherical plug with a round hole drilled through the center, mounted in the valve body. Ball valves have become the valve of choice over gate and globe valves in many applications for reasons of cost and performance.

Backflow prevention valves are usually called check valves and come in several types. The most common is the swing check. A flapper swings open to allow flow in one direction but closes if flow is reversed. This valve must be mounted so that gravity will assist in closing the flapper. A spring-loaded check valve includes a spring to assist in closing the flapper; consequently it has a higher resistance to flow. A lift check is arranged so that the flapper lifts off the seat to allow flow.

A pressure-reducing valve is an automatic control valve, usually a globe type with a diaphragm operator which acts to modulate flow through the valve to maintain a specified downstream pressure. For compressible fluids such as steam, air, or gas, maximum flow through the valve occurs at a ratio of downstream pressure to upstream pressure which is the critical pressure drop for the fluid, that is, 0.53 for steam. Thus, if a greater than 50 percent reduction is required, it is best to use two or three stages of pressure reduction for good control.

6.6.6 Pipe sizing

The principal criteria for sizing piping systems to serve a given flow rate are velocity in feet per second, and pressure drop in feet of water or pounds per square inch per 100 ft of pipe. The velocity is important because the turbulence due to velocity causes noise, and the noise due to high velocities may be unacceptable. The pipe may erode in turbulent high-velocity regions. The pressure drop in pumped systems becomes part of the pump head and is, therefore, a contributor to operating cost. The higher first cost of larger piping must be balanced against the increased operating cost of smaller piping.

Each design office has its target values of velocity and pressure drop for water, usually in the range of 3 to 4 ft/100 ft and 6 to 8 ft/s for large pipe to as low as 2 ft/s in small pipe. It will be found that the pressure drop governs in small pipe and the velocity governs in larger pipe.

Figure 6.10 shows flow versus head loss (pressure drop per 100 ft) and velocity in schedule 40 steel pipe. This is for water at 60°F, but the small error for warmer water gives conservative results. Figure 6.11 covers similar data for copper tubing. Data for plastic and PVC

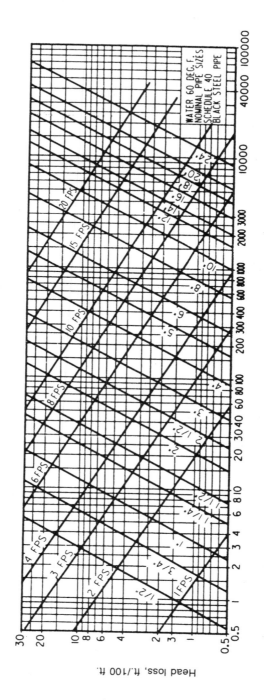

Figure 6.10 Friction loss for water in commercial steel pipe (schedule 40). (*Reprinted by permission from ASHRAE Handbook, 1997 Fundamentals, Chap. 33, Fig. 1, p. 33.5.*)

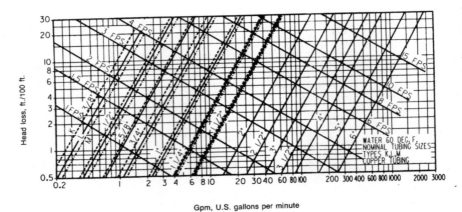

Gpm, U.S. gallons per minute

Figure 6.11 Friction loss for water in copper tubing (types K, L, M). (*Reprinted by permission from ASHRAE Handbook,* 1997 Fundamentals, *Chap. 33, Fig. 2, p. 33.5.*)

piping are available from the manufacturers. Reference 6 is an excellent resource for pipe sizing data.

Fittings — tees, elbows, valves — are allowed for on the basis of size and velocity, by using "equivalent length" values which have been determined empirically. Table 6.7 lists equivalent lengths for 90° elbows. Table 6.8 shows multipliers for Table 6.7 for various other fittings. A rule of thumb which is frequently used is to double the measured length of the piping system to allow for fittings. This will not be satisfactory for systems with a high fittings-to-length ratio, and in any case, it should be used only for preliminary estimates.

Steam line sizing is based on flow rate (pounds per hour) at a specified pressure and pressure drop. Figure 6.12 is a graph of flow rate in pounds per hour versus pressure drop in pounds per square inch per 100 ft and velocity in feet per minute. The figure is based on steam at 0 lb/in^2 gauge. The allowable pressure drop depends on the initial pressure and the acceptable pressure at the end of the system. For 15 lb/in^2 steam, typical pressure drops are in the range of 0.5 to 0.7 lb/in^2 per 100 ft. The maximum velocity should not exceed 10,000 ft/min in large pipes (10 to 12 in), dropping to 2000 ft/min in 2-in and smaller lines. This follows the old rule of thumb that "steam velocity in thousands should not exceed the pipe diameter in inches." Figure 6.13 provides velocity correction factors for pressures other than zero.

Condensate return lines, for gravity flow, can be sized by using Table 6.9. Wet return lines include no vapor. Dry lines include both steam and liquid — this is most common — and *vacuum* refers to a return line going to a vacuum pump and thus at a pressure below atmospheric. Return lines must slope downward in the direction of flow.

TABLE 6.7 Equivalent Length of Pipe for 90° Elbows, ft

Velocity, ft/s	\(\frac{1}{2} \)	\(\frac{3}{4} \)	1	\(1\frac{1}{4} \)	\(1\frac{1}{2} \)	2	\(2\frac{1}{2} \)	3	\(3\frac{1}{2} \)	4	5	6	8	10	12
									Pipe size						
1	1.2	1.7	2.2	3.0	3.5	4.5	5.4	6.7	7.7	8.6	10.5	12.2	15.4	18.7	22.2
2	1.4	1.9	2.5	3.3	3.9	5.1	6.0	7.5	8.6	9.5	11.7	13.7	17.3	20.8	24.8
3	1.5	2.0	2.7	3.6	4.2	5.4	6.4	8.0	9.2	10.2	12.5	14.6	18.4	22.3	26.5
4	1.5	2.1	2.8	3.7	4.4	5.6	6.7	8.3	9.6	10.6	13.1	15.2	19.2	23.2	27.6
5	1.6	2.2	2.9	3.9	4.5	5.9	7.0	8.7	10.0	11.1	13.6	15.8	19.8	24.2	28.8
6	1.7	2.3	3.0	4.0	4.7	6.0	7.2	8.9	10.3	11.4	14.0	16.3	20.5	24.9	29.6
7	1.7	2.3	3.0	4.1	4.8	6.2	7.4	9.1	10.5	11.7	14.3	16.7	21.0	25.5	30.3
8	1.7	2.4	3.1	4.2	4.9	6.3	7.5	9.3	10.8	11.9	14.6	17.1	21.5	26.1	31.0
9	1.8	2.4	3.2	4.3	5.0	6.4	7.7	9.5	11.0	12.2	14.9	17.4	21.9	26.6	31.6
10	1.8	2.5	3.2	4.3	5.1	6.5	7.8	9.7	11.2	12.4	15.2	17.7	22.2	27.0	32.0

SOURCE: Reprinted by permission from ASHRAE Handbook, *1997 Fundamentals*, Chap. 33, Table 6, p. 33.6.

TABLE 6.8 Iron and Copper Elbow Equivalents

Fitting	Iron pipe	Copper tubing
Elbow, 90°	1.0	1.0
Elbow, 45°	0.7	0.7
Elbow, 90° long turn	0.5	0.5
Elbow, welded, 90°	0.5	0.5
Reduced coupling	0.4	0.4
Open return bend	1.0	1.0
Angle radiator valve	2.0	3.0
Radiator or convector	3.0	4.0
Boiler or heater	3.0	4.0
Open gate valve	0.5	0.7
Open globe valve	12.0	17.0

*See Table 6.7 for equivalent length of one elbow.
SOURCE: Reprinted by permission from ASHRAE Handbook, *1997 Fundamentals,* Chap. 33, Table 7, p. 33.6.

6.6.7 Insulation

Pipe insulation is required whenever the temperature of the fluid in the pipe is significantly different from the ambient temperature around the pipe. This covers most heating and cooling applications. An exception is condensate return piping which may be left uninsulated to allow any flash steam to condense. In such a case, vagrant heat should be determined not to be a problem.

The required thermal resistance of the insulation is determined from applicable energy or building codes, the insulation manufacturer's recommendations, and calculations based on acceptable energy losses in the piping system.

Piping insulation is usually "preformed" to snap readily in place around the pipe. A protective jacket is required, usually canvas on hot lines and a vapor barrier type on cold lines. The vapor barrier specification must include a mastic seal at all joints and edges and should prohibit the use of staples. In areas where the pipe is exposed to damage, a metal or plastic jacket may be specified, including preformed jacket sections for fittings and valves. Insulation should be verified as acceptable for the application. Materials such as fiberglass, calcium silicate, foamed glass, and foamed plastic are all available.

6.6.8 Distribution piping configurations for water

Three basic system configurations are used: out-and-back, reverse-return, and loop.

The out-and-back system (Fig. 6.14) is common in large campus systems. The supply and return mains run in parallel from the central

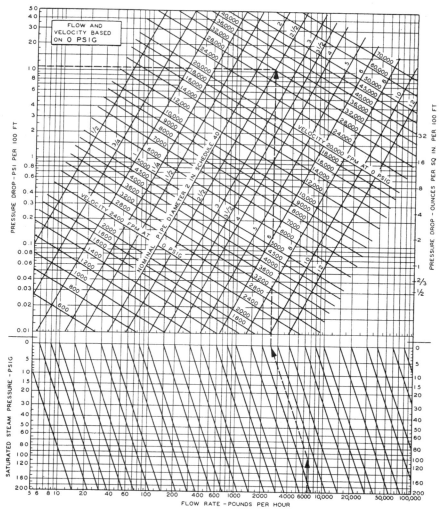

Figure 6.12 Basic chart for flow rate and velocity of steam in schedule 40 pipe. (*Reprinted by permission from ASHRAE Handbook,* 1997 Fundamentals, *Chap. 33, Fig. 10, p. 33.11.*)

plant to the points of use, with flow in opposite directions and with equal reduction in size as flow rates reduce. This means that the point with the lowest pressure difference from supply to return is at the end of the main, as shown on the pressure profile (Fig. 6.15). Then the branches nearest the central plant will require extra pressure reduction by means of balancing valves.

The reverse-return system (Fig. 6.16) is designed to provide pressure differences from supply to return which are similar at all

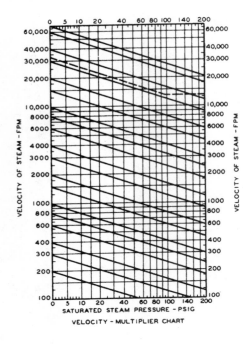

Figure 6.13 Velocity multiplier chart for Fig. 6.12. (*Reprinted by permission from ASHRAE Handbook,* 1993 Fundamentals, *Chap. 33, 1997 edition similar.*)

branches, as shown in the pressure profile in Fig. 6.17, so that the minimum amount of balancing is required. This is accomplished by beginning the return main at the location of the first takeoff from the supply and continuing the return main parallel to the supply main with flow in both pipes in the same direction. The return main increases in size as the supply main decreases. Finally, the return main goes back to the central plant.

The loop system (Figs. 6.18 and 6.19) consists of supply and return mains which are closed loops, with a constant pipe size. The loops are fed at one point from the central plant, flow goes in both directions on the loop, and branch takeoffs may be located at any point on the loop. At some point in the loop, there will be no flow, depending on the geometry and demand of the branches. Because the loop is closed, it is self-balancing; i.e., as branch takeoffs are added or removed, the no-flow point will move until balance is restored. For an overall pressure loss equivalent to that in a reverse-return system, the loop pipe size (diameter) should be about 40 to 50 percent of the main size from the central plant.

The pressure profiles for the three systems are based on full design flow. If flow is reduced, the profiles will change to less main slope and greater available head at each branch.

In comparing the three configurations, note that the system geometry, especially the relationships among the central plant and the var-

TABLE 6.9 Return Main and Riser Capacities for Low-Pressure Steam Systems, lb/h

Pipe size, in	$\frac{1}{32}$ lb/in² or ½ oz drop per 100 ft			$\frac{1}{24}$ lb/in² or ⅔ oz drop per 100 ft			$\frac{1}{16}$ lb/in² or 1 oz drop per 100 ft			$\frac{1}{8}$ lb/in² or 2 oz drop per 100 ft			¼ lb/in² or 4 oz drop per 100 ft			½ lb/in² or 8 oz drop per 100 ft		
	Wet	Dry	Vac.	Wet	Dry	Vac.	Wet	Dry	Vac.	Wet	Dry	Vac.	Wet	Dry	Vac.	Wet	Dry	Vac.
G	H	I	J	K	L	M	N	O	P	Q	R	S	T	U	V	W	X	Y
¾						42			100			142			200			283
1	125	62		145	71	143	175	80	175	250	103	249	350	115	350			494
1¼	213	130		248	149	244	300	168	300	425	217	426	600	241	600			848
1½	338	206		393	236	388	475	265	475	675	340	674	950	378	950			1,340
2	700	470		810	535	815	1,000	575	1,000	1,400	740	1,420	2,000	825	2,000			2,830
2½	1,180	760		1,580	868	1,360	1,680	950	1,680	2,350	1,230	2,380	3,350	1,360	3,350			4,730
3	1,880	1,460		2,130	1,560	2,180	2,680	1,750	2,680	3,750	2,250	3,800	5,350	2,500	5,350			7,560
3½	2,750	1,970		3,300	2,200	3,250	4,000	2,500	4,000	5,500	3,230	5,680	8,000	3,580	8,000			11,300
4	3,880	2,930		4,580	3,350	4,500	5,500	3,750	5,500	7,750	4,830	7,810	11,000	5,380	11,000			15,500
5						7,880			9,680			13,700			19,400			27,300
6						12,600			15,500			22,000			31,000			43,800
¾		48			48	143		48	175		48	249		48	350			494
1		113			113	244		113	300		113	426		113	600			848
1¼		248			248	388		248	475		248	674		248	950			1,340
1½		375			375	815		375	1,000		375	1,420		375	2,000			2,830
2		750			750	1,360		750	1,680		750	2,380		750	3,350			4,730
2½						2,180			2,680			3,800			5,350			7,560
3						3,250			4,000			5,680			8,000			11,300
3½						4,480			5,500			7,810			11,000			15,500
4						7,880			9,680			13,700			19,400			27,300
5						12,600			15,500			22,000			31,000			43,800

SOURCE: Reprinted by permission from ASHRAE Handbook, *1997 Fundamentals*, Chap. 33, Table 16, p. 33.15.

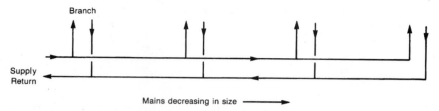

Figure 6.14 Out-and-back piping system.

ious points of use, will influence cost and efficiency greatly. In general, the out-and-back system will have the lowest first cost but the highest pumping cost. The loop system will have an intermediate first cost and the lowest pumping cost. The reverse-return system will have the highest first cost and an intermediate to low pumping cost. The loop system is preferable but only if the geometry is suitable.

6.6.9   System expansion and pressurization

The volume of the water in a piping system varies as its temperature changes. If not compensated for in some way, this can result in unacceptably high pressures or, at the other extreme, such low pressures that pump cavitation takes place due to flashing (boiling). The simplest way to accomplish this is to allow some water to leave the system through a pressure relief valve as the pressure rises, and to replace it through a makeup valve as the pressure falls. This is not a satisfactory solution, because the continuous addition of fresh water, with its entrained air, increases corrosion and requires venting of the air.

The purpose of the pressurization system is to maintain the pressure within allowable limits while limiting the addition of fresh water.

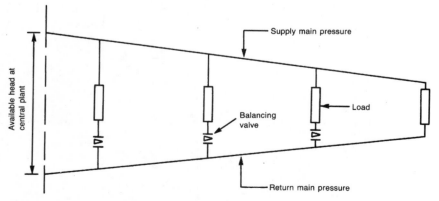

Figure 6.15 Pressure profile for out-and-back piping.

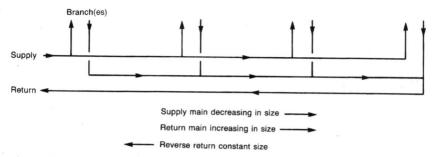

Figure 6.16 Reverse-return piping system.

The usual method of doing this involves an expansion tank (sometimes called a *compression tank*) (Fig. 6.20). This provides a place to store the excess volume of water created by expansion and includes a cushion of air or inert gas which will compress or expand to maintain pressure within limits. The gas cushion will be gradually absorbed by the water so that, unless it is replenished, the system will eventually become "waterlogged." Maintenance of the cushion can be manual or automatic. This problem can be alleviated by use of a diaphragm-type tank which has a watertight flexible diaphragm to separate the air and water.

Open expansion tanks are sometimes used (Fig. 6.21). Because this tank is open to air, the maximum water temperature is limited. The tank must be above the high point of the piping system, and the corrosion problem created by the air interface must be recognized. The possibility of freezing also exists if the tank is in an unheated space.

The point at which the expansion tank connects to the piping system is the point of no pressure change when the pump is started or stopped. It is preferable that this point be on the suction side of the

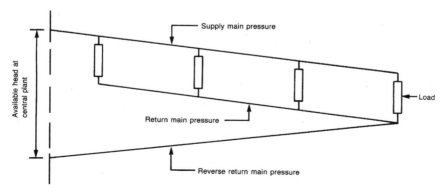

Figure 6.17 Reverse-return system pressure profile.

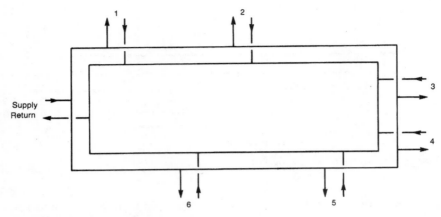

Figure 6.18 Loop piping system.

pump to minimize the total pressure. If the connection were on the discharge side of the pump, the total pressure would have to be greater in order to prevent cavitation at the pump inlet. An exception to this is an attic location for the tank, which would result in a smaller tank than one at the bottom of the system. This arises out of the pressure relationships in the sizing equation, as described below.

The size of a closed expansion tank is determined by (1) the volume of water in the system, (2) the range of water temperatures expected in normal operation, (3) the air pressure in the expansion tank when water first enters it, (4) the initial fill pressure (minimum operating pressure) at the tank, and (5) the maximum allowable operating pressure at the tank. The initial fill pressure is determined by the system static head at the tank location plus an allowance to maintain a net positive head at the pump suction under all operating conditions. Thus, if the tank is on the pump suction (Fig. 6.22), a few feet of gauge

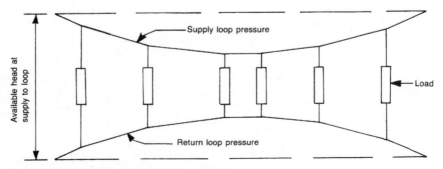

Figure 6.19 Pressure profile for loop piping system.

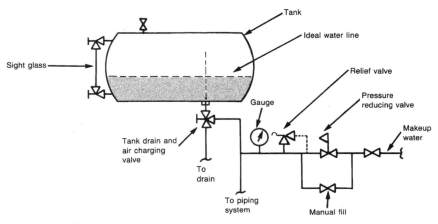

Figure 6.20 Closed expansion tank.

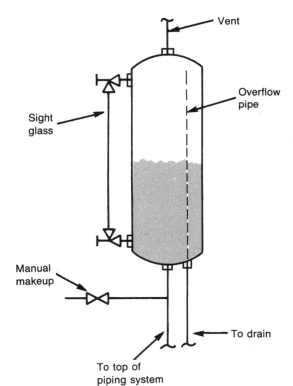

Figure 6.21 Open expansion tank.

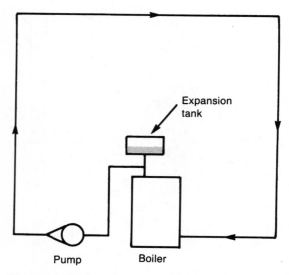

Figure 6.22 Tank at pump suction, pumping away from boiler.

pressure is sufficient. If the tank is on the pump discharge (Fig. 6.23), the minimum pressure must include the pump head plus a small margin. For an attic tank (Fig. 6.24), the extra static head (if the pump is in the basement or ground floor) will usually be sufficient. The maximum pressure allowable is determined from the lowest-rated device

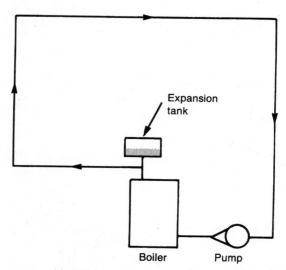

Figure 6.23 Tank at pump discharge, pumping into boiler. (Not recommended.)

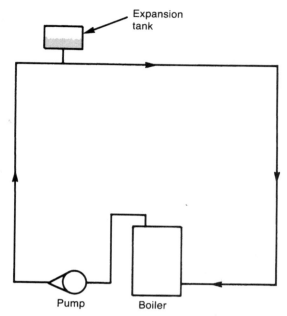

Figure 6.24 Tank at top of system.

in the system, most often the boiler or its pressure relief valve. Because of this, it is preferable to pump away from the boiler.

The following ASME formula for determining the size of a closed system expansion tank is valid for temperatures between 160 and 280°F.

$$V_t = \frac{(0.00047t - 0.0466)V_s}{P_a/P_f - P_a/P_o} \tag{6.2}$$

where V_t = minimum volume of expansion tank, gal
V_s = system volume, gal
t = maximum operating temperature, °F
P_a = pressure in expansion tank when water first enters (usually atmospheric), ftH$_2$O absolute
P_f = initial fill pressure at tank, ftH$_2$O absolute
P_o = maximum operating pressure at tank, ftH$_2$O absolute

For temperatures below 160°F, a simpler formula may be used:

$$V_t = \frac{E}{P_a/P_f - P_a/P_o} \tag{6.3}$$

where E = net expansion of the water in the system when heated from minimum to maximum temperature. Expansion E equals the system volume times the percentage increase indicated in the graph of Fig. 6.25. For chilled water, the minimum design temperature is used, combined with the maximum anticipated temperature during a summer shutdown.

Expansion tanks for high-temperature water systems are always provided with a cushion of inert gas (usually nitrogen) or high-pressure steam, which is continuously maintained by an automatic control system with rapid response. Maintenance of system pressure is critical to proper operation of the HTW system. HTW systems are usually large and have a wide temperature difference from supply to return. For details on handling expansion in HTW systems, see the relevant ASHRAE handbook chapter.[5]

6.6.10 Air venting

Some air is present in any piping system. Air in excessive amounts can cause noise. Air can also impede the flow of water in a closed piping system. It is desirable to remove as much air as possible. Air removal is based on two principles: (1) air will be entrained and carried along with the water stream at velocities in excess of 2 ft/s, and

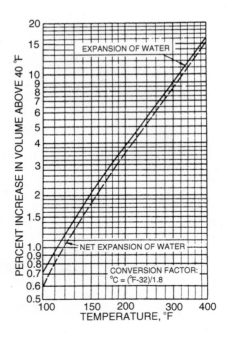

Figure 6.25 Expansion of water above 40°F. (*Reprinted by permission from ASHRAE Handbook*, 1987 HVAC Systems and Applications, *Chap. 13, p. 13.14.*)

(2) air tends to migrate to high points in the system when flow is stopped.

The second principle is employed in the installation of air vents at high points in the piping system. Air vents may be manual or automatic. Automatic vents use float valves or water-expansive materials to close the vent when water is present. Under low-pressure conditions, automatic vents may allow air to enter the system. Manual vents do not have this problem but do depend on regular operation by maintenance personnel. Always provide a drain line from each air vent, to prevent damage if water is carried over.

The first principle is utilized in air separation devices. The most common is the centrifugal separator (Fig. 6.26). This consists of a large vertical pipe section which the water enters tangentially near the bottom. The combination of centrifugal force and decreased velocity separates the entrained air, which is removed through a vent at the top center of the separator. The separator is usually vented to the expansion tank air cushion, although it can also be vented to atmosphere, manually or automatically.

An in-line air separator can be field-fabricated, by using a section of pipe large enough to reduce the flow velocity below 1.5 ft/s and with a length about 6 times the diameter (Fig. 6.27). This will allow some

Figure 6.26 Centrifugal air separator. (*Courtesy of ITT Fluid-Handling Division, Bell and Gossett.*)

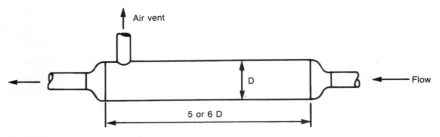

Figure 6.27 In-line air separator.

of the air to separate and to be carried off through a vent line. For HTW systems, all air vents must be manual because of the high pressures.

6.7 Pumps

Centrifugal pumps are used in HVAC systems for circulation of brine and chilled, hot, and condensing water. They are also used for pumping steam condensate and for boiler feed.

The operating theory of centrifugal pumps is exactly analogous to that of centrifugal fans, discussed in Chap. 5. The rotating action of the impeller (equivalent to the fan wheel) in a scroll housing generates a pressure which forces the fluid through the piping system. The pressure and volume developed are functions of pump size and rotational speed. For higher pressures, multistage pumps are used.

6.7.1 Pump configurations and types

The majority of the centrifugal pumps used in HVAC work have a backward-curved blade impeller (Fig. 6.28). For pumping hot condensate, a turbine-type impeller is used to minimize flashing and cavitation.

Most pumps are direct-driven at standard motor speeds such as 3500, 1750, and 1150 r/min. Typical arrangements include combinations of alternatives such as end or double suction, in-line or base-mounted, horizontal or vertical, and close-coupled or base-mounted (see Figs. 6.29 through 6.33). Vertical turbine pumps (Fig. 6.34) are used in sumps, i.e., in cooling-tower installations.

In general, in-line pumps are used in small systems or secondary systems, such as freeze prevention loops. Base-mounted pumps are used for most applications. Double-suction pumps are preferred for larger water volumes over 300 to 400 gal/min, because the purpose of the double-suction design is to minimize the end thrust due to water entering the impeller.

Rotation

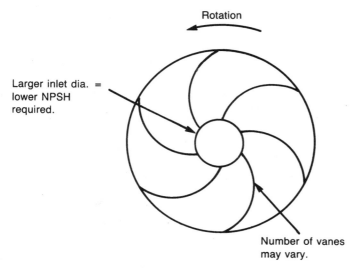

Larger inlet dia. =
lower NPSH
required.

Number of vanes
may vary.

Figure 6.28 Backward-curved pump impeller.

6.7.2 Performance curves

A typical pump performance curve (Fig. 6.35) is drawn with coordinates of capacity and head. The curves show the capacity of a specific
pump-casing size and design at a specific speed and with varying impeller diameters. The same impeller is used throughout, but when it
is "shaved" (machined) to reduce the outside diameter, the capacity is
reduced. This allows the pump to be matched to the design conditions.
The graph includes brake horsepower curves for standard-size motors,
based on water with a specific gravity of 1.0. For brines, or liquids
with other specific gravities, the horsepower must be corrected in direct proportion to the specific-gravity change. Also shown are efficiency curves.

Figure   6.29 In-line   circulator
pump. (*Courtesy   ITT   Fluid-
Handling   Division,   Bell   and
Gossett.*)

Figure 6.30 Closed-coupled end-suction pump. (*Courtesy of ITT Fluid-Handling Division, Bell and Gossett.*)

The point at which a pump curve intersects the no-flow line is the *shutoff head*. At this or a higher head, the pump will not generate any flow. If the pump continues to run under no-flow conditions, the work energy input will heat the water. The resulting temperature and pressure rise has been known to break the pump casing.

If the speed of the pump is varied, the result will be a family of curves similar to Fig. 6.36. These data are needed to evaluate a variable-speed pumping design.

6.7.3 Suction characteristics—NPSH

The condition of the liquid entering the pump can interfere with pump operation. If the absolute pressure at the suction nozzle approaches

Figure 6.31 Frame-mounted end-suction pump. (*Courtesy of Allis-Chalmers.*)

Figure 6.32 Base-mounted double-suction pump. (*Courtesy of Allis-Chalmers.*)

the vapor pressure of the liquid, vapor pockets will form and collapse in the impeller passages. This will be noisy and can cause damage to and destruction of the pump impeller. The condition is called *cavitation*. It is more likely to occur with warm water and at high flow rates. The pump performance curve of Fig. 6.35 includes an NPSHR (net positive suction head required) curve, which is a characteristic of this pump. This is the amount of pressure required in excess of the vapor pressure. The pumping system must be designed so that the available

Figure   6.33 Vertical   in-line pump. (*Courtesy ITT Fluid-Handling Division, Bell and Gossett.*)

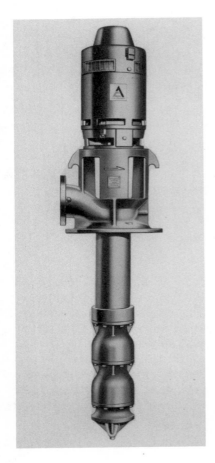

Figure 6.34 Vertical turbine pump. (*Courtesy of Allis-Chalmers.*)

NPSH exceeds the NPSHR. In a closed system, this can be done by increasing the initial fill pressure. In an open system, such as a cooling-tower sump or condensate return tank, the static head of the water column at the pump inlet represents the available pressure (less the friction loss between the sump and the pump inlet).[6]

6.7.4 Pump selection

To select a pump, it is necessary to calculate the system pressure drop at the design flow rate. Losses include pipe, valves, fittings, control valves, and equipment such as heat exchangers, boilers, or chillers. The design operating point or a complete system curve can then be plotted on a pump performance curve (Fig. 6.37). Usually several different pump curves will be inspected to find the best efficiency and

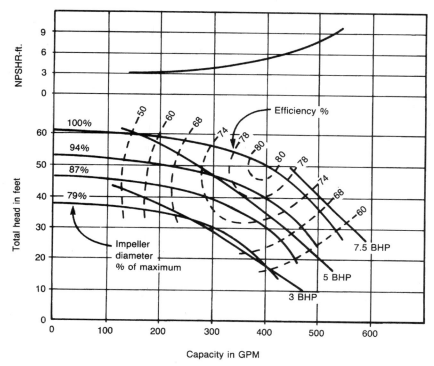

Figure 6.35 Pump performance curve.

lowest horsepower. In general, for large flows at low heads, lower speed pumps—1150 or even 850 r/min—will be most efficient. For higher heads and lower flow rates, 1750 or 3500 r/min will be preferable. Multistage pumps may be needed at very high heads. Always select a motor horsepower that cannot be exceeded by the selected pump at any operating condition; e.g., the horsepower curve should be above the pump curve at all points.

When two or more identical pumps are installed in parallel, the resulting performance is as shown in Fig. 6.38. The performance curve for two pumps has twice the flow of one pump at any given head. When the system curve is superimposed, the curve for one pump will intersect the system curve at about 70 percent of the design flow rate and about one-half of the design head. Similar curves can be drawn for three or more pumps in parallel.

Two or more identical pumps in series provide twice the head at any given flow rate, as shown in Fig. 6.39. Here the flow with one pump will be about 75 percent of design flow. However, unless a bypass is provided around the second pump, the system curve will change some-

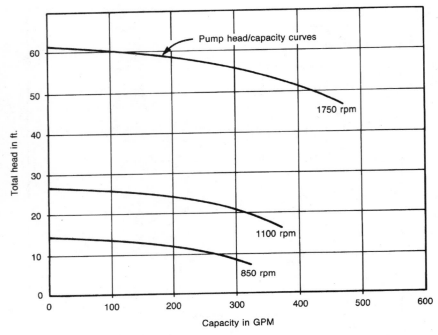

Figure 6.36 Pump speed versus capacity and head.

what with only one pump running, due to the pressure loss through the second pump. A bypass should be provided around both pumps to allow one pump to operate while the other is being repaired or replaced.

6.8 Refrigerant Distribution

The pumping of liquid refrigerant from condensing units to points of use is common practice in the cold-storage industry but is seldom used in HVAC work. To properly design such a system requires an extensive and specialized background in refrigerants and refrigeration machinery which is beyond the scope of this book.

6.9 Summary

The distribution of thermal energy by means of heated or cooled fluids from a central plant to points of use is a common HVAC practice. The process uses energy which contributes nothing to the final air-conditioning result; therefore the transport energy is said to be "parasitical." Thus, the transport system should be designed to use as little

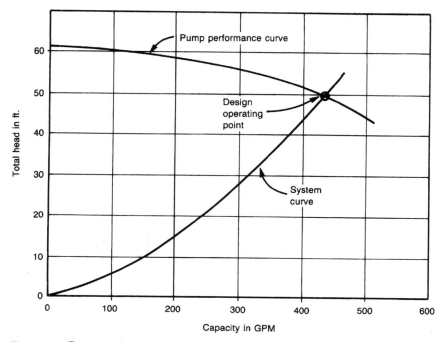

Figure 6.37 Pump performance versus system.

energy as possible. Some methods for accomplishing this include (1)
using a large ΔT for water systems to minimize flow rates, (2) minimizing water velocities and pressure drops (without greatly oversizing
piping), (3) using variable-speed or staged pumping, and (4) using secondary pumping for loops with higher heads than the other parts of
the system, preferably with variable-speed pumping to match loads.

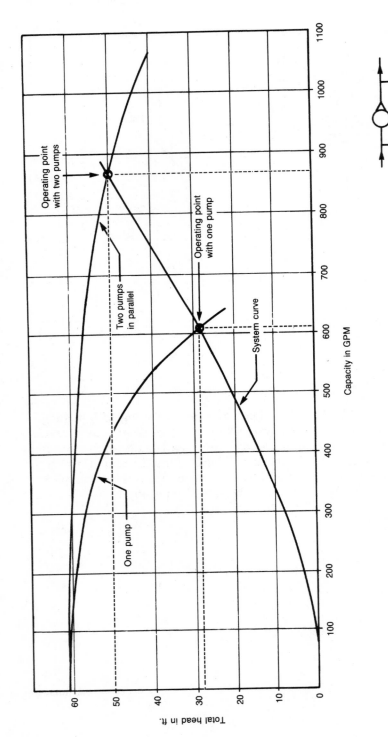

Figure 6.38 Operating with two pumps in parallel.

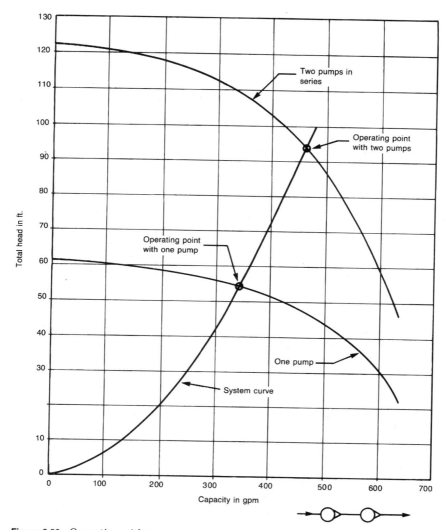

Figure 6.39 Operation with two pumps in series.

References

1. American Society of Mechanical Engineers (ASME), *Thermodynamic and Transport Properties of Steam,* 1967.
2. ASHRAE Handbook, *1997 Fundamentals,* Chap. 6, Table 3.
3. ASME, *Power Piping Code,* ANSI/ASME B31.1-1983.
4. ASHRAE Handbook, *1996 HVAC Systems and Applications,* Chap. 11, "District Heating and Cooling."
5. ASHRAE Handbook, *1996 HVAC Systems and Applications,* Chap. 14, "Medium- and High-Temperature Water-Heating Systems."
6. Ingersoll-Rand, *Cameron Hydraulic Data,* Woodcliff Lake, N.J., 1988.

7

Design Procedures: Part 5

Central Plants

7.1 Introduction

The design and construction of central plants for heating and cooling is one of the most challenging and interesting aspects of the HVAC design profession. Central plants range in size from small to very large, from residential to industrial utility scale. There are many areas of individual expertise and many levels of competence among designers. In this chapter we discuss several fundamental types of plants and aspects of plant design, still leaving much detail to literature and experience beyond the scope of this book. See Ref. 1 for additional discussion of the topics treated here.

7.2 General Plant Design Concepts

Independent of the service being produced, some concerns are common to central plants.

1. *Siting.* Central plants preferably are located in the middle of or adjacent to the loads they serve. Distribution piping costs may loom large if primary piping runs long distances to get to the service point. On the other hand, the combining of multiple service units into one plant is the act which achieves the economy of scale and the convenience of operation, so distance is a tradeoff, but the central location is still a favored point to start. For large plants serving congested campuses, a remote or peripheral location may be preferred. This allows better access to the plant and removes plant, traffic, noise, and emissions from the more densely populated areas.

For high-rise buildings, there is the question of the basement, roof, or in between. On-grade locations have the best access. Sometimes buildings are occupied from the ground up during extended construction, suggesting a low-level site. Where water systems are involved, pressures may become very high at lower building levels. This is less of a problem with chillers than with boilers. Systems with boilers often take the equipment to the roof, partly for pressure considerations, partly to eliminate the problem of taking the flue up through the building, partly for emission dispersion. Cooling towers need to be near the chiller served if possible, to reduce the cost of piping, but the cooling-tower vapor plume can be a problem in cool weather if it impacts the building (window cleaning, condensation on structure, etc.). A vapor plume is a cold-weather visual problem in year-round operation and may cause a local "snow" effect in cold climates.

2. *Structure.* The enclosure and support for major plant equipment should be strong enough to withstand vibration, to support equipment and piping, to contain yet accept expansion and contraction, to enclose and subdue noise, and to support maintenance through access and hoist points.

In some environments, plant structures are fully enclosed by heavy masonry. In the industrial environment, in mild climates, plant structures may be open, offering only a roof and access, possibly a sound enclosure. Some well-designed plants may take on an aesthetic aspect including large expanses of glass and careful lighting. It is a fun experience to sculpt in pipe and equipment for all to see. This can be accomplished with little premium construction cost, but it takes more design time and an artist's inclination. Some feel that a plant that looks good may work better, since more time is given to function and layout than in the "quick and dirty" arrangements so often encountered. Well-arranged plants usually are more easily maintained, given the space associated with form and symmetry.

As a general note, reinforced-concrete floors and below-grade walls have proved to be durable. Steel-frame superstructures with intermediate floors of concrete and steel work very well. Steel members with grating for walkways are very popular.

Plant enclosures should allow for future equipment replacement or addition, with wall openings and possibly roof sections which can be removed and replaced.

3. *Electrical Service.* Many plants, particularly those with chillers or electric boilers, comprise a major electrical load for the facility. Proximity to the primary electrical service is a cost concern. The electrical service should be well thought through, and should allow for any projected plant expansion, if not in present gear, at least in space and concept. Since the plant environment may be coarse (although

cleanliness is a virtue), electrical equipment is often housed in a separate room with filtered, fan-forced ventilation. Some electronic gear needs to be in an air conditioned space.

Where many motors are involved in a plant, *motor control centers* (MCCs) are preferred to individual combination starters. Large plants may have several MCCs to reduce the length of wiring runs.

The electrical service should have a degree of redundancy. Hospitals and other critical-care facilities require access to at least two independent utility substations. This carries into the large plant in the form of multiple transformers and segmented switch groups with tie breakers. Standby power generation may be included in plant design in addition to backup power for life safety issues.

4. *Valving.* In central plants there is no substitute for isolation valves for every piece of equipment. Multiple high-pressure steam boilers require double valving with intermediate vent valves to protect workers inside a unit that is down for maintenance. Valves should be installed in accessible locations.

7.3 Central Steam Plants

Some general concepts of steam distribution were presented in Chap. 6. Steam plants require considerations of siting, structure, and electrical service, as described in this chapter. Boilers are the primary component of steam plants and are supported by a host of auxiliary components such as boiler feed pumps, deaerating feedwater heaters, condensate holding tanks, water softeners, blowdown heat recovery systems, water treatment systems, flue gas economizers, fuel-handling equipment, etc. See Fig. 7.1.

Each component of the steam system is available in a range of quality and performance characteristics. Selection depends on duty and on the sophistication of the plant operation. Equipment for a smaller school will be of a different character than for a campus or an industrial plant. With all the subjective differences, the technical calculations are similar.

Because condensate originates in heat exchange devices as a fluid without pressure, it must drain by gravity to a collection point. If a steam plant can be located at the low point of the served system, the entire condensate return line may flow by gravity. Otherwise, intermediate collection points and booster pumps may be required.

An important aspect of a steam plant is the condensate storage vessel. When a boiler fires up after a time of setback or at the onset of a peak heating load, a significant amount of feedwater will be evaporated and sent out into the system with a time lag before any of the condensate will get back to the plant. The storage tank must hold

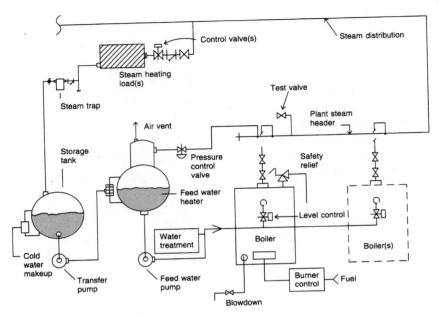

Figure 7.1 Steam plant diagram.

enough water to sustain the initial demand, and then it must have enough "freeboard" or residual capacity to accept the returning condensate after an evening load shutdown. Failure to provide adequate storage is observed through storage tank overflow, with high makeup water rates and high treatment costs.

Small plants often use the feedwater heating tank as a combination storage-and-preheat vessel.

Most steam plants use a version of a feedwater heater to remove dissolved oxygen by bringing the feedwater to the boiling point. This also tempers the water to reduce the potential for damaging the boiler with a shot of cold water.

Feedwater makeup to boilers is accomplished with feedwater pumps. If feedwater is heated to near the boiling point, the pumps must have a low net positive suction head (NPSH) to avoid cavitation. Small plants often have a dedicated pump for each boiler with a level control on the boiler drum which cycles the pump on a call for more water. Larger plants usually have a continuously running pump for several boilers with modulating valves and automatic level controls to maintain a constant level in the boiler steam drum.

7.3.1 Steam plant controls

In a very small steam system, a space thermostat may cycle the boiler on and off, and the steam drum-level control will activate the feedwater pump. In a more complex system, the boiler(s) will maintain a constant steam pressure in the main header, and a pressure control will modulate the fuel input to match the load. For multiple-boiler operation, there may be a plant master control which will apportion the load to the several boilers on a proportional or a programmed basis.

7.3.2 Flue gas economizers

Flue gas economizers are often used on steam boilers to pick up an additional 3 to 7 percent of combustion efficiency. Reclaimed heat from the economizers may be used for combustion air preheating or feed-water preheating. In either case, care must be taken to keep the ex-iting flue gas above the water vapor condensation temperature, and for feedwater heating, there must be adequate flow to avoid steaming in the economizer.

7.3.3 Boiler testing

It is often desirable or necessary to test steam boiler performance. To this end, a valve to open for discharge to atmosphere is included in the plant design.

7.4   Central Hot Water Plants

Some general concepts of heating water distribution were discussed in Chap. 6. Chapter 10 discusses boilers and some other pieces of heat-ing plant equipment.

Low-temperature water (LTW) heating systems (150 to 250°F) are simple in design. They include boiler(s), pump(s), and secondary com-ponents such as water treatment, air eliminators, and expansion tanks. See Fig. 7.2. The simplicity of these systems is compelling. They become so automatic and reliable that even in larger sizes they are often taken for granted.

Most hot water plants serve loads of varying magnitude. If constant-flow systems were common in the past, variable-flow systems are be-coming more common because of the reduced pumping energy which can be obtained at lower loads.

Water heating plants are usually designed for a heating differential of 20 to 40°F through the boilers. Return water temperatures below 140°F to the boiler should be avoided in most cases out of concern for flue gas vapor condensation and for "cold shock" of the boiler itself.

Multiple hot water boilers are almost always piped in parallel. See Figs. 7.3 and 7.4. Where two boilers are selected, it is common to size each for 60 percent of the peak load, to allow one boiler to keep the

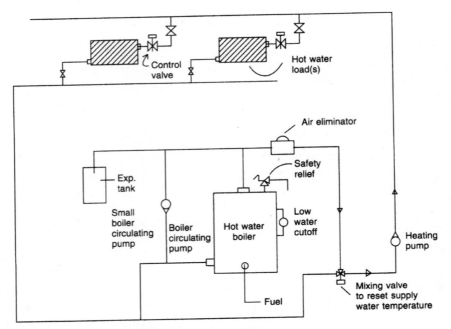

Figure 7.2   Elementary heating water system diagram.

system "alive" if the other boiler fails. For a three- or four-boiler or more system, boilers are usually sized so that the entire load can be carried even if the largest boiler fails. There is usually a smaller boiler sized to the summer load. Care must be taken not to underestimate the peak summer demand. Undersizing the small boiler forces the use of a larger boiler, losing the benefit of the smaller selection.

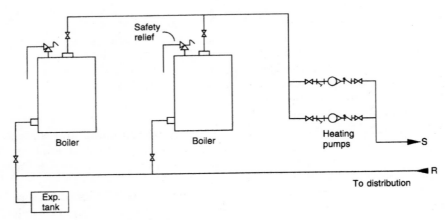

Figure 7.3   Central heating plant, multiple boilers/common pumps.

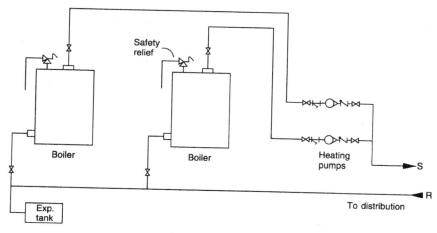

Figure 7.4 Central heating plant, multiple boilers/individual pumps.

Water heating plants usually have a means of introducing an oxygen scavenging chemical with corrosion inhibitor to the system. Soft water is often used for fill water. Heating water systems should be quite tight, requiring little makeup water. Where glycol solutions are used for freeze protection, a means of introducing the glycol-water mixture must be included in the plant. This often takes the form of a holding tank with a feed pump. Glycol solutions require attention to materials in the system. Some elastomers are sensitive to some petroleum-derived glycols.

7.5 High-Temperature Hot Water Plants

High-temperature water (HTW) plants usually have supply water temperatures between 350 and 450°F. This discussion also includes plants with a supply temperature between 250 and 350°F because the principles are similar. These systems became popular in the post-World War II era, as an alternative to steam plants for large campus and military-base central heating systems. The advantage is related to the ability of water with high temperature differential to move large quantities of heat in smaller distribution pipe. Pumping and control may be simplified. System design pressures are similar as for high-pressure steam, but must be handled carefully to avoid flashing related to changes in elevation across the facility.

The detailed design of HTW plants is a specialty beyond the scope of this book. There are few definitive works on the technology and only a few design offices across the country, with personnel having HTW experience. Chapter 14 of the 1996 ASHRAE Handbook, *HVAC Systems and Equipment,* discusses the topic.

A designer working in or with an HTW plant will find recognizable components. High pressure boilers are the heart of the plant. Almost any fuel can be accommodated. HTW boilers are almost always circulated with constant flow independent of the load, to avoid hot spots and steaming on the heat transfer surfaces. Some plants use a large drum with a steam cushion to accept the wide fluid expansion and contraction episodes encountered in large systems. An alternative and now more common practice is to use an expansion drum pressurized with nitrogen in a manner similar to a conventional lower-temperature heating plant.

HTW plants usually serve variable-flow secondary systems (the loads have control valves which meter the supply water to match the load) and therefore benefit from variable-speed control for the system pumps.

To protect the plant from power outage, most HTW plants have standby power generation capability. To protect from sudden water loss due to rupture in the distribution system, quick-closing valves on the piping in and out of the plant are recommended.

HTW plants usually look for return water temperatures ranging from 200 to 250°F. If the water comes back warmer than the design value, it becomes difficult to load the fixed-circulation-rate boilers. Building system designers working with HTW should recognize that steam generation at pressures above 15 lb/in^2 is not a good load for an HTW system. Since the HTW leaving the steam generator must be above the steam saturation temperature, it is impossible for a steam generator to get the return water temperature down to the plant design inlet condition. Large HTW flows are required, and this wastes distribution system capacity. This problem can be relieved by cascading the steam generator HTW return into a lower-grade heating service; but, in general, high-pressure steam requirements should be accommodated with an independent boiler.

7.6 Fuel Options and Alternative Fuels

A nice feature of central heating plants is that if the load requirements are not extreme, almost any fuel source can be utilized to make steam or hot water. Coal, oil, and gas (natural, liquefied, manufactured) are traditional fuels. But wood refuse, combustible by-product, and municipal and industrial waste and industrial process exhaust streams are all candidates for central plant heating sources. Under some conditions, electricity may be used as an energy input for a central plant.

High-temperature geothermal waters or steam can be used through heat exchangers for central heating. Direct use of many geothermal resources incurs problems with corrosive and precipitate aspects of the waters.

Solid fuels present design challenges related to delivery, handling and storage of the input fuel, and the collection, storage, and disposal of the residual matter.

7.7   Chilled Water Plants

Central chilled water plants for HVAC systems have evolved to a combination of factory-built chillers and pumps in a variety of piping and pumping configurations. Since the cooling effort may require a large amount of energy to drive the process, much attention is given to schemes which reduce energy use. In some office space cooling services, the cooling function may be considered noncritical and subject to a low initial and operating cost design concept. In other applications such as computer rooms and electronics manufacturing, the product may have such high value and the quality of product may be so sensitive to environmental conditions that no expense will be spared to provide reliable cooling.

Interestingly, systems which have low operating cost may be quite reliable because it takes better equipment and better arrangements to operate with less energy input, assuming proven technology in the equipment design.

There are several key factors in designing a quality chilled water plant:

- Well-configured chiller(s)
- Efficient pumps
- A good piping scheme with ample valving
- A good control concept
- Good access for maintenance and replacement

Chillers as a piece of equipment are discussed in Chap. 9. Pumps are discussed in Chap. 6, as are several piping schemes.

There is an old saying: "Pump out of a boiler and into a chiller." While this is not a hard-and-fast rule, it has some basis in good practice. Pumping out of a boiler places the boiler at the pump suction, which is the lowest pressure point in the system. This allows any dissolved air to work its way out at that point. It lets the boiler be designed for and work at no more than the fill pressure of the system.

Pumping through the chiller makes the chiller, which typically has a relatively high (10- to 20-ft) pressure drop, the first pressure-drop device in the system. This reduces the remaining pressure throughout the system. Chiller heat exchangers (tube bundles) are usually rated for 150 lb/in² gauge working pressure and are not threatened by the condition.

7.7.1 Central plant piping configurations for water

The design challenge in the central plant arrangement is to deliver service to the distribution system while operating the plant as efficiently as possible. Variations in load have more impact on chillers than on boilers, so the following discussion will concentrate on chiller plants. Most that is said also applies to heating plants.

- In an elementary system with one chiller and one or more air-handling unit coils, the layout shown in Fig. 7.5 works best. The goal is to provide essentially constant flow through the chiller while

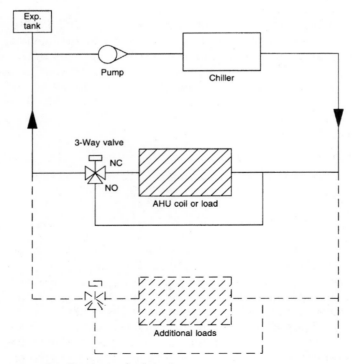

Figure 7.5 Constant volume system.

modulating flow through the AHU coils. The three-way control valves accomplish this. The typical chiller requires that flow be maintained within perhaps ± 10 percent of some base rate. In addition, most chillers can modulate capacity down to about 20 percent of a base rating. When the load or flow rate is too small, the chiller will shut down. In the system of Fig. 7.5, the design water temperature difference across the chiller may be 6 to 20°F. As the three-way valves modulate to match a decreasing load, the temperature difference will decrease; and when this difference is about 2°F, the chiller controls will sense a near no-load condition and shut down. The potential chiller cycling can be offset to some extent by raising the supply water temperature as the load decreases, but too great a rise might make it impossible to maintain a desired humidity level in the building. The pump and chiller must run continuously as long as cooling is required. Hot gas bypass is sometimes used to force a chiller to stay on-line in low-load conditions.

- In any system with two or more chillers, the situation becomes more complex, with potential operating difficulties, but with greater opportunities for energy conservation. Figure 7.6 shows one possible arrangement. The first necessary step is to use two-way valves for control of the AHU coils. If three-way valves are used, then the system requires constant flow through all pumps and chillers and no energy conservation is possible. Either the flow rate or temperature control varies with the number of chiller-pump combinations on-line. With two-way valves, the distribution system flow can vary in proportion to the load. To maintain adequate flow through the chillers at reduced system loads, it is necessary to bypass some flow. The bypass valve is controlled to maintain a constant pressure differential between supply and return mains, sufficient to serve the most remote AHU. This valve is frequently located in the central plant but may be located at the end of the distribution system or at any convenient place. It must be sized for the nominal flow rate of one chiller. As the load and distribution flow decrease to the point where the valve is fully open (or nearly so), a limit switch and alarm light are provided to inform the plant operator that one chiller and pump can be taken off-line. When the bypass valve is fully closed (or nearly so), a similar alarm signals the operator to start a pump and chiller. This avoids the operating cost of unnecessary equipment and allows the on-line equipment to operate more efficiently. The annual saving in pumping cost alone can be very significant. The distribution system can be any of the three types discussed in Chap. 6.

- With the advent of low-cost, reliable, variable-speed pumping capability, a chilled water plant scheme has developed which is becom-

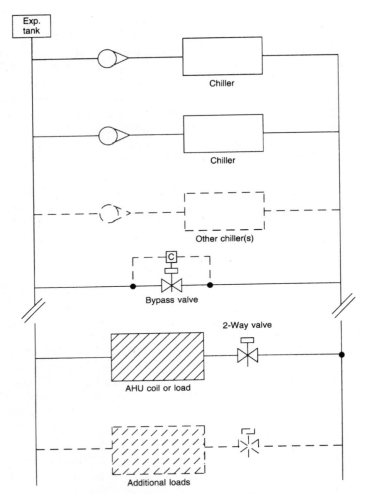

Figure 7.6 Multiple chiller plant with pressure bypass.

ing a favorite in the industry. The concept is illustrated in Fig. 7.7. The plant is set up in a loop with one or more chillers circulated, with a low-head pump for each chiller. The chillers and chiller pumps are staged on and off based on system demand. Coming off the plant supply line, pumps with variable-speed control deliver supply water to the system loads. *Every* load has a two-way modulating valve to meter chilled water to match the load. Load-side coils are preferably oversized to obtain highest reasonable temperature differentials, which also allows the plant supply temperature to be raised in moderate weather conditions. Metering water to each load also obtains the greatest possible system diversity.

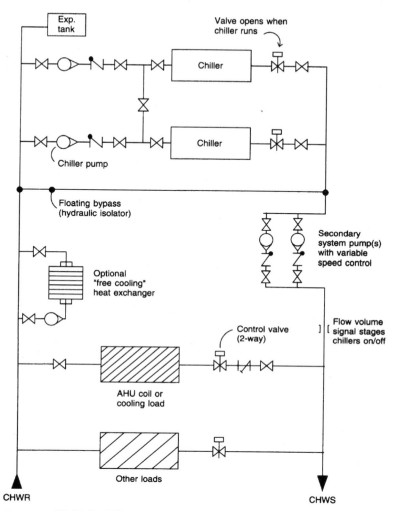

Figure 7.7 Multiple chiller plant with floating bypass, secondary pumping.

This system can utilize chillers of different capacities and easily allows any one of several machines to operate in a standby mode. "Free" winter cooling can be incorporated as part of a chiller or can be added into the system, preferably precooling the system return water. The direction of flow in the floating-bypass line combined with flow measurement in the secondary supply line of the system pumps yields the signal to add or remove chillers from service. Note that a one-to-one relationship between pump and chiller means that failure of either element makes the combination unusable.

The reliability of these systems can be increased by using an additional header between the pumps and the chillers (Fig. 7.8). This allows any combination of pumps and chillers to be used. With either arrangement, two-position manual or automatic control valves must be provided to isolate off-line chillers. Pumps are isolated by the check valve.

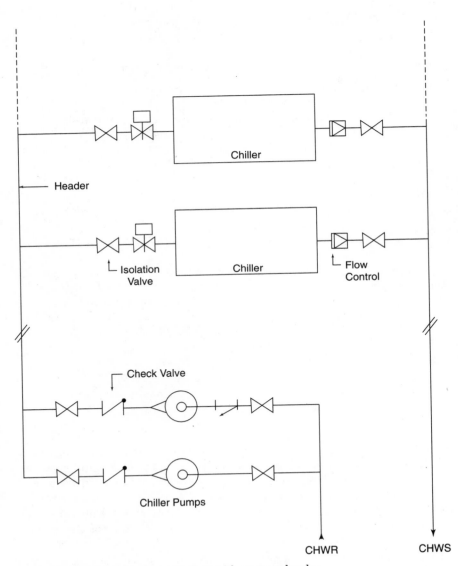

Figure 7.8 Multiple pumps and chillers with common header.

In large campus systems, the piping-head losses in buildings may vary greatly from one building to another. To avoid penalizing the distribution pump, it has been common practice to provide a secondary pump (or pumps) at each building or at least at the high-head-loss buildings. Figure 7.9*a* shows a common method of interfacing the building system to the distribution system. A hydraulic isolator *A-B* is required to prevent the building system pressure variations from affecting the distribution system. The building distribution system can be any of the three basic types. Flow between the building and distribution systems can be modulated by the control valve shown, which responds to building load as it affects the secondary water supply and return temperature. The central plant pump must bear the responsibility for the pressure drop to the building and through the pump.

A disadvantage of secondary pumping is the high parasitic pumping cost which may ensue. Secondary pumps are sized for maximum pressure drop at maximum flow. Taking advantage of load diversity is a key to energy cost control. If secondary pumps are used, they should be fitted with variable-speed control related to secondary system load.

Figure 7.9*b* shows a secondary pumping scheme which has been used by many, but should be avoided. When the three-way valve goes to the full-demand condition, it places the secondary pump in series with the central plant pumps and in hard-coupled parallel with all other secondary pumps. At times of peak demand or at a time when there is insufficient plant capacity on line to satisfy all secondary systems, the secondary pumps start fighting each other for water and the strong units win while the weak units suffer. A common consequence of hard-coupled secondary pumps is a pressure reversal in the system where the central return line pressure rises above the supply line pressure, usually to the consternation of all involved. Decoupling the secondary pumps is a quick fix, but energy savings follow with variable-speed control for all pumps.

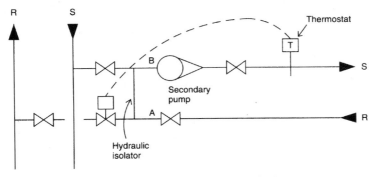

Figure 7.9*a* Secondary pumping with hydraulic isolation.

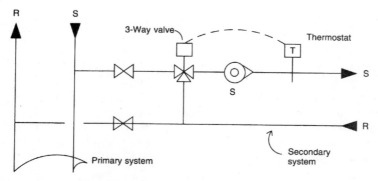

Figure 7.9b Secondary pumping with 3-way valve (not recommended).

Condensing water piping between chillers and cooling towers can be arranged with pump, chiller condenser, and tower cell in a one-to-one relationship (see Fig. 7.10) or, for greater reliability, can be headered and cross-connected so that each chiller can relate to two or more pumps and two or more cooling towers. See Fig. 7.11.

Some systems have been designed to maintain constant flow on even a large scale by putting chillers in series. This allows chillers to stage on and off, but incurs the high cost of constant-flow pumping.

7.8   Thermal Storage Systems

An important variation of the central chilled water plant scheme includes thermal storage. There are several reasons to create a bank of passive cooling capacity, most focused on saving energy cost, but including system reliability in case of chiller failure or power curtailment.

The basic concept of thermal storage which may be applied to chilled or hot water, but most often is used with chilled water, is to use cooling capacity at off hours to reduce the peak electrical demand of the complex. This is meaningful since an electrically driven chiller and pumps may represent 20 to 30 percent or more of the total building demand. Further, many utilities offer financial incentives to owners who will incorporate demand shifting concepts into building system designs. The utility benefit is that a kilowatt of shifted or deferred demand is equivalent to a kilowatt of new generating capacity. Note that there is little actual energy saving in thermal storage. It takes almost as much energy — or more energy — to make chilled water at night as during the day, particularly where storage systems use colder primary supply water temperatures.

The thermal storage media are usually water in large-volume tanks; ice as part of the circulating chilled water system or indirectly made

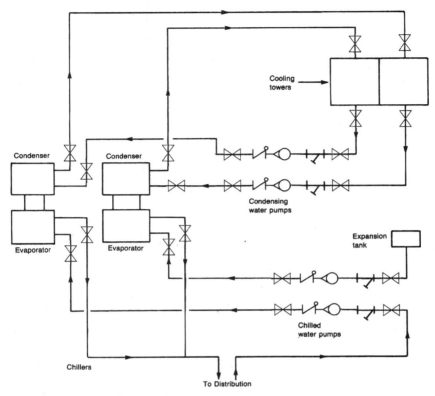

Figure 7.10 Central plant, one-to-one arrangement.

by dropping the chilled water temperature to below freezing (requires a glycol-brine chilled water solution) and building ice on a heat exchange surface; or blocks of encapsulated eutectic salts, where the salts undergo a phase change in the middle of the chilled water supply/return temperature range. The blocks are stacked and involved in the chilled water circulating pattern.

It can be reasoned that ice requires the least total storage volume since the latent heat of freezing is 144 Btu/lb compared to 1 Btu/(lb · °F) (from 40° to 60°F = 20 Btu/lb) for water, with the eutectic salt thermal capacity somewhere in between. In reality, the gross space required for ice storage is about one-third to one-half that required for water. One advantage of ice storage is that it is self-stratifying. Obtaining uniform stratification and full access to storage capacity is a major challenge in water storage systems.

Thermal storage can be easily incorporated into a chilled water system, as indicated in Fig. 7.12. By placing the storage in place of the conventional floating-bypass line, the system pumps can be shut off

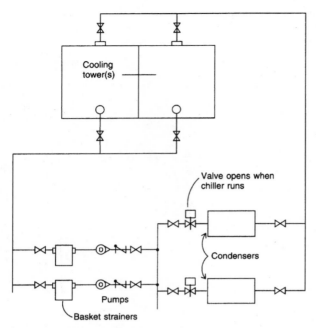

Figure 7.11 Condensing water system for multiple chillers, multiple pumps, and multiple cooling towers.

and the chillers run to load the storage. When the system pumps are on, capacity can be taken from chillers, storage, or a combination of both. The suggested scheme can use water or ice or eutectic salts as a storage medium. If ice is used, the entire system must use a glycol solution, or else the ice chiller and ice storage must be a separate loop from the chilled water.

There are so many variations of production, storage, and use that not all are discussed here. One obvious variation is to make ice with direct-expansion refrigeration. Another is to make ice on an evaporator sheet and then periodically slough it off or scrape it off into a tank which is part of the circulated chilled water system.

Many water-based and some ice-based thermal storage systems wind up being "open," or unpressurized. This creates problems of water treatment, air venting, and possibly increased pumping head.

7.8.1 Sizing of thermal storage systems

Thermal storage is typically sized on the basis of ton-hours. That is, the system will provide so many tons of cooling for so many hours of

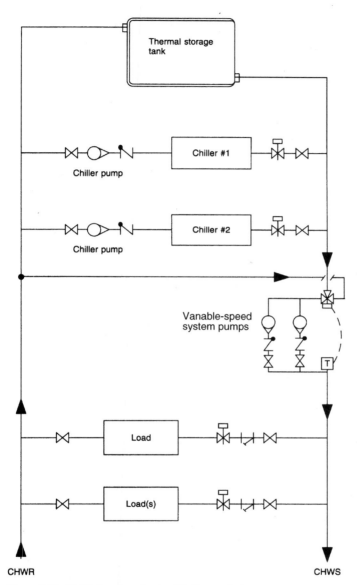

Figure 7.12 Chilled water plant with thermal storage.

time. The converse is that it will take a chiller of so many tons of capacity so many hours to recharge the storage.

For an ice storage system, assume a desired capacity of 2000 ton · h (200 tons for 10 h). The amount of ice required will be (heat of fusion alone)

$$\frac{2000 \text{ ton} \cdot \text{h} \times 12,000 \text{ Btu/(ton} \cdot \text{h)}}{144 \text{ Btu/lb}} = 166,700 \text{ lb ice}$$

If the ice water can be utilized to a 60°F return water temperature, the storage volume can be reduced somewhat:

$$\frac{2000 \text{ ton} \cdot \text{h} \times 12,000 \text{ Btu/(ton} \cdot \text{h)}}{144 \text{ Btu/lb} + (60°F - 32°F) \times 1 \text{ Btu/(lb} \cdot °F)} = 139,500 \text{ lb ice}$$

For a water storage system of the same capacity, assuming water stored at 40°F and utilized to 60°F,

$$\frac{2000 \text{ ton} \cdot \text{h} \times 12,000 \text{ Btu/(ton} \cdot \text{h)}}{(60°F - 40°F) \times 8.3 \text{ lb/gal} \times 1 \text{ Btu/(lb} \cdot °F)} = 14,500 \text{ gal}$$

The benefit of highest possible return water temperature is immediately obvious whether for ice or water storage. Higher return temperatures are obtained with greater heat exchange surface at the load, with counterflow as a required condition. Larger system temperature differentials also allow reduced flow rates which can reduce pumping energy as well as initial piping cost.

A difficult challenge in water storage system design is stratification. Tanks are designed to have the coldest water on the bottom with warmer water returned to or taken from the top. Failure to keep the cold and warm water segregated results in loss of storage capacity. Some early designs used a membrane to physically provide this separation. Sometimes multiple tanks in series can be used to create compartmental segregation. More common is an effort to design intake and outlet structures that yield laminar flow at the top and bottom of the tank while letting the natural difference in the density of the cold supply and warmer return provide the stratification. At best, there is still an interface region between chilled water and return water which must be discounted in calculating the net storage capacity. See Ref. 2 for a further discussion of this thermal storage topic.

7.9 Heat Recovery Plants

The recognition that a machine (chiller) used to generate chilled water rejects low-grade heat that is normally thrown away, but could be used for heating in certain applications, gives rise to a variation in plant design which may have economic desirability. The *coefficient of performance* (COP) of the heating side of a centrifugal or helical-gear chiller is on the order of 6:1 or more:

$$COP_H = \frac{12,000 \text{ Btu/ton} + 0.7 \text{ kW/ton} \times 3410 \text{ Btu} \cdot \text{h/kWh}}{(0.7 \text{ kW/ton})(3410 \text{ Btu} \cdot \text{h/kWh})} = 6.0$$

This means that for every 1 unit of purchased power, 6 units, more or less, of heat can be delivered at near condensing temperatures, given an adequate resource to the evaporator. For many systems, the resource can be the internal heat of a building, recovered by the cooling coils. The reclaimed heat can be used for heating ventilation air, for reheat, for offsetting envelope heat losses, and perhaps for heating domestic water. Well water, lake water, stream waters, cooling-tower waters, etc., have all been successfully used as a heat recovery resource.

The coefficient of performance indicates that electricity, which is often too valuable to use for direct resistance heat, can be applied for less cost than more conventional fossil fuels such as oil and natural gas.

In counterpoint, the fuel utilities are promoting the use of gas or oil in fuel-fired absorption chillers. These chillers require 10,000- to 20,000-Btu input for every ton of effective cooling, and the heating COP can be calculated as

$$COP_H = \frac{12,000 \text{ Btu/(ton} \cdot \text{h)} + 10,000 \text{ Btu/(ton} \cdot \text{h)}}{10,000 \text{ Btu/(ton} \cdot \text{h)}} = 2.2$$

The operating-cost comparison, then, is whether the cost of electricity divided by the electric COP is less than the cost of fuel divided by the absorber COP, with the fuel consumption adjusted for combustion efficiency. Note that there will still be capital cost factors to consider.

There are still other factors to consider in the economic analysis. Some chillers are fitted with double-bundle condensers to segregate a heating loop from a cooling-tower loop. Chillers designed for higher condensing temperatures may not be as efficient in the conventional cooling range. Centrifugal chillers must be carefully selected to avoid surging at reduced capacity in the higher-condensing-temperature range.

Heat recovery plants have the general design characteristics of conventional chiller plants. They can be married to thermal storage, both heating and cooling, and they can be combined with fuel-fired plants for almost any operating mode. One consequence of heat recovery system design is that heating water temperatures range from 90 to 115°F. Radiant and natural convection heating methods don't work well at these temperatures. Fan forced-circulation schemes are required with coils similar to cooling coils for rows and fin spacing.

Design hint: Keep a high degree of separation between the heating loop and the cooling loop. Good intent notwithstanding, cross-connections between hot and cold circuits, even with controlled valving, often fail with consequent blending of hot and cold flows, which places false loads on the central system.

Heat recovery plants may be small or large. Factory-packaged, air-to-air equipment is available, starting at 3 to 5 tons up to 50 tons or more. Chiller-based systems have been designed from 50 tons up to several thousand tons. The technology for such systems is more than fifty years old. See Fig. 7.13 for a simplified diagram of one variation of a chiller-based heat recovery system that uses deep wells as a heat source and as a heat sink.

7.10 Central Plant Distribution Arrangements

The schematic plant arrangements in Figs. 7.1 through 7.8 are representative of common practice for heating or cooling plants in the HVAC industry. One or two units are shown, but the principles apply

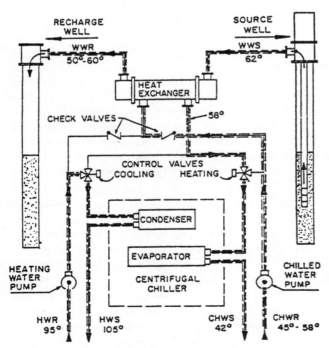

Figure 7.13a Well water based, chiller heat recovery system, heating mode.

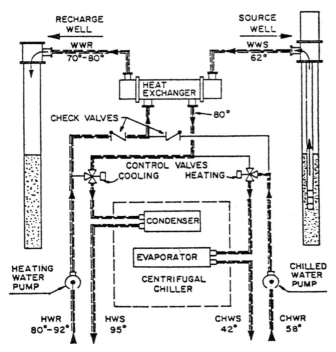

Figure 7.13b Well water based, chiller heat recovery system, cooling mode.

to any number of units. The interface to the distribution system was discussed in Sec. 6.6.8.

All the above figures (Figs. 7.1–7.8) suggest four-pipe systems, in which chilled water and hot water or steam are separated. A two-pipe distribution system is designed to provide either hot or chilled water (but not both at the same time) to air-handling units, usually individual room fan-coil units. Figure 7.14 shows a typical arrangement. The chiller and boiler are interfaced through heat exchangers to a common distribution system, so that either hot or chilled water is distributed, depending on the season (sometimes depending on the time of day). Special thermostats are required for control in the AHUs. Heat exchangers, or specially controlled changeover valves, are needed to prevent thermal shock at the boiler or chiller when changeover occurs, although these valves are not always provided. This type of system is not satisfactory in spring and fall, when changeover may be needed twice each day. A great deal of energy is wasted at each changeover.

A compromise is the three-pipe system (Fig. 7.15), which uses a hot supply pipe, a cold supply pipe, and a common return. Special valves are required at the AHUs and fan-coil units. The single coil in the

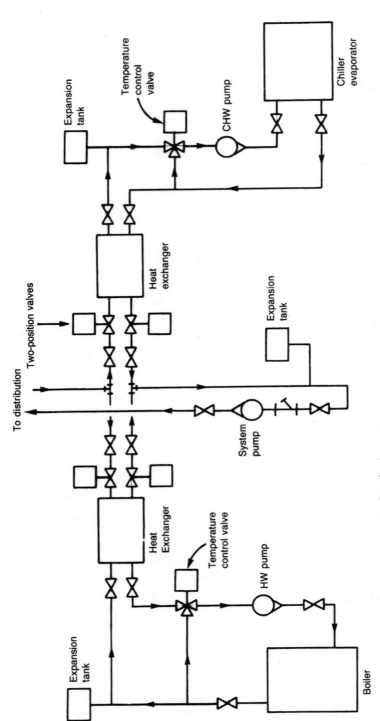

Figure 7.14 Central plant serving a two-pipe distribution system.

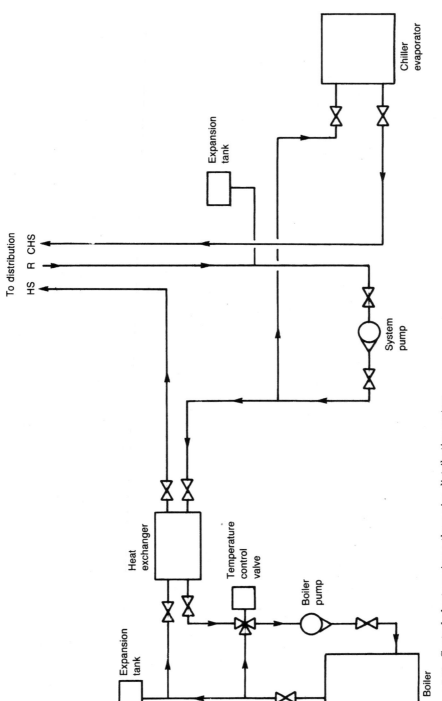

Figure 7.15 Central plant serving a three-pipe distribution system.

219

AHU is used for heating or cooling. Again, there is much wasted energy due to the common return. The purpose, and only benefit, of both the two- and three-pipe systems is to decrease first cost.

Another type of central system provides a neutral temperature water source to individual heat pumps throughout the building. See Fig. 7.16. The design philosophy is that when the heat pumps are in the heating mode, the return water will be colder and heat will be supplied from the boiler. When cooling is needed, the return water will be warmer and will be cooled by the evaporative water cooler. Much of the time, some heat pump units will be heating while others are cooling, in effect transferring energy between those building areas, and very little primary heating or cooling will be needed. There is no outside air or economizer cycle, so some energy must be expended to cool interior zones all year long.

7.11   Cogeneration Plants

Cogeneration implies the simultaneous production of power combined with a beneficial use of the rejected heat from the power production or, inversely, the production of power from the waste heat from a thermal process, often an industrial furnace.

A rather common system uses a reciprocating engine to drive a generator, with the rejected engine heat at 180 to 220°F used for hot water heating functions. A heat exchanger may be used to recover additional heat from the engine exhaust gases. Gas- or oil-fired turbines are also utilized to drive generators, with boilers to recover the available exhaust gas heat. Engines produce more power per unit of fuel than do turbines, but have higher maintenance costs.

Where high-pressure steam is available, a steam turbine driving a generator or piece of equipment may be used as a pressure-reducing device with almost one-to-one conversion of heat to power. This assumes that the lower-pressure steam can be efficiently used.

Simple sketches of the systems described are given in Figs. 7.17, 7.18, and 7.19. These are simple block diagrams to illustrate the concepts. Design of complete plants requires the development of a fully configured flow diagram (P&ID) for the proposed system with subsequent design for each component. Detailed plant design is beyond the scope of this book.

Cogeneration plants require a substantial feasibility study, for they involve major capital expenditure which will be recovered only over time under favorable economic conditions. Absolute commitment of the owner to qualified operation and planned maintenance is required. Conservative projections of fuel and power price conditions should be used. The opportunity to use reject heat should be real. The very best

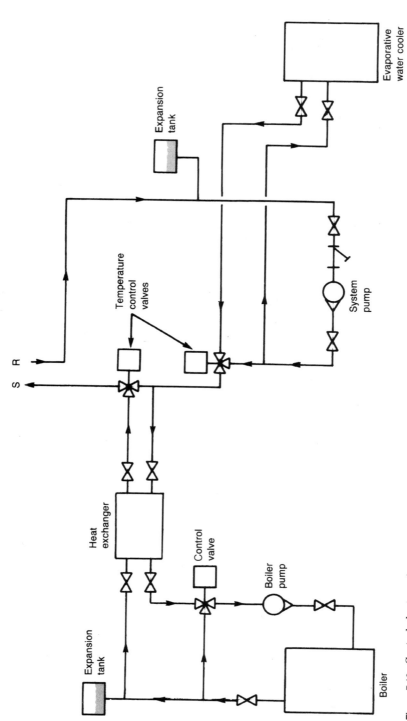

Figure 7.16 Central plant serving a water to air heat pump system.

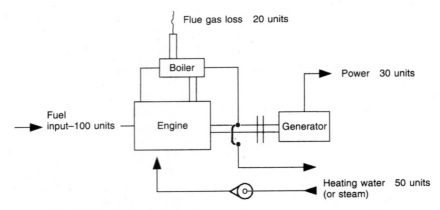

Figure 7.17 Engine driven cogeneration.

systems in economic terms are the ones that can use all the power, and all the heat, all the time. Beware of the cost of standby systems. If power delivery is critical, standby production must be available. If the local utility is used for standby production, it usually has a significant cost factor. Cogeneration implies being a personal utility company, at least to some degree. Except on a large scale. or under unusual circumstances, it is difficult to be more cost-effective than a utility company. Utility companies usually respond to anomalies in rate schedules which favor cogeneration by changing the rate schedules. As an example, remember an earlier day when power companies sold electricity for heating at discount rates. The low rates lasted until demand caught up with supply.

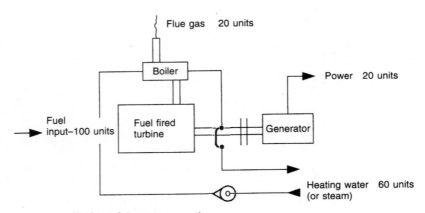

Figure 7.18 Turbine driven cogeneration.

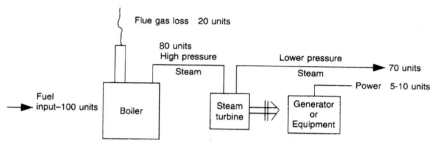

Figure 7.19 Steamboiler/topping turbine cogeneration.

7.12 Summary

Central plants are the heart of the systems which produce heating and cooling capacity for buildings and complexes. They range from small to large, simple to complex. They involve a variety of inputs, manipulations, and outputs, each suited to the needs and resources of the specific project. Smaller plants should generally stay with simple schemes and standard equipment, while large systems can drift toward more sophisticated concepts and more tailored installations. Economic evaluations should support design concepts, and long-term ease of operation and maintenance should be established as a key factor.

This chapter identifies plants for heating, cooling, and combinations of both. There are many variations of the basic ideas offered. There are additional plants related to compressed air, refrigeration, and water treatments which have energy implications, but are left to another discussion. The HVAC systems designer will recognize plants as potential areas of specialty experience as assignments and interest allow.

References

1. ASHRAE Handbook, *1996 HVAC Systems and Equipment,* Chap. 7, "Cogeneration Systems"; Chap. 8, "Applied Heat Pump and Heat Recovery Systems"; Chap. 10, "Steam Systems"; Chap. 11, "District Heating and Cooling"; Chap. 12, "Hydronic Heating and Cooling System Design"; and Chap. 14, "Medium and High Temperature Water Heating Systems."
2. ASHRAE Handbook, *1995 HVAC Applications,* Chap. 40, "Thermal Storage."

Design Procedures: Part 6

Automatic Controls

8.1 Introduction

HVAC systems are sized to satisfy a set of design conditions, which are selected to generate a maximum load. Because these design conditions prevail during only a few hours each year, the HVAC equipment must operate most of the time at less than rated capacity. The function of the control system is to adjust the equipment capacity to match the load. Automatic control, as opposed to manual control, is preferable for both accuracy and economics; the human as a controller is not always accurate and is expensive. A properly designed, operated, and maintained automatic control system is accurate and will provide economical operation of the HVAC system. Unfortunately, not all control systems are properly designed, operated, and maintained.

The purpose of this chapter is to discuss control fundamentals and applications, concisely but completely. For a yet more detailed discussion, see the references at the end of the chapter. The diagrams shown are "generic" and use symbols defined in Fig. 8.68 at the end of the chapter.

Control systems for HVAC do not operate in a vacuum. For any air conditioning application, first, it is necessary to have a building suitable for the process or comfort requirements. The best HVAC system cannot overcome inherent deficiencies in the building. Second, the HVAC system must be properly designed to satisfy the process or comfort requirements. Only when these criteria have been satisfied can a suitable control system be applied.

8.2 Control Fundamentals

All control systems operate in accordance with a few basic principles. These must be understood as background to the study of control devices and system applications.

8.2.1 Control loops

Figure 8.1 illustrates a basic control loop as applied to a heating situation. The essential elements of the loop are a sensor, a controller, and a controlled device. The purpose of the system is to maintain the controlled variable at some desired value, called the *set point*. The process plant is controlled to provide the heat energy necessary to accomplish this. In the figure, the process plant includes the air-handling system and heating coil, the controlled variable is the temperature of the supply air, and the controlled device is the valve which controls the flow of heat energy to the coil. The sensor measures the air temperature and sends this information to the controller. In the controller, the measured temperature T_m is compared with the set point T_s. The difference between the two is the *error signal*. The controller uses the error, together with one or more gain constants, to generate an output signal that is sent to the controlled device, which

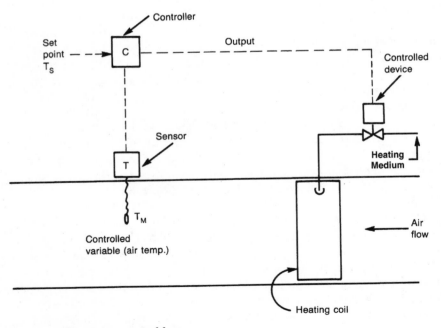

Figure 8.1 Elementary control loop.

is thereby repositioned, if appropriate. This is a closed loop system, because the process plant response causes a change in the controlled variable, known as *feedback,* to which the control system can respond. If the sensed variable is not controlled by the process plant, the control system is open loop. Alternate terminology to the *open-loop* or *closed loop* is the use of *direct* and *indirect* control. A directly controlled system causes a change in position of the controlled device to achieve the set point in the controlled variable. An indirectly controlled system uses an input which is independent of the controlled variable to position the controlled device. An example of a direct control signal is the use of a room thermostat to turn a space-heating device on and off as the room temperature varies from the set point. An indirect control signal is the use of the outside air temperature as a reference to reset the building heating water supply temperature.

Many control systems include other elements, such as switches, relays, and transducers for signal conditioning and amplification. Many HVAC systems include several separate control loops. The apparent complexity of any system can always be reduced to the essentials described above.

8.2.2 Energy sources

Several types of energy are used in control systems. Most older HVAC systems use pneumatic devices, with low-pressure compressed air at 0 to 20 lb/in^2 gauge. Many systems are electric, using 24 to 120 V or even higher voltages. The modern trend is to use electronic devices, with low voltages and currents, for example, 0 to 10 V dc, 4 to 20 mA, or 10 to 50 mA. Hydraulic systems are sometimes used where large forces are needed, with air or fluid pressures of 80 to 100 lb/in^2 or greater. Some control devices are self-contained, with the energy needed for the control output derived from the change of state of the controlled variable or from the energy in the process plant. Some systems use an electronic signal to control a pneumatic output for greater motive force.

8.2.3 Control modes

Control systems can operate in several different modes. The simplest is the two-position mode, in which the controller output is either on or off. When applied to a valve or damper, this translates to *open* or *closed.* Figure 8.2 illustrates two-position control. To avoid too rapid cycling, a control differential must be used. Because of the inherent time and thermal lags in the HVAC system, the operating differential is always greater than the control differential.

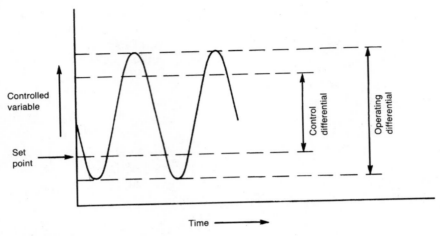

Figure 8.2 Two-position control.

If the output can cause the controlled device to assume any position in its range of operation, then the system is said to *modulate* (Fig. 8.3). In modulating control, the differential is replaced by a throttling range (sometimes called a *proportional band*), which is the range of controller output necessary to drive the controlled device through its full cycle (open to closed, or full speed to off).

Modulating controllers may use one mode or a combination of three modes: proportional, integral, or derivative.

Proportional control is common in older pneumatic control systems. This mode may be described mathematically by

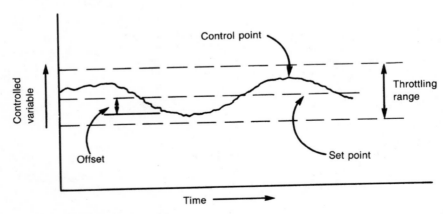

Figure 8.3 Modulating control.

$$O = A + K_p e \qquad (8.1)$$

where O = controller output

A = constant equal to controller output with no error signal

e = error signal

K_p = proportional gain constant

The gain governs the change in the controller output per unit change in the sensor input. With proper gain control, response will be stable; i.e., when the input signal is disturbed (i.e., by a change of set point), it will level off in a short time if the load remains constant (Fig. 8.4).

However, with proportional control, there will always be an *offset*—a difference between the actual value of the controlled variable and the set point. This offset will be greater at lower gains and lighter loads. If the gain is increased, the offset will be less, but too great a gain will result in instability or *hunting,* a continuing oscillation around the set point (Fig. 8.5).

To eliminate the offset, it is necessary to add a second term to the equation, called the *integral mode:*

$$O = A + K_e e + K_i \int e \, dt \qquad (8.2)$$

where K_i = integral gain constant and $\int e \, dt$ = integral of the error with respect to time.

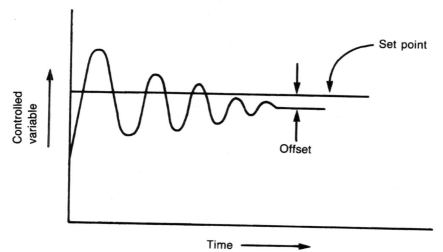

Figure 8.4 Proportional control, stable.

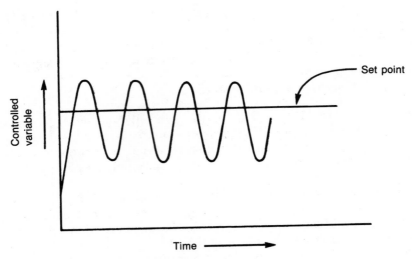

Figure 8.5 Proportional control, unstable.

The integral term has the effect of continuing to increase the output as long as the error persists, thereby driving the system to eliminate the error, as shown in Fig. 8.6. The integral gain K_i is a function of time; the shorter the interval between samples, the greater the gain. Again, too high a gain can result in instability.

The derivative mode is described mathematically by $K_d \, de/dt$, where de/dt is the derivative of the error with respect to time. A control mode

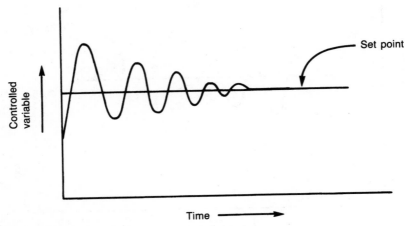

Figure 8.6 Proportional plus integral control.

which includes all three terms is called *PID mode.* The derivative term describes the rate of change of the error at a point in time and therefore promotes a very rapid control response—much faster than the normal response of an HVAC system. Because of this it is usually preferable to avoid the use of derivative control with HVAC. Proportional plus integral (PI) control is preferred, and will lead to improvements in accuracy and energy consumption when compared to proportional control alone.

Most pneumatic controllers are proportional mode only, although PI mode is available. Most electronic controllers have all three modes available. In a computer-based control system, any mode can be programmed by writing the proper algorithm.

8.3 Control Devices

Control devices may be grouped under the four classifications of sensors, controllers, controlled devices, and auxiliary devices. The last group includes relays, transducers, switches, and any other equipment which is not part of the first three principal classifications.

8.3.1 Sensors

In HVAC work, the variables commonly encountered are the temperature, humidity, pressure, and flow.

8.3.1.1 Temperature sensors. The most common type of temperature sensor is the bimetallic type (Fig. 8.7). The element consists of two strips of dissimilar metals, continuously bonded together. The two metals are selected to have very different coefficients of expansion. When the temperature changes, one metal expands or contracts more than the other, creating a bending action which can be used in various ways to provide a two-position or modulating signal. A widely used configuration of the bimetal sensor is in the form of a spiral (Fig. 8.8), allowing greater movement per unit temperature change. Another bimetal type is the rod-and-tube sensor (Fig. 8.9), usually inserted into a duct or pipe. The rod and tube form the bimetal.

The bulb-and-capillary sensor (Fig. 8.10) utilizes a fluid contained within the bulb and capillary. Various liquids and gases are used, each suitable for a specific temperature range. The bulb may be only a few inches long, for spot sensing, or it may be as long as 20 ft, for averaging across a duct. A special application is the low-temperature safety sensor which uses a refrigerant with a condensing temperature of about 35°F. Whenever any short portion of the long bulb is exposed

Movement when heated

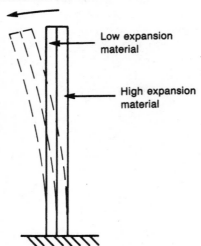

Low expansion
material

High expansion
material

Figure 8.7 Bimetal temperature
sensor.

to freezing temperatures, the refrigerant in that section condenses, causing a sharp drop in the sensor pressure. This can open a two-position switch to stop a fan and to prevent coil freeze-up.

The sealed bellows sensor (Fig. 8.11) operates on the same principle as the bulb-and-capillary sensor. It is usually vapor-filled.

The one-pipe bleed-type sensor (Fig. 8.12) is widely used in pneumatic systems. Control air at 15 to 20 lb/in² gauge is supplied through a small metering orifice. A flapper valve at a nozzle is modulated by one of the previously described temperature sensors or by sensors for flow, pressure, or humidity. As the valve varies the nozzle airflow, pressure builds up or reduces in the branch line to the controller. By add-

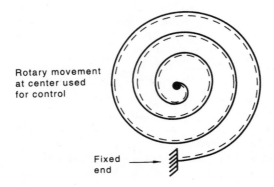

Rotary movement
at center used
for control

Fixed
end

Figure 8.8 Spiral bimetal temperature sensor.

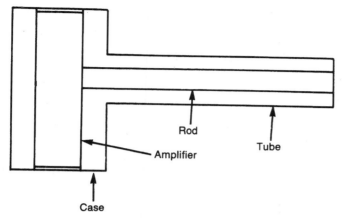

Figure 8.9 Rod-and-tube temperature sensor.

ing appropriate springs and adjustments, this device can also be used directly as a proportional controller.

Modern electronic control systems use some form of resistance or capacitance temperature sensor. Widely used is the thermistor, a solid-state device in which the electrical resistance varies as a function of temperature. Most thermistors have a base resistance of 3000 Ω (or more) at 0°C and a large change in resistance per degree of temperature change. This makes the thermistor easy to apply, because the resistance of wire connections (leads) is small compared to that of the thermistor. Thermistor response is very nonlinear, but circuitry can be added to provide a linear signal. The principal objections to thermistors are (1) their tendency to drift out of calibration with time (although this can be minimized with proper factory burn-in) and (2) the

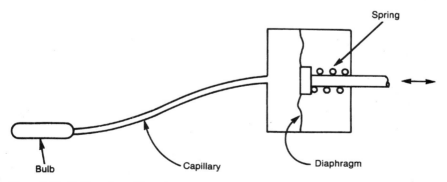

Figure 8.10 Bulb-and-capillary temperature sensor.

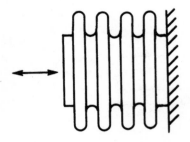

Figure 8.11 Bellows temperature sensor.

problem of matching a replacement to the original thermistor (manufacturers will provide "replaceable" devices at extra cost).

Resistance temperature detectors (RTDs) are made of fine wire wound in a tight coil. The resistance to electric current flow varies as a function of temperature. Various alloys are used. One alloy, with the tradename Balco, has a base resistance of 500 Ω at 0°C. The best RTDs are made of platinum wire. The platinum RTD has a low base resistance — 100 Ω at 0°C — so three- or four-wire leads must be used. Platinum RTDs are very stable, showing little drift with time. Another type of RTD is made by thin-film techniques, with a platinum film deposited on a silicon substrate. Resistance varies with temperature, and high base resistance can be obtained; for example, 1000 Ω at 0°C. All these electronic sensors can be obtained in several configurations, for room or duct or pipe mounting.

8.3.1.2 Humidity sensors. Many hygroscopic (moisture-absorbing) materials can be used as relative-humidity sensors. Such materials absorb or lose moisture until a balance is reached with the surrounding

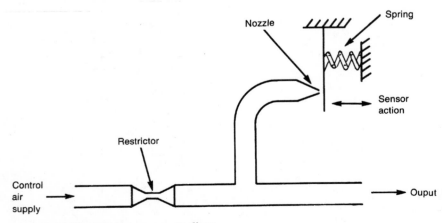

Figure 8.12 Bleed-type sensor controller.

air. A change in material moisture content causes a dimensional change, and this change can be used as an input signal to a controller. Commonly used materials include human hair, wood, biwood combinations similar in action to a bimetal temperature sensor, organic films, and some fabrics, especially certain synthetic fabrics. All these have the drawbacks of slow response and large hysteresis effects. Their accuracy tends to be questionable unless they are frequently calibrated. Field calibration of humidity sensors is difficult.

A different style of absorption-type dew-point sensor uses a tape impregnated with lithium chloride and containing two wires connected to a power supply. As moisture is absorbed by the lithium chloride, a high-resistance electric circuit is created, which heats the system until the system is in balance with the ambient moisture. The resulting temperature is interpreted as the dew point. This device is accurate when maintained at regular, frequent intervals; dirt in the system diminishes its accuracy.

Thin-film sensors are now available which use an absorbent deposited on a silicon substrate such that the resistance or capacitance varies with relative humidity. These are quite accurate—± 3 to 5 percent—and have low maintenance requirements.

The more accurate dew point sensor is the chilled-mirror type shown in Fig. 8.13. A light source is reflected from a stainless-steel mirror to a photocell. The mirror is provided with a small thermoelectric cooler. When it is cooled to the dew point, condensation begins to form on the

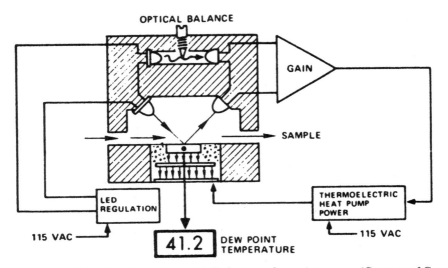

Figure 8.13 Principle of operation of chilled-mirror dew-point sensor. (*Courtesy of General Eastern Instruments, Watertown, Massachusetts.*)

mirror face, the reflectivity changes, the fact is noted, and the mirror temperature is read as the dew point temperature. The dew-point temperature establishes the moisture content of the air. The dew-point temperature combined with the ambient condition yields the relative humidity (RH) by calculation. This system has a high degree of accuracy, within ±1°F, which allows calculation of relative humidity to an accuracy of ±2 to 3 percent. Maintenance consists of occasionally cleaning the mirror. The device can be obtained for duct or wall mounting and with circuitry to provide a dew-point temperature or a relative-humidity signal.

8.3.1.3 Pressure sensors. For sensing differential pressure, some type of diaphragm is used (Figs. 8.14 and 8.15). The diaphragm separates the two halves of a closed chamber, with one of the two pressures introduced on each half, or one-half may be open to atmosphere as a reference. Diaphragm materials may be a flexible elastomer or thin metal, depending on the use and pressure range. Sensors are available for pressures ranging from a few inches of water to several thousand pounds per square inch gauge. The flexing of the diaphragm as the pressures change is amplified in various ways to provide a modulating signal, which can be used for modulating or two-position control. One special diaphragm application, called a *piezometer,* utilizes a crystalline structure in which the electric current flow varies as the crystal is deformed.

A *bellows* (Fig. 8.16) is a corrugated cylinder which expands or contracts linearly as the pressure changes. The input pressure is always compared to atmospheric pressure.

A *bourdon tube* (Fig. 8.17) is a closed semicircular tube, which tries to straighten as the pressure is increased. This is the sensing element in most dial pressure gauges, but it is seldom used in control systems.

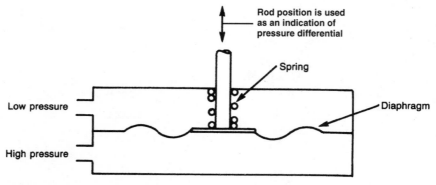

Figure 8.14 Diaphragm pressure sensor.

Figure 8.15 Diaphragm pressure sensor, PMND pneumatic transmitter. (*Courtesy of TSBA Controls, Inc.*)

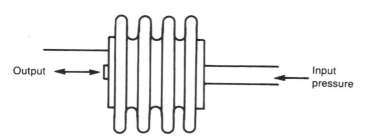

Figure 8.16 Bellows pressure sensor.

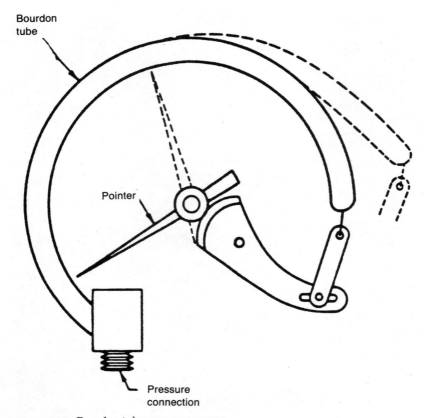

Figure 8.17 Bourdon tube pressure sensor.

Note that the air velocity measured with a hot-wire anemometer (Sec. 8.3.1.4) can be used to measure an air pressure difference between two adjacent spaces, with pressure being indicated as a function of the airflow velocity.

8.3.1.4 Flow sensors. In HVAC work, it is often necessary or desirable to measure or detect the flow of air, water, steam, or other gases. Several methods are available.

A *sail switch* is a two-position device for detecting airflow or no flow. It consists of a lightweight sail mounted in the duct and connected to close a switch when the sail is displaced by air movement. A similar device is used in piping to detect liquid flow. These devices have a tendency to "stick" in the open or closed position or to oscillate between open and closed, providing a false signal. Some engineers believe that a more reliable flow/no-flow indication can be obtained by reading the pressure differential across the fan or pump.

The *hot-wire anemometer* airflow sensor (Fig. 8.18) includes a small electric resistance heater and a temperature sensor. The air flowing over the heater has a cooling effect, and the air velocity is proportional to the amount of electric energy required to maintain a reference temperature. A variation of the concept imposes a fixed voltage across the resistance, with circuitry to read the current flow which changes with sensed temperature and with flow. Hot-wire anemometers require a reference to neutralize the effect of changing temperature on the output signal.

The *pitot tube* (Fig. 8.19) is a double tube, installed in a duct or pipe so that the tip points directly into the fluid flow and therefore measures the total pressure (TP). Openings in the outer tube face at right angles to the flow and measure static pressure (SP) only. When these two pressures are conveyed to a differential pressure sensor or a manometer, the difference between them can be read and is equal to *velocity pressure (VP)* from the equation

$$VP = TP - SP \qquad (8.3)$$

The velocity V can then be determined from

$$V = C\sqrt{VP} \qquad (8.4)$$

where, for HVAC work, V is in feet per minute and VP is in inches of

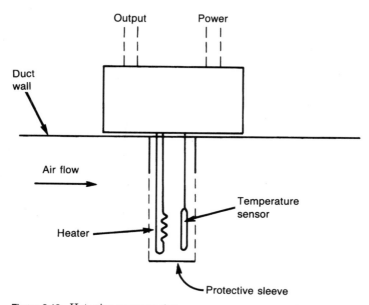

Figure 8.18 Hot-wire anemometer.

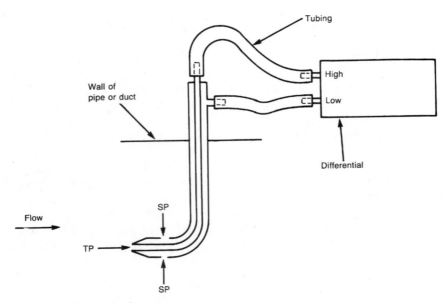

Figure 8.19 Pitot-tube flow sensor.

water; C is a constant related to the density of the fluid. For standard air, C is equal to 4005. For other than standard air, the value of C is corrected by dividing 4005 by the square root of the new air density ratio. For example, the air density ratio at 5000-ft elevation (with respect to standard air) is 0.826. The value of C at 5000 ft is therefore

$$C_{alt} = 4005/\sqrt{0.826} = 4400$$

The pitot tube can be used for measuring the velocity pressure of any fluid with a known density. It is often used for measuring water flow. The accuracy depends on the accuracy of the device being used to measure the differential pressure.

The *orifice plate* (Fig. 8.20) is used for measuring the flow of all types of fluids. In HVAC systems, it is used primarily for water and steam flow but can also be used for airflow. The reduction of the conduit cross section causes an increase in fluid velocity and velocity pressure, thereby reducing the static pressure. The change in static pressure can be measured and used to determine total flow rate

$$Q = CA\sqrt{H} \qquad (8.5)$$

where Q = flow rate
     A = cross-sectional area of conduit
     H = static-pressure change

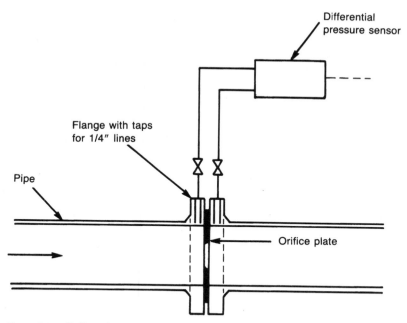

Figure 8.20 Orifice plate flow measurement.

C = constant relating to orifice and conduit areas and fluid density

All values must, of course, be in consistent units.

The orifice plate is simple and relatively inexpensive, and therefore it is widely used. It creates a dynamic loss because of the abrupt contraction and expansion. Its accuracy falls off rapidly as flow decreases below about 20 percent of design. (The turndown ratio is, therefore, about 5:1.)

The dynamic losses can be eliminated or decreased by using a *venturi* (Fig. 8.21), in which the area changes are gradual. The flow equation for the venturi is

$$Q = C\sqrt{H} \tag{8.6}$$

where C is a constant determined by the manufacturer.

A turbine flow meter is a small propeller mounted in the fluid stream and connected by a gear train to a measuring/totalizing mechanism. The propeller speed is proportional to the fluid velocity. Because of hysteresis and friction in the gear train, the accuracy of measurement will vary in a nonlinear way and must be corrected for. The

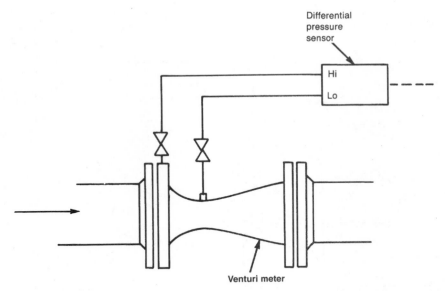

Figure 8.21 Venturi meter flow measurement.

rotating-vane anemometer, used for measuring airflow, is also a turbine type of meter.

The *paddle wheel* type (Fig. 8.22) is turned by the flowing fluid, with speed being proportional to velocity. This device is connected magnetically to the measuring mechanism and is accurate over a wide range of flow.

There are other ways of sensing flow rates, but those described above are representative of HVAC use. Note that all the modulating sensors may also be used for two-position flow detection.

8.3.2 Controllers

Controllers may be classified by the type of control action and type of energy used for the control signal. Control action may be two-position or modulating, with modulating control utilizing proportional, integral, or differential modes or some combination of these. Control energy sources include pneumatic, hydraulic, electric, electronic, and self- or system-generated. The continued development of transistor technology into microchip technology with programmable control adds the words *analog* and *digital* to the vocabulary. *Analog* suggests continuous modulation of a signal by voltage or current flow, while *digital* implies discrete or on/off. However, digital signals can approximate analog signals by stepping a controlled device toward the open or

Figure 8.22 Paddle wheel flow sensor. (*Courtesy of Signet Industrial.*)

closed position. In fact, the accuracy of digitally controlled circuits and devices is coming to be state of the art.

8.3.2.1 Two-position controllers. Perhaps the simplest two-position control is derived from the bimetal sensor, as in Fig. 8.23. The bending action of the bimetal is used to make or break an electrical contact. The set point is adjusted by moving the fixed contact in or out. To provide a "snap-action make-and-break," a small magnet is used. This system has been largely superseded by one consisting of a bimetal helical coil with a mercury switch mounted at its center (Fig. 8.24). The switch mounting is flexible so that when the switch tilts, the mercury running to one end will cause a snap action.

All the modulating-type sensors can be connected to two-position switches of the mechanical or mercury type.

8.3.2.2 Modulating controllers. The methods which controllers use to determine the value of the output signal have already been discussed. To be considered here are the various energy types used by HVAC controllers.

By far the most common, historically, are pneumatic devices. The principle of the nonbleed relay-type pneumatic controller is shown in

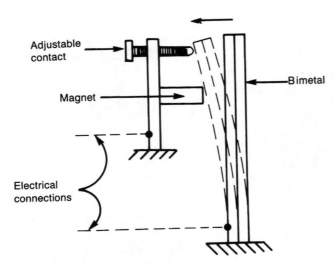

Figure 8.23 Bimetal two-position controller.

Fig. 8.25. When the sensor causes a downward movement against the lever, the air supply valve opens and air pressure increases in the chamber and in the output. The flexible diaphragm pushes upward against the sensor action until the air supply valve closes at some new balance point. This is internal feedback. When the sensor action is upward, the exhaust valve opens and some air bleeds out, reducing the pressure to a new balance point. The gain or sensitivity to change in the sensed variable is adjusted by varying the length of the lever arm. The set point is adjusted by varying the spring tension at the

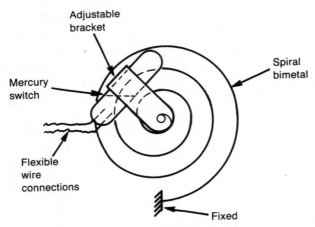

Figure 8.24 Spiral bimetal mercury switch controller.

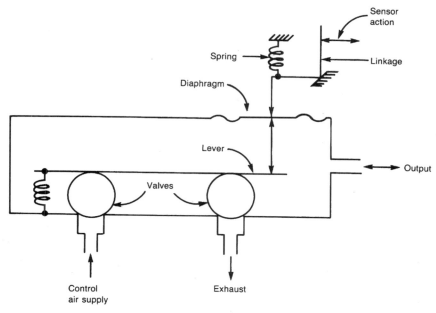

Figure 8.25 Nonbleed relay-type controller.

sensor action. This is the classical method of building a pneumatic controller.

The relay-type controller as shown in Fig. 8.25 uses only proportional mode. It is possible to add reset functions and integral-mode operation, although this makes a more complex device.

The bleed-type pneumatic sensor already discussed (Fig. 8.12) can become a proportional controller if the air supply orifice is adjustable—for gain adjustment—and the sensor-nozzle combination is provided with a set point adjustment.

A common method of obtaining an electric modulating output employs a rheostat—a variable resistance. The rheostat may be circular (Fig. 8.26) or linear. It forms part of an electric circuit, with current flowing in at one end and out through the moving-arm contact. As the arm moves in response to a modulating sensor, the amount of resistance varies; therefore, the output voltage varies. The gain is a function of resistance per unit length and speed of arm travel in relation to change in sensor input. The set point is adjusted by changing the starting point of the moving arm.

The *Wheatstone bridge* (Fig. 8.27) is used in some form in most electric and electronic controllers. The principle of the bridge circuit is that in a balanced bridge all four resistances are equal. When power is applied, the voltages at the two output terminals are equal, and a

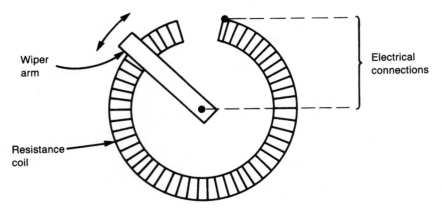

Figure 8.26 Circular rheostat.

meter placed across those terminals shows a zero difference in potential. If one of the resistances is variable (as indicated by the arrow across it) and is, in fact, varied, then there will be a difference in voltage between the output terminals that is proportional to the change in resistance. A basic bridge controller (Fig. 8.28) includes a change-in-resistance type of sensor, an adjustable set point resistor, and a calibrating resistor.

Electric controllers generally operate with power supplies of 24 V ac or more. An electronic controller usually needs a 24-V dc power supply and provides an output signal in a range of 0 to 5 or 0 to 10 V or 4 to 20 or 10 to 50 mA. The preferred ranges are 0 to 10 V or 4 to 20 mA. Most electronic controllers are designed around a device called

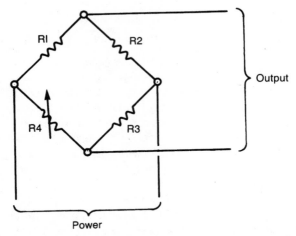

Figure 8.27 Wheatstone bridge.

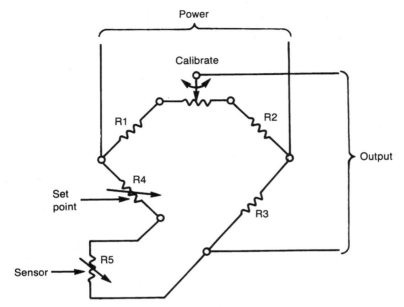

Figure 8.28 Bridge circuit with calibration and set point.

an *op amp* (operational amplifier). This solid-state device provides an almost infinite amplification of an input signal. By means of appropriate circuitry, the op amp can add or subtract signals and provide proportional, integral, or derivative functions. These may be combined as shown in Fig. 8.29 to make a PID controller or any desired com-

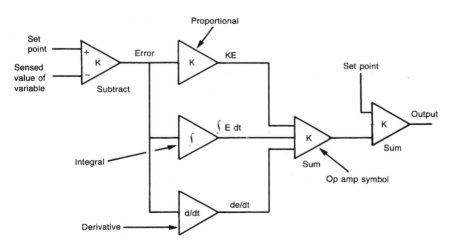

Figure 8.29 PID controller using op amps.

bination thereof. For further discussion, consult a good electronics text or Refs. 1 and 3 at the end of this chapter. For the use of a computer as a controller, see Sec. 8.7.

The sensor and controller are often combined in one package, although they serve separate functions. This package is referred to as a *stat*—thermostat, humidistat, pressurestat, and flowstat. The most common arrangement is wall-mounted in a room, but duct-mounted stats are also frequently encountered. Electronic control systems often use room sensors wired to remotely located control boards.

8.3.3 Controlled devices

In HVAC work, controlled devices are usually valves, dampers, or motors. As previously noted, these devices may be controlled in two-position or modulating modes.

8.3.3.1 Control valves. A control valve (Fig. 8.30) includes a body, within which are passages for fluid flow; a seat; and a plug. The plug is connected to a stem, which in turn is connected to an operator. Where it penetrates the body, the stem is provided with some kind of seal or packing to prevent loss of fluid. Many different materials are used for the various elements to suit the requirements of temperature, pressure, and fluid characteristics.

The three principal plug types encountered are shown in Fig. 8.31*a*, *b*, and *c*. *Plug lift* refers to opening the valve by lifting the plug off the seat. The flat or quick-opening plug is suitable only for two-position control, as shown in Fig. 8.32; a very small lift results in a

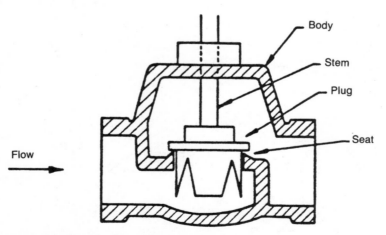

Figure 8.30 Straight-through (two-way) control valve.

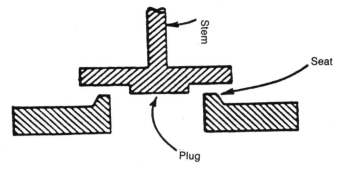

Figure 8.31a Quick-opening (flat seat) valve.

large change in the design flow rate. The linear plug is notched or tapered in a straight line; this results in a linear or near-linear response to lift. The equal-percentage plug is notched or tapered in a curved shape, resulting in an exponential response. The origin of the curves in Fig. 8.32 is not at zero flow. This is because the valve must be constructed with some clearance between plug and port to prevent sticking in the closed position; when the valve is cracked open, some minimum flow occurs. The amount of clearance, and therefore the minimum flow rate, is a function of the valve design and quality. The ratio of this value to 100 percent flow is known as the *turndown ratio*. For a typical commercial-quality valve, this minimum flow is about 5 percent, making the turndown ratio 100:5 or 20:1. Ratios of 50:1, 100:1, or even 200:1 are available, but 20:1 is acceptable for most HVAC work.

The equal-percentage plug is used for modulating control of water flow because it has a lower ratio of flow increase to lift increase in the

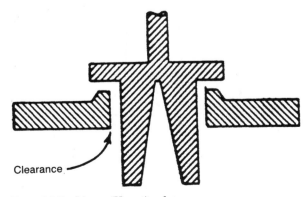

Figure 8.31b Linear (V-port) valve.

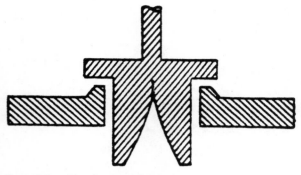

Figure 8.31c Equal-percentage valve.

region near to closure. This tends to offset the effect shown in Fig. 8.33. This figure shows the output of a heating coil as a function of the flow rate of hot water through the coil. At lower flow rates, the water temperature drop increases, with a resultant increase in heating capacity per pound. Thus, a flow rate decrease of 80 percent may result in an output decrease of only 20 to 30 percent. The shape of the curve is determined by the temperature difference between the entering water and the entering air; as the difference decreases, the curve

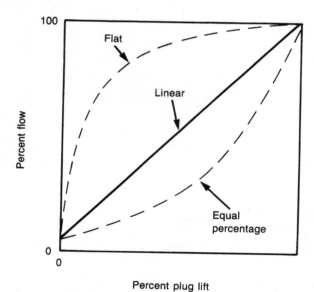

Figure 8.32 Flow versus plug lift in a control valve.

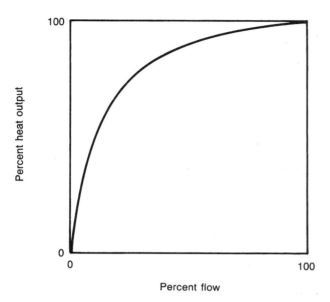

Figure 8.33 Coil heating capacity versus hot water flow rate.

becomes flatter. A similar effect, although not so pronounced, is obtained in a cooling coil.

In Fig. 8.30 the fluid flow direction is indicated as tending to open the valve, i.e., to raise the plug off the seat. With flow in the opposite (wrong) direction, as the valve plug approaches the nearly closed position, the velocity pressure of the fluid will tend to completely close it — there is sufficient hysteresis in the operator linkage to allow this. When the valve closes and fluid flow stops, the velocity pressure goes to zero. This allows the valve to crack open, flow is restored, and the process is repeated. The result is called *chatter,* which results in *water hammer,* which can sometimes destroy a valve or piping system. Each control valve has an arrow cast into the outside of the body to show the flow direction. Three-way valves are made for mixing or diverting service (Fig. 8.34*a* and *b*) and must be used only in the designated arrangement. Because the fluid pressure differential tends to open the two-way valve, the valve operator must be designed to provide tight closure at the shutoff pressure differential expected, which may be as high as the "shutoff" head of the pump or the static pressure of the system.

The appropriate valve size is determined by means of the valve constant C_V. This number is furnished by the manufacturer and is used to relate capacity to pressure drop or head:

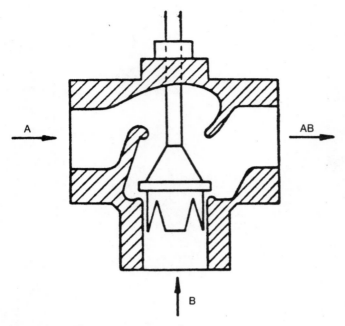

Figure 8.34a Three-way mixing valve.

$$Q = C_V\sqrt{H}$$

where Q = flow rate, gal/min for water, lb/h for steam
        H = head loss (pressure drop), usually in lb/in^2
        C_V = determined with valve wide open and decreases as valve
               modulates closed

Selection of the control valve is made by determining the flow rate
needed, assuming a head loss, determining a C_V, and selecting a valve
with that C_V from the manufacturer's catalog. Because the calculated
C_V seldom equals a catalog C_V rating, most valves are chosen to be
somewhat oversized. Shutoff pressure, plug type, and valve materials
are also part of the selection criteria.

The design head loss is very important in determining the effectiveness of the control valve. Figure 8.35 shows a subsystem with a
water-type heating or cooling coil, branch piping from supply and return mains, and a control valve. This subsystem could easily have a
pressure loss of 20 to 25 ft without the control valve. For the valve to
be effective, its wide-open pressure drop must be a significant fraction
of the total subsystem drop, say 35 to 50 percent. In this case, a 10–15-
ft or greater valve head loss would be required. A point sometimes

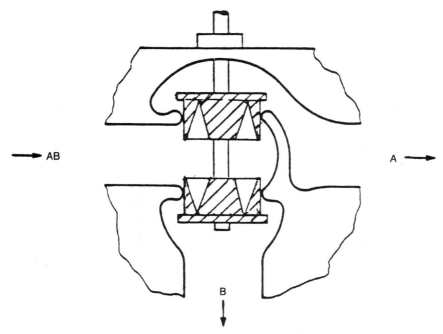

Figure 8.34b Three-way diverting valve.

overlooked in selecting a control valve for steam (or any gas) is the phenomenon of *critical flow.* The control valve acts as an orifice, and there is some minimum ratio of downstream to upstream pressure below which flow will not increase—therefore, there is a maximum flow through that valve at any given inlet pressure. For steam, the

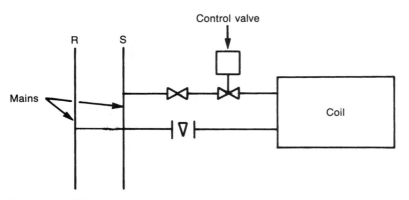

Figure 8.35 Coil, valve, and piping subsystem.

ratio is about 0.53. Most sizing charts and tables include this critical-pressure ratio.

Valve operators (or motors) may be pneumatic, electric, electronic, or hydraulic. Hydraulic operators are used for large valves (8 to 10 in and larger), where the higher control fluid pressures (80 to 100 lb/in^2) allow the use of smaller operators. Electric operators are two-position or modulating mode, the latter using a motor/gear system which takes 30 s to 2 min to drive from one extreme to the other. Electronic operators may be similar to electric, or they may use a stepping motor, which is driven a fractional distance by a short electric pulse; the number and duration of the pulses determine the total distance traveled.

Pneumatic operators are simple pistons, driven by air pressure to one extreme and returned by a spring when the pressure is reduced. There are double-acting pistons driven by air in both directions for some applications. For accurate operation when modulating action is required—to overcome hysteresis and friction—a positive or pilot positioner is recommended (Fig. 8.36). This device is an "amplifier" which uses a small controller output signal change to provide full control air pressure or exhaust to move the piston. A built-in feedback signal limits the travel to an amount proportional to the change in the controller signal.

The pneumatic operator has, at this writing, advantages of simplicity, low maintenance, and low cost, compared with electric and electronic operators. A general exception is a control system with only a few valves and no available compressed air supply.

There are new electrically driven actuators coming to market to work with the *direct digital control* (DDC) systems. But the simplicity and benefits of compressed-air-driven operators still support their use in many instances.

Different from all these is the *self-contained valve,* which utilizes the change of state of the controlled variable such as temperature or pressure to provide operating power. The device includes the sensor, controller, and operator in single package. Self-contained valves typically use the hydraulic power of a bulb sensor or the electric power of a thermocouple to initiate action.

8.3.3.2 Air control dampers. Most control dampers are the parallel or opposed multiblade type, square or rectangular (Fig. 8.37), although single-blade round dampers are also used. In parallel-blade dampers, all blades rotate in the same direction. In opposed-blade dampers, contiguous blades rotate in opposite directions. This results in different control characteristics, as shown in Figs. 8.38 and 8.39. Note that to approach "linear" response, the parallel-blade damper must have a

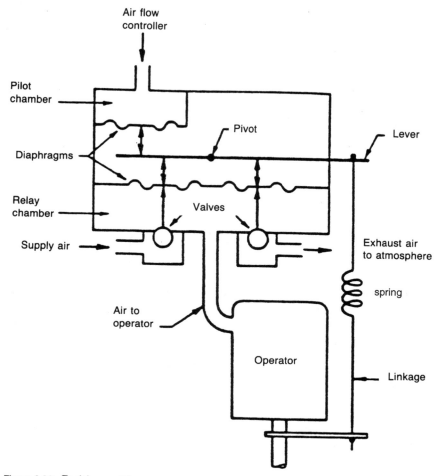

Figure 8.36 Positive positioner.

wide-open pressure drop of 30 to 50 percent of the total system pressure loss, while the opposed-blade damper can do the same at 15 to 20 percent. Because the total air system pressure drops without dampers are often in the range of 2 to 3 in H_2O or higher, very few dampers are actually sized to provide good control. A typical multiblade damper will have a wide-open pressure drop of about 0.1 in H_2O at 1000 ft/min (thus 0.4 in H_2O at 2000 ft/min, 0.9 in at 3000 ft/min, etc.). Because typical duct air velocities range from 1000 to 2000 ft/min, a properly selected control damper could be much smaller than the duct. If a blank-off plate is used to accommodate the smaller damper, there will be turbulent losses due to the orifice effect, which may be greater

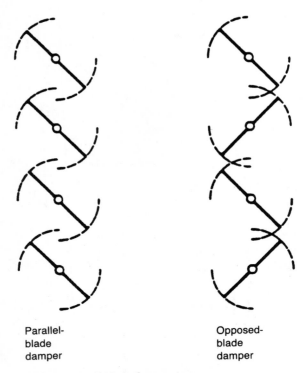

Parallel-
blade
damper

Opposed-
blade
damper

Figure 8.37 Multiblade damper types.

than the effect of the damper. Ideally, then, the duct size reduction and increase should be gradual, as in a venturi (Sec. 8.3.1.4), although the available space may not allow this.

For two-position operation, parallel-blade dampers are preferred, because they are less expensive than the opposed-blade type.

For use in outside air and relief and exhaust air openings, low-leakage dampers are preferred to minimize the possibility of freezing coils.

Damper operators are similar in construction concept to operators used on valves; but the mountings and linkages and sometimes the body designs are different.

8.3.3.3 Motors. Motor control in HVAC systems includes start/stop and variable-speed control. Included in this section is the control of devices which vary the speed of driven equipment while allowing the motor to run at constant speed.

Start/stop control of small motors is easily accomplished directly from the controller, or if the current-carrying capacity of the controller

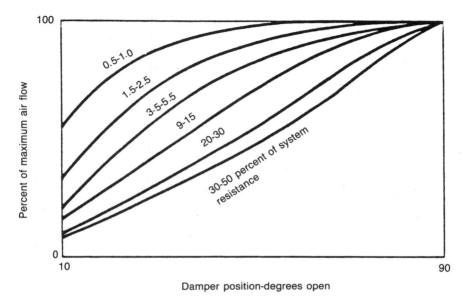

Figure 8.38 Parallel-blade damper flow characteristics. (*Reprinted by permission from ASHRAE PDS-2 Workbook.*)

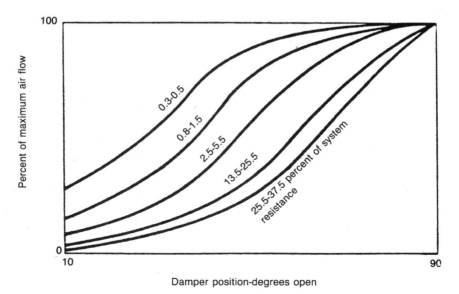

Figure 8.39 Opposed-blade damper characteristics. (*Reprinted by permission from ASHRAE PDS-2 Workbook.*)

is not adequate, a relay or an independent motor starter can be used. The control signal may come from a source such as a temperature or pressure switch, a programming clock, a limit switch, a flow switch, or another motor. A typical control circuit for an across-the-line motor starter is shown in Fig. 8.40*a* and *b*. Larger motors may require star-delta, or auto transformer, or reduced-voltage (Fig. 8.41) starting to limit the current drawn during start-up. This control circuit includes a time delay during which voltage and current are low during the initial acceleration period, then are higher during a final stage at full voltage to complete the acceleration to operating speed. Several different methods are used; consult a motor starter manufacturer's catalog for details.

Two-speed motor control is also used, with manual or automatic change from fast to slow. For equipment with large inertial forces, such

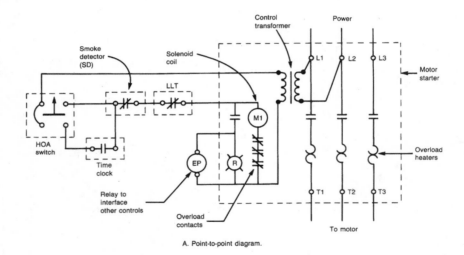

A. Point-to-point diagram.

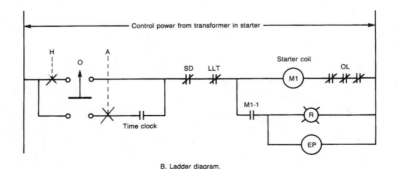

B. Ladder diagram.

Figure 8.40 Across-the-line motor starter.

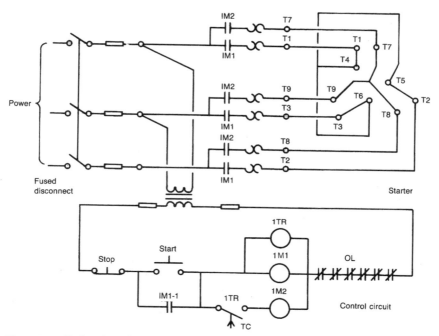

Figure 8.41 Reduced-voltage (part-winding) starter.

as large fans, it is best to provide time delays for both acceleration—to minimize starting current—and deceleration from fast to slow speeds—to minimize dynamic braking effects, which can seriously damage motors.

Mechanical fan or pump speed control can be accomplished through variable-speed transmission systems or by varying the motor speed. Mechanical belt-and-pulley systems in which the pulley diameter is changed to control the speed are suitable for small, low-horsepower fan systems (less than 1 hp). Magnetic or hydraulic clutches have been very popular for both fan and pump applications on larger systems. Recent developments in variable-frequency motor speed controls have improved the reliability and lowered the costs of these devices until they are now preferred for most applications. A variable-frequency speed controller (Fig. 8.42) is a solid-state device which takes the standard 50- or 60-cycle input power and varies the output frequency in response to an input signal from an electronic or pneumatic controller. Because the speed of a standard squirrel-cage motor is a direct function of frequency, this provides very accurate speed control down to about 30 percent of nominal speed. The device has a minimum speed adjustment; recommended settings are from 30 to 50 percent. The speed control takes the place of the conventional motor starter and

Figure 8.42 Variable-frequency drive (VFD). (*Courtesy of Toshiba International Corp.*)

can be adjusted to limit the starting current to provide a "soft start," in which acceleration varies as a function of speed. This reduces start-up stress on both the motor and the drive transmission (belts or couplings). Refer to Chap. 12 for a more detailed discussion of variable speed drives.

8.3.4 Auxiliary control devices

This classification includes relays, transducers, switches, and timers.

8.3.4.1 Relays and transducers. The dictionary speaks of a *relay* as a device for amplifying, modifying, or passing on a signal. Relays may be pneumatic, hydraulic, electric, electronic, or optical and may have many different functions. A *transducer* is a special kind of relay which deals with modulating signals and changes a signal from one form of energy to another or which amplifies the signal in the same energy form.

The most common traditional form of a relay is the *electromechanical relay* (Fig. 8.43*a* and *b*), a two-position device which utilizes an electromagnetic coil, called a *solenoid,* to move a plunger or an armature to open and/or close electrical contacts. The *time-delay relay* (Fig. 8.44) is similar, but the movement of the plunger is restricted by, in this illustration, a diaphragm enclosing an air chamber from which air bleeds slowly through a leak port. The time interval is adjusted by varying the leak-port orifice. In modern practice, most time-delay re-

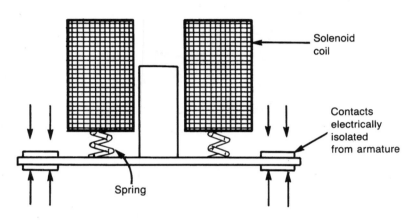

A. Plunger type

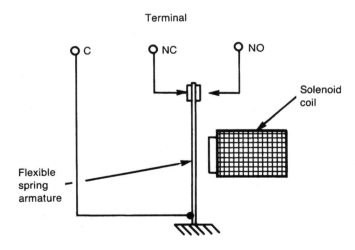

B. Reed type

Figure 8.43 Electromechanical relays.

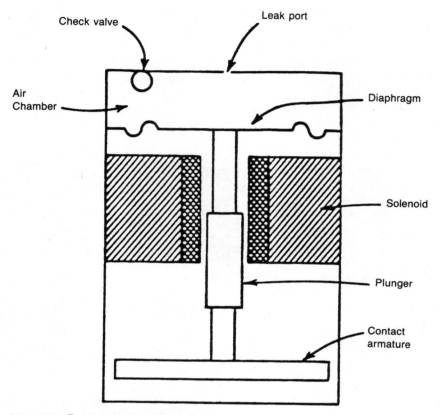

Figure 8.44 Pneumatic time-delay relay.

lays are electronic, typically variable-set-point, multifunction, solid-state devices. In all these, the control voltage (to the solenoid or the solid-state counter) and the load voltage (on the contacts) are usually different. The contact "rating" will be specified as the *inductive* or *resistive* maximum current at a stated voltage. The inductive rating relates to opening the contact, which creates a tendency to arc across the opening if current is too high. This causes contact burning and premature failure.

Pneumatic systems use several different types of "relays." The *electropneumatic* (EP) *relay* is a three-way air valve operated by an electric solenoid. Its normal function is to provide or bleed off control air as a fan or pump is started or stopped. The typical arrangement is shown in Fig. 8.45. The *pneumatic-electric* (PE) *relay* is a pressure switch which trips at some desired control air pressure. The set point is usually adjustable. This is used most often to provide control of an

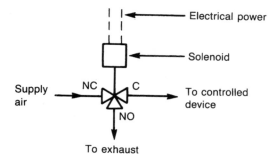

Electrical power

Solenoid

Supply air → NC   C → To controlled device

NO

To exhaust

Figure 8.45 EP relay arrangement.

electric circuit in response to a pneumatic sensor or controller. A reversing relay provides a modulating output which is the reverse of the input—a change from direct- to reverse-acting or vice versa (Fig. 8.46). An *averaging* relay provides a modulating output which is the average of several inputs. A *selector* relay provides a modulating output signal equal to the higher or lower of two inputs. A *discriminator* relay does the same for several inputs, up to 20 in some designs. Virtually all these functions are also available in electronic devices. Further, electronic devices are surpassing the older electromechanical devices in adaptability and function.

The discriminator function is widely used as an energy-saving concept in multizone and dual-duct systems. It is often used to provide reset of set point signals for the hot-and-cold duct temperature controllers, for the purpose of selecting the zones with the greatest demand for heating and cooling. The use of the discriminator is somewhat controversial because its usefulness depends entirely on the

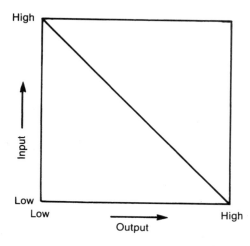

High

Input →

Low

Low   High
Output →

Figure 8.46 Input versus output in a reversing relay.

validity of the incoming signals. A bad sensor or a zone thermostat set much lower than those in the other zones may drive the discriminator to the exclusion of all other signals. Some authorities feel that discriminators should not be used. The discriminator function can be used in a computer-based controller. Here, the program can include a procedure for eliminating questionable signals.

Transducers are used in HVAC work to change a modulating pneumatic signal to a modulating electric signal, or vice versa. The electric signal may be current or voltage—generally 4 to 20 mA or 0 to 10 V. A good voltage- or current-to-pneumatic (I/P or V/P) transducer is critical to the success of hybrid control systems using electronic or computer-based controllers and pneumatic valve and damper operators. One method of constructing such a transducer is shown in Fig. 8.47. Varying the voltage or current on the solenoid coil will cause the valve to modulate, thus varying the output pressure.

8.3.4.2 Switches. A *switch* is a device for opening or closing a circuit or for transferring a load from one circuit to another. A *circuit* can be electric or pneumatic. A switch may be operated manually or automatically.

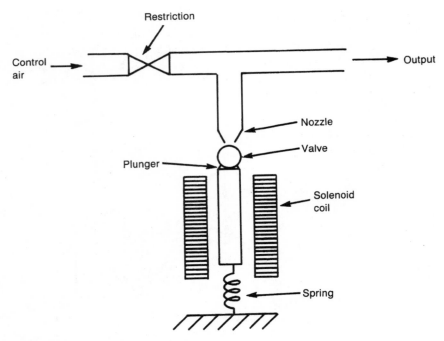

Figure 8.47 I/P (or V/P) transducer.

Figure 8.48 SPST switch.

In its simplest form, a switch is single-pole, single-throw (SPST) (Fig. 8.48). This is an on/off function, with the switch open or closed. Electrically, a contact is closed (on) or open (off). Pneumatically, a valve is open (on) or closed (off). A single-pole, double-throw (SPDT) switch transfers load. Electrically, as one contact opens, another closes. Pneumatically, as one port in a three-way valve opens, another closes (Fig. 8.49). Multiple contacts and ports and multiple "throw" operations are available, in many configurations. An electric relay is a switch. Both pneumatic and electric switches are available with electric solenoid operators or pneumatic (pressure) operators.

A *limit switch* (Fig. 8.50) is operated by the movement of some item of equipment. For example, a valve or damper may be provided with a limit switch which closes or opens at some point in the valve or damper operating cycle. Programming or sequencing switches make use of cam-operated limit switches.

A "hand gradual" switch is actually an adjustable pressure-reducing valve used in pneumatic control systems as a set point adjuster or to provide a signal such as a minimum outside-air damper setting.

A pushbutton switch may be either a maintained-contact or a momentary-contact type. A momentary-contact pushbutton requires an

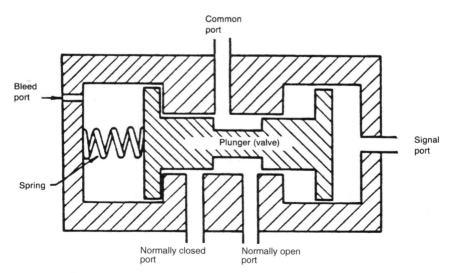

Figure 8.49 Pneumatic switching relay (SPDT).

Figure 8.50 Limit switch. (*Courtesy of Micro Switch, a division of Honeywell, Inc.*)

auxiliary "holding" or "latching" contact in the device or relay being controlled (Fig. 8.51). This contact closes when the relay is energized by pushing the start button and maintains the circuit after the start button has been released. The stop button breaks the circuit, allowing the holding contact to open.

8.3.4.3 Timers. Timing devices include clocks, program timers, and time-delay relays. Historically, a *time clock* (Fig. 8.52) is a clockwork mechanism with a rotating dial on which trip tabs can be mounted.

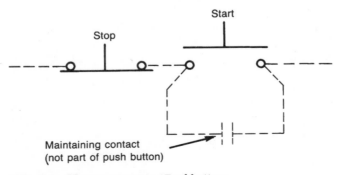

Figure 8.51 Momentary-contact pushbuttons.

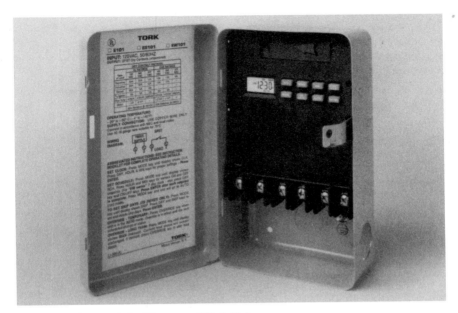

Figure 8.52 Time switch. (*Courtesy of Tork Co.*)

As the dial rotates, the tabs trip a switch at preset times. Both 24-h and 7-day programs are available.

A program timer contains several switches which are opened and closed in a preset sequence. The timer is started manually or automatically, and either will stop when the cycle is complete or will repeat the cycle indefinitely. Historically clocks and program timers were driven by a clockwork mechanism. Modern practice uses solid-state techniques or special-purpose computers.

8.4 Instrumentation

Although an "instrument" can sometimes be used for sensing or control, instrumentation is generally understood to refer to devices which monitor, i.e., measure, indicate, and record status. The purpose of instrumentation is to provide information for the guidance of the plant operator in analyzing, troubleshooting, and improving the operation of the HVAC system and controls.

The monitoring process is similar to the control process and uses similar sensors and controllers—or sometimes the same sensors and controllers. The controller output is used by an indicator to show status. Status may be analog, as shown on a pressure or temperature gauge (Fig. 8.53), or digital, usually shown by a pilot light or lights.

Figure 8.53 Pressure gauge. (*Courtesy of Dwyer Instruments, Inc.*)

Status may also be recorded, where desired, on a circular or strip-chart recorder or on computer storage media.

The importance of good instrumentation cannot be overemphasized. To be useful, the indicating devices must be accurate, reliable, and easily monitored. Otherwise they will be ignored. One advantage of computer-based systems is the amount of data available; the problem is to limit and clearly present the data to avoid overwhelming the operator.

8.5 Typical Control Systems

The following systems represent good basic practice. They are included to show the interaction among devices and loops in real situations. For a much more complete discussion of systems and subsystems, see Refs. 1 and 2 at the end of this chapter.

Most packaged equipment includes control systems designed and installed by the manufacturer and requiring a minimum of field installation. This is particularly true for chillers, boilers, and self-contained units.

The diagrams which follow are generic. The devices and their interconnections are shown, but specifics are omitted, because each manufacturer handles a given function in a different way. The symbols used are specific to this text because the industry has been slow to respond to attempts to use standard nomenclature. A symbol legend can be found in Fig. 8.68 (p. 287).

8.5.1 Control of outside air quantity

Almost all air-handling systems utilize a mixture of outside air and recirculated air. The quantity of outside air needed is that required to replace exhausted air or that required for ventilation to maintain a desired indoor air quality (IAQ). In some industrial applications the exhaust makeup will govern, but in most commercial and institutional applications the IAQ will govern. In most places the ventilation rate will be specified by the local code authorities. Typically the codes are based on ASHRAE Standard 62, the newest version of which will be available about the time this edition is published. The outside air volume control methods are similar for all types of air-handling systems. The essential point here is that the outside air flow must be measured and recorded to prove compliance with these codes.

8.5.1.1 Where exhaust makeup requirements govern, the outside air quantity will be fixed and controlled by a two-position damper which opens whenever the supply fan runs. This may provide 100 percent outside air or a fixed percentage of total air quantity. With 100 percent outside air it is usually recommended that the fan start-up sequence provide for opening the outside air damper before the fan starts. Low static pressure in the intake plenum may cause a collapse of that plenum.

8.5.1.2 The traditional "economy cycle" control of outside air (Fig. 8.54) operates as follows: When the supply fan is off, the outside air and relief dampers are closed. When the fan is started, these dampers open to minimum position, as determined by the setting of the minimum position switch through selector relay R2. The return air damper will be partially closed to match. If the outside air temperature is below the set point of controller C1, this controller will adjust the damper positions to maintain the required minimum mixed air temperature as sensed by T2. As the outside air temperature increases,

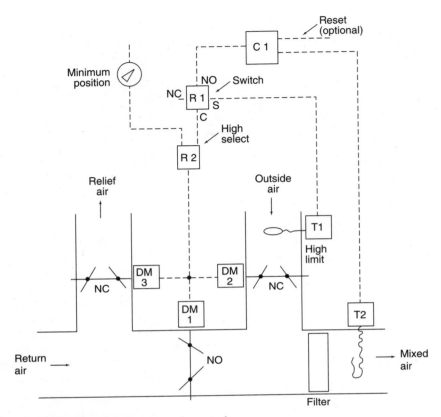

Figure 8.54 Classical economy cycle control.

the dampers will modulate and finally reach full open for outside and relief and closed for return. As the outside air temperature continues to increase, it will reach the set point of high-limit thermostat T1 (typically about 75°F). This will cause switch R1 to block the signal from C1, allowing the dampers to return to minimum outside air position.

The purpose of this control system is to use outside air for cooling when possible and to minimize outside air use when refrigeration is required. Note that there are three control loops involved: the main loop with T2 and C1, another loop with T1 and R1, and a third loop with the minimum position switch.

8.5.1.3 With variable air volume (VAV) systems the economy cycle is no longer satisfactory, since it provides a fixed percentage of outside air which will result in a decrease in the quantity of outside air as the total air volume decreases. To maintain a constant minimum volume of outside air, it is necessary to use other methods. Many methods have been proposed. A few of these are:

The *outside air fan* uses a fixed-volume fan to force outside air into the mixed air plenum of the air-handling system. Some variations include economy cycle control as well.

Mixed air plenum pressure requires that the outside and return air dampers be controlled to provide a fixed pressure in the mixed air plenum. This pressure is assumed to be enough lower than the outside air pressure to provide the required outside air quantity. There is no provision for changes in outside pressure due to wind velocity and direction, nor for stack effect in the building, and these effects will change the outside air volume.

One hundred percent outside air with heat reclaim systems use a total heat reclaim system between outside air and exhaust air. This system is effective and economical when using modern heat reclaim equipment in mild climates.

A more complex system is shown in Fig. 8.55. This resembles the configuration of the traditional economy cycle but is controlled in an entirely different manner. The primary sensor is the outside air flow

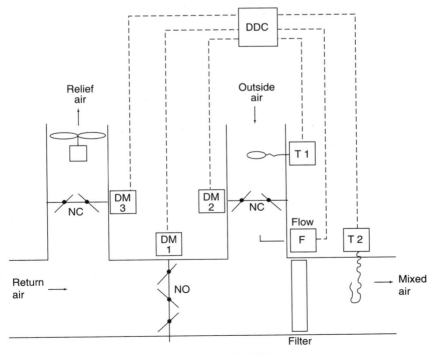

Figure 8.55 Control system for outside air with VAV.

velocity sensor F. There is also a low-limit mixed air temperature sensor T2 and an outside air temperature sensor T1. Note that the outside air (OSA), return air (RA), and relief air (REL) dampers are controlled individually. Control is provided by a DDC, programmed as follows: The velocity corresponding to the minimum OSA flow rate is the set point. At start-up the REL damper is closed, the RA damper is open, and the OSA damper is modulated to obtain the minimum OSA flow rate. When the OSA damper is fully open, the RA damper must be modulated toward closed to maintain the OSA flow rate. The low-limit sensor T2 can be allowed to reset the OSA flow rate set point *upward* so long as the mixed air temperature is above a minimum. The high-limit sensor returns the flow rate set point to minimum whenever the OSA temperature is too high for economy. The REL damper and fan come into action whenever the OSA flow rate exceeds the minimum.

8.5.2 Single-zone AHU

The single-zone air-handling unit serves one zone or a group of contiguous zones with similar loads (Fig. 8.56).

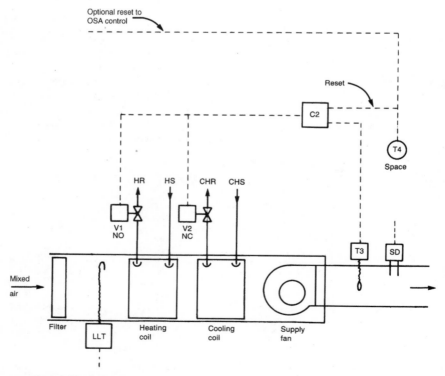

Figure 8.56 Single-zone air-handling unit.

The space temperature control loop operates as follows. Controller C2 controls valves V1 (heating) and V2 (cooling) in sequence. Valve V1 is normally open and closes as the output of C2 increases. When V1 is fully closed, V2 (normally closed) begins to open; and when C2 is at maximum output, V2 will be fully open. The output of C2 depends on the supply air temperature as sensed by T3. The set point of C2 is reset from the space temperature sensor T4.

The signal from T4 can also be used to reset the set point of the outside air controller of Fig. 8.54, for the purpose of minimizing heating energy use in mild weather.

The fire or smoke detector is required by most codes and will stop the supply fan if fire or smoke is detected. The low-limit temperature switch will stop the supply fan if freezing air is detected. Although the control loops are independent of one another, there is some interaction through the airstream in the AHU. Sometimes additional zoning is needed and can be provided by duct reheat (Fig. 8.57) or perimeter reheat by convectors, fin pipe, or baseboard radiation.

8.5.3 Multizone AHU

The multizone AHU (Fig. 8.58) provides both hot and cold air and therefore allows simultaneous heating and cooling in contiguous zones. The traditional multizone unit had only two ducts—hot and cold—and was therefore a reheat system. This is now not allowed by

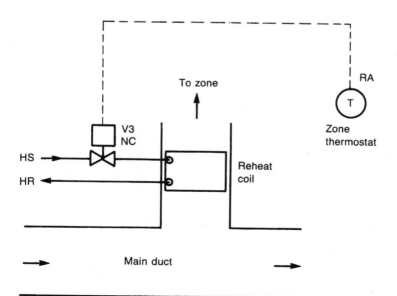

Figure 8.57 Zone reheat coil.

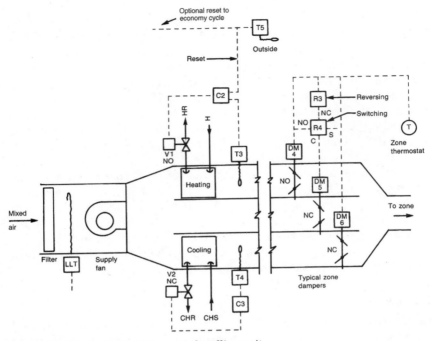

Figure 8.58 Three-duct multizone air-handling unit.

many energy codes, and the three-duct or "Texas" multizone system shown is designed to satisfy the codes and still provide the control advantages of the multizone unit. The economy cycle is the same as described in Sec. 8.5.1 except that C1 may be reset from the outside air temperature—higher when the outside air temperature is lower. The hot plenum is maintained at a temperature which is reset as a function of outside air, decreasing as the outside air temperature increases. The cold plenum is maintained at a constant temperature, usually somewhat above the low-limit mixed-air temperature to minimize refrigeration requirements. The zone supply air temperature is controlled in response to the zone temperature as follows. When maximum heating is required, the hot damper is fully open and bypass and cold dampers are closed. As less heating is required, the hot damper modulates toward closed and the bypass damper modulates toward open. When the hot damper is fully closed and the bypass fully open, switching relay R2 transfers so that the bypass damper signal comes through the reversing relay R1. Now, as the zone needs cooling, the bypass damper closes and the cold damper opens. The sequence is shown graphically in Fig. 8.59.

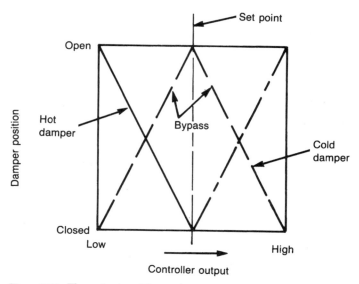

Figure 8.59 Three-duct multizone damper sequence.

8.5.4 Dual-duct air-handling system

The dual-duct system (Fig. 8.60) operates on the same principles as the multizone system, but the hot and cold plenums are extended through the building, with a "mixing box" provided at each zone. A common arrangement (not shown) has a single fan supplying both ducts. That arrangement has the same reheat problems as the two-duct multizone system. The two-fan system shown in Fig. 8.60 overcomes most problems and provides very economical operation. The warm return air goes almost entirely to the hot duct, minimizing the use of heating energy. Outside air goes almost entirely to the cold duct, minimizing the use of cooling energy. Each fan is provided with volume control, based on maintaining a constant static pressure in its related duct. Exterior zones, which may need some heating, are provided with mixing boxes. Interior zones, which need only cooling, are provided with VAV boxes (see below).

This system may be made even more economical by the use of variable volume mixing boxes (Fig. 8.61*a* and *b*). The VAV mixing box utilizes separate operators for the hot and cold dampers, allowing them to function independently and sequentially as follows. At maximum heating, the hot damper is fully open, the cold damper closed. As the heating load decreases, the hot damper modulates toward closed, and air volume decreases. When it is about 60 percent closed, the cold damper starts to open and total air volume remains stable.

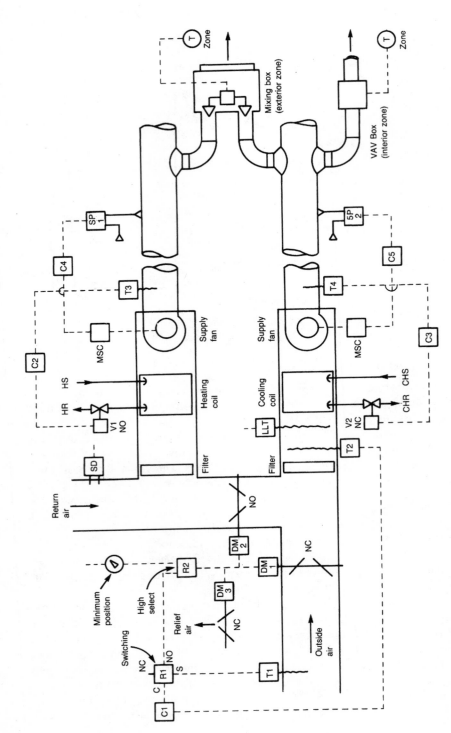

Figure 8.60 Double-duct two-fan air-handling system.

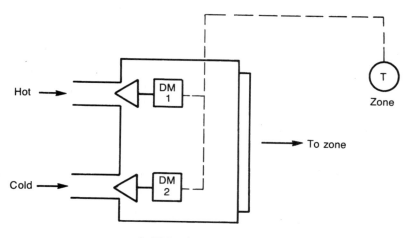

A. Mixing box arrangement.

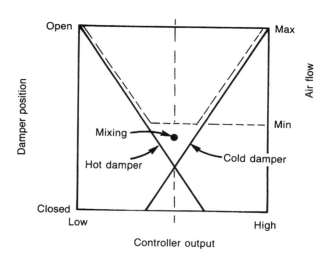

B. Damper operating sequence.

Figure 8.61 Double-duct VAV mixing box.

The hot damper continues to close and is fully closed when the cold damper is 40 percent open. The cold damper can continue on to fully open, if required, with a corresponding increase in air volume.

When variable volume is used, the outside and return air dampers must be controlled to maintain minimum flows as described in Sec. 8.5.1.

8.5.5 Variable-volume air-handling system

The VAV system (Fig. 8.62) is based on the principle of matching the load by varying the air volume supplied to each zone rather than varying the temperature, with the intent of saving fan work energy as compared with a constant-volume system. As the individual VAV boxes modulate in response to zone demands, the total system volume will vary. If the fan volume is not controlled, the static pressure in the duct system will increase, resulting in noise, lack of control at the boxes, and the possibility of duct blowout. To overcome this, several methods of volume control are available. In many cases, the controlled variable is the static pressure in the duct at some point selected to provide sufficient pressure at the most remote VAV box. An old rule of thumb is to locate the sensor two-thirds to three-fourths of the distance from the fan to the most remote box. In reality, the best location is from the inlet duct to the most remote box. If this point is satisfied, all other points in the system will be satisfied. Complete building DDC systems allow every box condition to be monitored, which allows the total cfm requirements to be summed and the fan speed adjusted accordingly. Volume control methods include:

- A modulating damper at the fan discharge. This makes the fan "ride up its curve" and saves little, if any, fan work energy.

- A bypass from supply to return, with a modulating damper. This means that the fan is working at constant volume at all times, while system volume varies. Good control is obtained, but there is no energy savings.

- Inlet vane dampers. As these dampers modulate, they change the operating characteristic of the fan (see Chap. 4) and energy is saved. Inlet vanes pose an energy penalty as an added resistance to airflow.

- Mechanical variable-speed drives. As described in Sec. 8.3.3.3, these systems save energy because of the physical law which states that the fan (or pump) horsepower varies as the cube of the speed. This savings is not fully realized because of mechanical losses in the systems.

- Electronic (solid-state) variable-frequency speed controllers. As described in Sec. 8.3.3.3, variable-frequency drives are used for fans, pumps, and chiller compressors. One of the largest of such systems is used with a 5000-hp motor driving a chiller compressor at the Dallas-Fort Worth airport. Energy use varies as the cube of the speed, although there are some losses in the electric circuits.

The single-duct VAV system provides only cooling, so supplemental heating is required in perimeter zones. This may be provided by

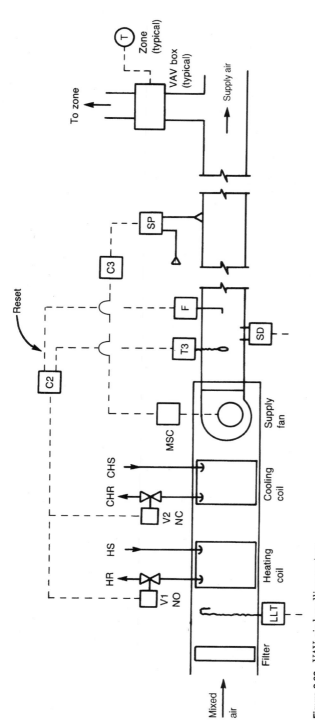

Figure 8.62 VAV air-handling system.

perimeter radiation, fan-coil units, or reheat coils in the supply duct. Heating is controlled in sequence, so that it is used only when the air supply is at its minimum.

8.5.6 Central plant control systems

Plant control systems range from simple on/off switches for residential-scale equipment to elaborate computer-based *supervisory control and data acquisition* (SCADA) systems for large institutional plants. A few brief notes are offered to suggest the concepts.

- *Steam boilers* are usually controlled by system steam pressure. A pressure sensor feeds a signal back to the boiler fuel control system. There are safety overrides to defend against firing beyond the operating condition. Water is fed to the boiler in response to a water-level sensor on the upper steam drum. Multiple boilers firing together may be modulated in common, or one boiler (or more) may be set at a fixed rate with another boiler modulating with load variations. Controls for water level at the deaerator and at the storage/transfer tank are usually self-contained. System pumps may be manually or automatically engaged. Much of the function of an automation system is related to data gathering and report generation.

- *Water heating boilers* are controlled from leaving hot water supply temperature. While boilers are best kept at a consistent temperature, the water supply temperature to the system may be controlled by blending return water with the heated primary supply water. Pumps and boilers may be turned on to match the load. As with steam, an automation system may have some control algorithms, but will offer value in data acquisition and report generation.

- *Water chillers* typically have a factory-designed, self-contained control system where the unit can be turned on or off by an operator at the local panel or from a remote interfaced control panel. The chiller control panel will allow local or remote definition of the chilled water temperature set point. In remote control, reset of the control point based on load or outside-air conditioning is possible. Again, an automated system can provide data acquisition and report generation over and above the basic system control functions.

  One aspect of chilled water system control which has been improved through DDC technology is the condensing water temperature control. A common control sequence for water temperature from cooling tower to chiller condenser was to cycle the cooling-tower fans to maintain the condensing water back to the chiller at a fixed condition, say, 80 to 85°F, based on a summer wet-bulb design temperature of 70 to 80°F. The fixed condition was energy wasteful much of the year since the tower could make cooler water at other than peak load times, and the chiller could operate with less energy input, given reduced condensing conditions. DDC technology now allows a reset of the cooling water set point related to load, outside air conditions, and chiller limitations.

8.6   Electrical Interfaces

It has been noted that motors are controlled devices whose operations (start, stop, and speed adjustment) must often be governed by the HVAC control system. However, the power and control wiring for the motors is usually furnished by the electrical contractor, based on design work by the electrical engineer. Because neither of these people understands the needs of the HVAC control system, the HVAC designer must communicate this information to the electrical designer in language that the electrical designer can understand.

The simplest and most effective way to do this is by means of the schematic or ladder diagram (Fig. 8.63). The format is that commonly used by electrical control designers. The diagram shows the control interlocks and logic. Power wiring is implied; wherever a motor starter solenoid coil and overload contacts are shown, the accompanying power contacts and wiring are required, as shown in Fig. 8.40. Auxiliary contacts in each starter are indicated by reference number; for example, M2-1 is in starter M2, etc. Numbers on the right edge of the diagram indicate the lines on which auxiliary contacts for that starter are to be found. At each contact there is a reference to the line number on which the starter solenoid occurs. On a simple diagram such as this, these reference numbers may seem unnecessary, but on a large, complex diagram, they are essential. Many logic diagrams include relay solenoid coils and contacts which are shown in a way similar to the motor starter coils, but without overload contacts.

8.7   Computer-Based Controls

The use of computers as controllers in HVAC systems began about 1969 with the introduction of computer-based "supervisory" systems. These systems provided central monitoring and recording of local activities and, in addition, such control programs as programmed start-stop and reset of local loop controller set points to decrease energy use. The systems are given a variety of names, the most common being *EMCS* (energy management and control system), and *BAS* (building

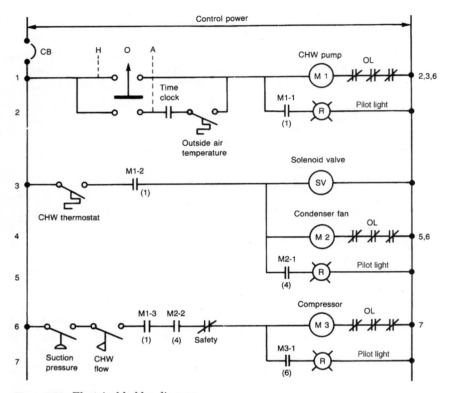

Figure 8.63 Electrical ladder diagram.

automation system). In industry there is reference to a SCADA system (supervisory control and data acquisition system). The principal justification for these systems is the energy savings and system management data which may be obtained.

8.7.1 Energy management and control system

As noted above, EMCS is also known by other names, including BAS and SCADA. The concept applies to a great many existing systems but in new HVAC work has been largely superseded by the DDC system (see Sec. 8.7.2).

Figure 8.64 is a generalized schematic of an EMCS. The central console includes a computer, software (programs), display monitor, printer, and the keyboard through which the operator communicates with the system. The group consisting of the monitor, printer, and keyboard is sometimes called the *operator-machine interface* (OMI).

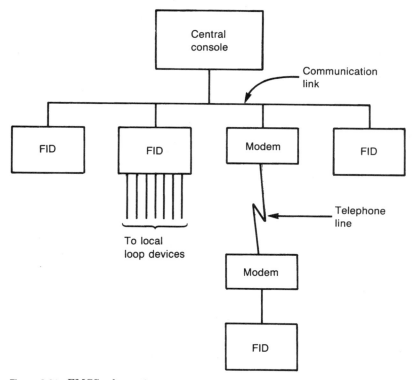

Figure 8.64 EMCS schematic arrangement.

The communication link is usually a coaxial cable or a "twisted pair" of wires. Information is transmitted on the link by means of digital "words" (Fig. 8.65), which include the address of the destination, the message, and a method of checking for accuracy. The *field interface device* (FID) serves as an interface between the communication link and the local loop devices. In many systems, the FID is "intelligent" (IFID); i.e., it contains a microcomputer and can make many of the decisions otherwise made at the central console. This minimizes traffic on the communication link and results in what is called *distributed intelligence*. The general arrangement of an IFID is shown in Fig. 8.66.

Figure 8.65 Digital message.

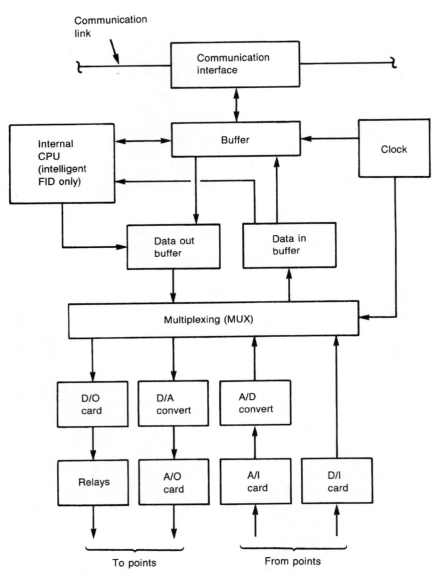

Figure 8.66 Field interface device.

The most important activity of the IFID is *signal conditioning* — interpreting the information received from the sensors, translating analog data to digital form (A/D conversion), and doing the reverse (D/A conversion) for outgoing commands. Digital information, both incoming and outgoing, requires interpretation but not conversion. The IFID does not replace any of the local loop control devices. Each device connected to the IFID is called a *point*.

8.7.2 Direct digital controller

A direct digital controller is a small computer which takes the place of one or more local loop controllers, together with many of the switches and relays associated with those controllers. Sensors and controlled devices remain and must be connected to the direct digital controller through a real-time interface so that the controller can deal with events as they happen. The direct digital controller is shown schematically in Fig. 8.67. Note that it is similar to an IFID and can function in a manner similar to an IFID when connected to a central console. The direct digital controller contains most of or all the software formerly associated with the central console, so that the central console is no longer needed. All the direct digital controllers in a system, building, or complex can be connected to a *common bus* through which they can communicate with one another. A central console can also be connected to the bus and can be used for data acquisition,

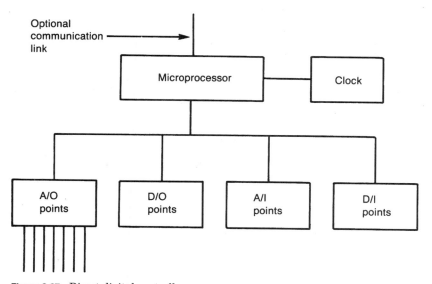

Figure 8.67 Direct digital controller.

monitoring, and display and for operator interface. In some systems, set points and other software variables can be reset from the central console. The variables can also be reset at the direct digital controller.

Most direct digital controllers are special-purpose microcomputers, designed for control purposes. A typical controller can connect to 12 to 24 points, although some are larger. Local display and access varies greatly, depending on the manufacturer.

8.7.3 Software

In a computer-based system, all the control functions are algorithms in software or firmware. The software must include a database, which is a list of all connected points together with information about each point. Typical data include the point name, address, type, conversion units, engineering units for display and recording, related algorithms, and other pertinent characteristics. The control algorithms can be written for P, PI, or PID control modes. All control gains are therefore numbers in software.

Software has become very sophisticated. Graphics — schematic displays on the central monitor — are widely used. Most graphics displays are interactive; i.e., they show the current status or value of the points in the system, updating the data as conditions change. Most programs are menu-driven, to assist the operator. Some new developments are in the area of *adaptive* systems. An adaptive program allows the computer to learn by studying the results of its actions. One valuable function involves adjusting gain constants in the control algorithm to compensate for changes in loads or system gains, thereby providing optimal performance under all conditions. It has been suggested that eventually the monitoring computer will be programmed to consider energy consumption efficiency as the controlled variable, while still maintaining the required comfort or process environment. This would be energy management of a high order.

8.7.4 Compatibility

One of the historical problems with computer-based systems has been the lack of compatibility among the various manufacturers. This condition is closely related to the long-term marketing benefit of having a proprietary (incompatible) system. Hardware parts of the systems are very similar and often interchangeable, but software is not. Generally, software protocol varies from system to system, with little or no compatibility. Thus additions to the original system usually must be made by the original supplier.

To overcome this, ASHRAE has developed an interface standard called BACNET (Building Automation and Control Network), which

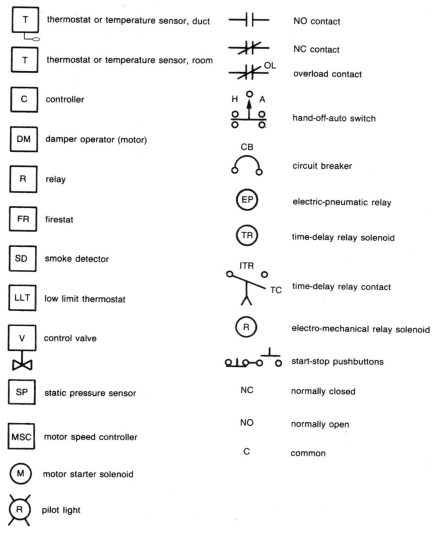

Figure 8.68 Legend for control drawings.

specifies methods of interfacing noncompatible protocols through a common communication bus system. For details, contact ASHRAE.

8.8 Control Symbols

Figure 8.68 shows the control symbols used in this book. There is no industry standard for control symbols, except that of the Instrument Society of America (ISA), which is seldom used in the HVAC world.

Therefore, any symbols may be used in design documents as long as they are carefully defined and consistent.

8.9 Summary

This chapter has necessarily been brief, and the reader is urged to use the references for further study. Good control systems must be as simple as possible for the application. The "real-world" rule is that a complex system will be simplified by operators to suit their understanding, whether the modified system meets the original design intent or not. Good documentation and operator training are essential.

References

1. Roger W. Haines and Douglas C. Hittle, *Control Systems for Heating, Ventilating and Air Conditioning,* 5th ed., Chapman and Hall, New York, 1993.
2. ASHRAE Handbook, *1995 Applications,* Chap. 42, "Automatic Control."
3. Ralph Thompson, "Electronic Controllers," *Heating/Piping/Air Conditioning,* July 1986, p. 123.
4. Roger W. Haines, "Ventilation Air, the Economy Cycle and VAV," *Heating/Piping/Air Conditioning,* October 1994, p. 71.

Equipment: Part 1

Cooling

9.1 Introduction

Cooling means the removal of heat. In HVAC, a cooling process is usually identified as one which lowers the temperature or humidity (or both) of the ambient air. The *effective temperature* includes not only the temperature and humidity of the ambient air but also radiant effects and air movement. Some cooling processes, i.e., evaporative cooling, do not actually remove any heat, but create a sensation of cooling by lowering the sensible temperature of the air.

9.2 Refrigeration Cycles

A refrigeration cycle is a means of transferring heat from some place where it is not wanted (heat source) to another place where it can be used or disposed of (heat sink). The necessary components are (1) two or more heat exchangers (one each at source and sink), (2) a refrigerant, (3) a conduit for conveying the refrigerant, (4) mechanical and/or heat energy to move the refrigerant through the system, and (5) devices to control the rate of flow, to control temperature and pressure gradients, and to prevent damage to the system.

There are several basic refrigeration cycles. The two most common—two-phase (vapor compression) mechanical, and single- and double-effect absorption—are discussed below. Steam-jet refrigeration has historical importance but is not used in modern practice. Noncondensing (one-phase) mechanical cycles are used primarily in aircraft where light weight and simplicity are important. Thermoelectric refrigeration utilizes thermocouples working in reverse: When an electric current is impressed on a thermocouple, a cooling effect is obtained. These are small systems for specialized applications and are comparatively expensive to install and operate.

9.2.1 The mechanical two-phase vapor compression refrigeration cycle

The most common cooling source in HVAC is mechanical two-phase vapor compression refrigeration. In this cycle (Fig. 9.1), a compressor is used to raise the pressure of a refrigerant gas. Work energy (Q_W) is required, usually provided by an electric motor or steam turbine or fuel-fired engine. The compression process raises the temperature of the gas. The high-pressure gas flows through piping to a condenser where heat is removed by transfer to a heat sink, usually water or air. The refrigerant is selected with properties which allow it to condense (liquefy) at the temperature and pressure in the condenser. The high-pressure liquid is passed through a pressure-reducing device to the evaporator. At the lower pressure, the liquid tends to evaporate, re-

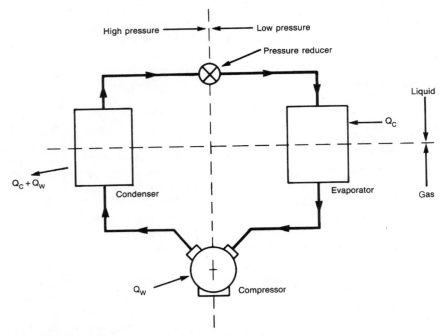

Figure 9.1 Mechanical two-phase refrigeration cycle.

moving the heat of vaporization (Q_C) from its surroundings (the evaporator—technically, the heat source). The cold, low-pressure vapor is then returned to the compressor to be recycled. Note that the heat removed in the condenser is equal to the sum $Q_C + Q_W$. One index of compressor effectiveness is its *coefficient of performance* (COP)

$$COP = Q_C/Q_W \qquad (9.1)$$

The refrigeration cycle can also be shown on a graph of the properties of a specific refrigerant. The graph in Fig. 9.2 is a *pressure-enthalpy* or *p-h diagram* with the basic coordinates of pressure and enthalpy. The four stages of the cycle include compression (with a rise in temperature and enthalpy due to work done), condensing (cooling and liquefying at constant pressure), expansion (at constant enthalpy) and vaporization at constant pressure. The use of the *p-h* diagram allows the selection of the most effective refrigerant for the pressures and temperatures appropriate to the process. To minimize the work energy required, the temperature difference between the heat source and heat sink should be minimized.

9.2.2 Absorption refrigeration cycle

An absorption refrigeration cycle involves a refrigerant-absorbent pair, where the refrigerant is moved from the low-pressure evaporator region to the high-pressure condensing region as an "absorbed" gas on the back of the absorbent. Common refrigerant-absorbent pairs include ammonia-water and water–lithium bromide. In each case there is a strong affinity of each compound for the other. Energy is given up in the absorption process and energy is required to separate the pair. As in vapor compression refrigeration, the beneficial cooling effect is obtained from evaporation of the refrigerant in the low-pressure region of the system.

Figure 9.3 is a schematic diagram of a two-shell lithium bromide water chiller using steam as the heat source. The saturated "strong" solution in the generator is heated to drive off water in vapor form. The resulting unsaturated "weak" solution flows by gravity to the absorber, where the solution absorbs water vapor from the evaporator and is then pumped back to the generator. The water driven off in the generator is condensed in the condenser, flows by gravity to the evaporator and is evaporated there, with the heat of vaporization being extracted from the chilled water. The condensing water is used first to cool the solution in the absorber and then to condense the refrigerant water. The evaporation and regeneration processes also create a pressure differential between the upper and lower shell, and restrictors

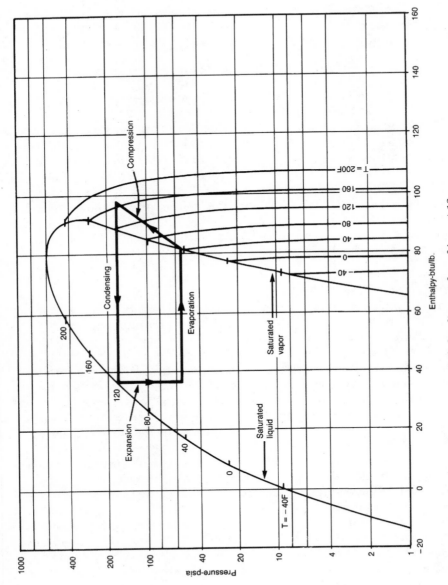

Figure 9.2 Refrigeration cycle on pressure–enthalpy diagram for refrigerant-12.

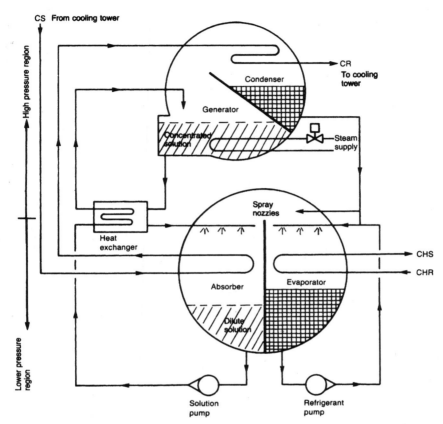

CS From cooling tower

CR
To cooling
tower

Condenser

Generator

Concentrated
solution

Steam
supply

High pressure region

Spray
nozzles

Heat
exchanger

CHS

CHR

Absorber

Evaporator

Dilute
solution

Lower pressure
region

Solution
pump

Refrigerant
pump

Figure 9.3 Two-shell absorption refrigeration cycle.

(not shown) are used in the pipelines to help maintain this pressure gradient. Heat rejection to the condensing water is roughly twice the refrigeration effect. In addition to the heat supplied to the generator, the chemical process of absorption creates some heat.

An absorption refrigeration system has a low coefficient of performance compared to a mechanical refrigeration system. Normally absorption can be justified economically only when plenty of comparatively low-cost or "waste" heat is available. Solar heat has been used, although in most areas the cost of its collection makes solar energy too expensive to compete with conventional fuels.

9.3 Compressors

In the refrigeration cycle, a compressor is a pump, providing the work energy to move the refrigerant from the low-pressure region to the high-pressure region through the system. Compressors come in two general types: positive displacement and centrifugal. Positive displacement compressors include reciprocating, rotary, scroll, and helical rotary (screw) types.

9.3.1 Reciprocating compressors

Reciprocating compressors are usually the single-acting piston type. Figure 9.4 shows one possible arrangement. The volume swept by the piston is the displacement. The remaining volume under the cylinder head is the clearance. The theoretical volumetric efficiency is a function of the ratio of these two volumes together with the compression ratio of the system. Higher compression ratios result in lower volumetric efficiencies. The actual volumetric efficiency will be somewhat less due to pressure drops across valves and other inefficiencies. The capacity of the compressor is a function of the volumetric efficiency, the properties of the refrigerant, and the operating pressures.

A compressor is always designed for a specific refrigerant at some narrowly defined range of operating pressure.

A compressor may have from 1 to 12 cylinders. Some older machines had as many as 24 cylinders. Most compressors for comfort cooling are direct-driven by electric motors. Historically, compressors were belt-driven, often by steam engines and at slow speeds.

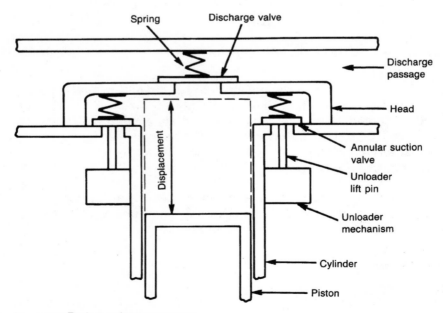

Figure 9.4 Reciprocating compressor.

Sizes range up to as high as 200 tons or more in one compressor, although units over 100 tons are rare. If larger capacities are needed, two or more compressors are used in parallel. Small compressors — to 5 or 7½ tons — are capacity-controlled by cycling the unit on and off. Larger units usually have unloaders on all but one or two cylinders. The unloader is activated electrically or pneumatically to lift the suction valve off its seat so that no compression takes place. Unloaders may be activated in stages so that two or more steps of capacity control may be obtained. Hot gas bypass is sometimes used with reciprocating compressors to maintain stable operation at reduced loads. Hot gas bypass does incur a power cost penalty.

Reciprocating compressors require from slightly less than 1 hp to as much as 1.5 hp/ton of actual refrigeration capacity at the maximum design temperatures and pressures typical of comfort-cooling processes. The horsepower per ton increases as the suction pressure and the temperature decrease. Therefore, it is more energy-efficient to operate at the highest suction pressure compatible with the needs of the application.

Compressors are lubricated by force-feed pumps or, in small units, by splash distribution of oil from the sump. Lubricating oils are selected to be miscible with the refrigerant and are carried throughout the piping system, which must be designed to ensure return of the oil to the compressor.

9.3.2 Rotary compressors

Rotary compressors are characterized by continuous circular or rotary motion. The two common types are shown in Figs. 9.5 and 9.6. In the

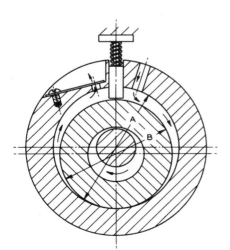

Figure 9.5 Rolling piston-type rotary compressor. (*Reprinted by permission from ASHRAE Handbook,* 1996 HVAC Systems and Equipment.)

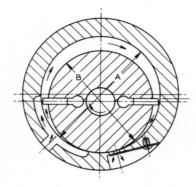

Figure 9.6 Rotating-vane type of rotary compressor. (*Reprinted by permission from ASHRAE Handbook,* 1996 HVAC Systems and Equipment.)

rolling-piston type, the rotor turns on an eccentric shaft, continuously sweeping a volume of space around the cylinder. Suction and discharge ports are separated by a vane which slides in and out against the cylinder wall. In the rotating-vane type of compressor, two sliding vanes are mounted in the rotor to form a compression chamber. The performances of the two types are similar.

In HVAC work, rotary compressors are seldom used in other than small sizes, up to about 10-hp capacity. Unloaders are not used on these small machines.

A special kind of rotary compressor has become prominent in the marketplace — the *scroll compressor.* The compression element consists of two interlocking spiral vanes, one stationary and the other rotating. The vanes are arranged so that low-pressure gas enters at the periphery and is compressed toward the center, where the gas flows out of an annular discharge port. As of this writing, these compressors are available in sizes up to 60 tons.[1]

9.3.3 Helical rotary compressors

Helical rotary or screw compressors are made in single-screw and twin-screw types. The single-screw compressor (Fig. 9.7) consists of a

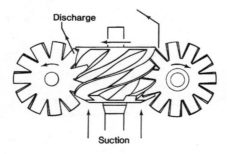

Discharge

Suction

Figure 9.7 Principle of operation of a single-screw compressor. (*Reprinted by permission from ASHRAE Handbook,* 1983 Equipment.)

helical main rotor with two star wheels. The enclosure of the main rotor has two slots through which the star wheel teeth pass; these teeth, together with the rotor and its enclosure, provide the boundaries of the compression chambers. The twin-screw compressor (Fig. 9.8) has two meshing helical gears and works much as a gear pump, with the helical shape forcing the gas to move in a direction parallel to the rotor shaft. These machines typically are direct-driven at 3600 r/min and are usually oil-flooded for lubrication and to seal leakage paths. Capacity control is obtained by means of a sliding or rotating slotted valve.

9.3.4 Centrifugal compressors

Centrifugal compressors belong to the family of turbomachines, which includes fans and centrifugal pumps. Pressures and flows result from rotational forces. In HVAC work, these compressors are used primarily in package chillers where the compressors provide large capacities. Typical driven speed is 3600 r/min or more, for electric motors or steam or gas turbines. Standard *centrifugal* chillers (Fig. 9.9) range in capacity from 100 to 2000 tons, although some special units have been built with capacities as great as 8500 tons. Capacity control is obtained by varying the driven speed or by means of inlet vanes, similar to those used on centrifugal fans. Noncondensing air-cycle systems, such as used on commercial aircraft, use high-speed gas-turbine drives at up to 90,000 r/min.

9.3.5 Hermetic compressors

Compressors may be built in either hermetic or open configurations. A hermetic unit has a casing which encloses both the compressor and the drive motor, minimizing the possibility of refrigerant leakage. Motors are specially constructed and are normally cooled with suction gas or with liquid refrigerant. In an open machine, the drive motor or turbine is separate from the compressor. Shaft seals must be provided to prevent refrigerant leakage. Standard drives—direct, gear, or belt—may be used. Semihermetic units have separate casings for the

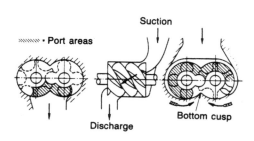

Figure 9.8 Helical rotary twin-screw compressor. (*Reprinted by permission from ASHRAE Handbook,* 1983 Equipment.)

Figure 9.9 Centrifugal compressor. (*Reprinted by permission of York International Corporation.*)

compressor and the motor with matching flanges for gas-tight assembly. Open drives have the advantage of removing the motor heat from the refrigeration cycle, thereby improving chiller performance. This advantage is lost if the motor heat is picked back up into the cooling load indirectly.

9.4 Chillers

The term *chiller* is used in connection with a complete chiller package — which includes the compressor, condenser, evaporator, internal piping, and controls — or for a liquid chiller (evaporator) only, where the water or brine is cooled.

Liquid chillers come in two general types: flooded and direct-expansion. There are several different configurations: shell-and-tube, double-tube, shell-and-coil, Baudelot (plate-type), and tank with raceway. For HVAC applications, the shell-and-tube configuration is most common.

9.4.1 Flooded chillers

A typical flooded shell-and-tube liquid chiller is shown in Fig. 9.10. Refrigerant flow to the shell is controlled by a high- or low-side float valve or by a restrictor. The water flow rate through the tubes is defined by the manufacturer but it generally ranges from 6 to 12 ft/s. Tubes may be plain (bare) or have a finned surface. The two-pass arrangement shown is most common, although one to four passes are available. The chiller must be arranged with removable water boxes so that the tubes may be cleaned at regular intervals, because even a small amount of fouling can cause a significant decrease in the heat exchange capacity. The condenser water tubes are especially subject to fouling with an open cooling tower. Piping must be arranged to allow easy removal of the water boxes.

9.4.2 Direct expansion (DX) chillers

In the DX liquid chiller (Fig. 9.11), the refrigerant is usually inside the tubes with the liquid in the shell. Baffles are provided to control the liquid flow. The U tube configuration shown is typical and less expensive than the straight-through tube arrangement but can lead to problems with oil accumulation in the tubes if refrigerant velocities are too low. Refrigerant flow is controlled by a thermal expansion valve.

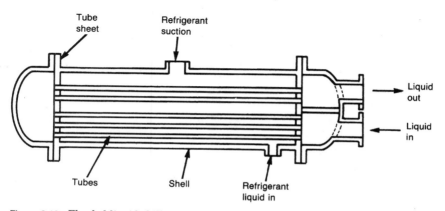

Figure 9.10 Flooded liquid chiller.

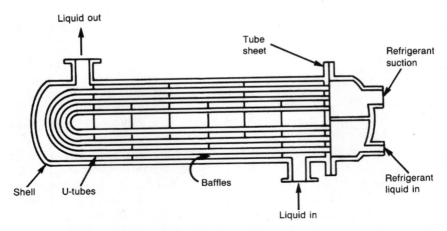

Figure 9.11 Direct-expansion chiller (U-tube type).

9.4.3 Package chillers

A complete package chiller will include compressor, condenser, evaporator (chiller), internal piping, and operating and capacity controls. Controls should be in a panel and include all internal wiring with a terminal strip for external wiring connections. In small packages — up to about 100 tons — motor starters are usually included. In larger chillers, unit-mounted starters are an option. Some units with air-cooled condensers are designed for outdoor mounting; freeze prevention procedures must be followed. Units with water-cooled condensers require an external source of condensing water.

Chillers with reciprocating compressors (Fig. 9.12) are found mostly in the 5- to 200-ton range. Although larger units are made, economics usually favor centrifugal compressor or screw chillers (Fig. 9.13), in sizes of 150 tons or more. Screw compressor systems are made in a growing range of sizes, by a growing number of manufacturers. With larger chillers or with high-voltage motors (2300 V, 4160 V), motor starters are usually separate from the centrifugal or screw chiller mounting frame and require field wiring of power and control circuits. Centrifugal compressor packages may be turbine-driven, occasionally engine-driven, but most often are driven by electric motors. The typical system is direct-driven at 3600 r/min. Wye-delta motors are often used for reduced-voltage starting. In larger units of 1000 tons or more, it is not unusual to use high-voltage motors; the lower current requirements allow smaller wire sizes and across-the-line starting. An unusual drive system evolved on one of the 8500-ton chillers at a major international airport. The utility plant manager replaced an original steam-turbine driver with a 5000-hp 4160-V variable-speed, variable-

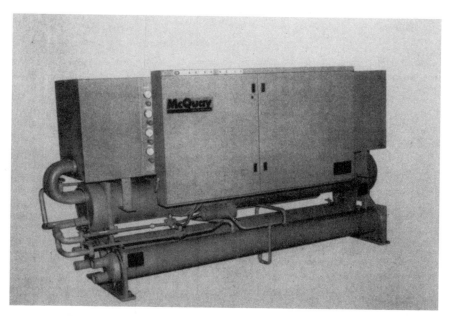

Figure 9.12 Reciprocating chiller package. (*Courtesy of McQuay Air Conditioning.*)

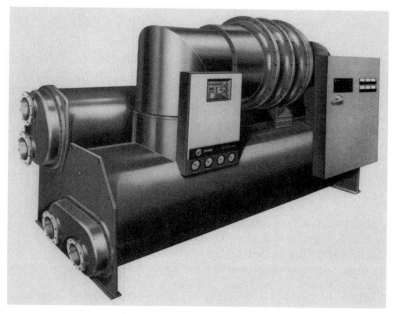

Figure 9.13 Centrifugal chiller package. (*Reproduced by permission of the Trane Co., Lacrosse, Wis.*)

frequency electric drive. The chiller capacity was reduced to 5500 tons, more in line with the actual load.

9.5 Condensers

The purpose of the condenser in a two-phase refrigeration cycle is to cool and condense the hot refrigerant gas leaving the compressor discharge. It is, then, a heat exchanger, of the shell-and-tube, tube-and-fin, or evaporative type. The heat sink is air, water, or a process liquid. Typically, small package systems use air-cooled condensers. Large built-up systems use water-cooled or evaporative condensers, although large air-cooled condensers are sometimes employed. The contrasting criteria here are the lower first cost of air-cooled condensers compared with the improved efficiencies obtained with water-cooled or evaporative condensers. The improvement in efficiency comes about because of the lower condensing temperatures achieved with water-cooled or evaporative condensers. The condenser capacity should match as closely as possible the capacity of the compressor in the system, although oversizing is preferable to undersizing if compressor efficiency is to be maximized.

9.5.1 Air-cooled condensers

An air-cooled condenser is usually of the finned-tube type (Fig. 9.14), with the refrigerant in the tubes and air forced over the outside of the tubes and fins by a fan. Capacity control, if used, is accomplished by cycling the fan, using a multispeed fan, or modulating airflow by means of dampers. Refrigerant flow velocities must be designed to prevent oil traps in the tubes. Capacities are based on square feet of coil face area, fan airflow rate, desired condensing temperature, and design ambient dry-bulb (db) temperature.

9.5.2 Water-cooled condensers

Water-cooled condensers are typically of the shell-and-tube type, with the water in the tubes and the refrigerant in the shell (Fig. 9.15). Chiller capacity control is not normally related to condenser water temperature. However, the water temperature may vary if it is supplied from a cooling tower, and most chiller manufacturers prefer that condensing water not be taken below 65 to 70°F because the oil may get held up in the condenser. Most code authorities do not allow direct use (and waste) of domestic water for condensing purposes.

Figure 9.14 Air-cooled condenser. (*Courtesy of McQuay Air Conditioning.*)

9.5.3 Evaporative condensers

An evaporative condenser (Fig. 9.16) includes a bare (no fins) tube coil which contains the refrigerant, a system for spraying water over the coil, a "casing" enclosure around coil and sprays, and a fan to force or draw air through the enclosure across the coil and sprays. The spray system includes a sump with a float valve for makeup water and a

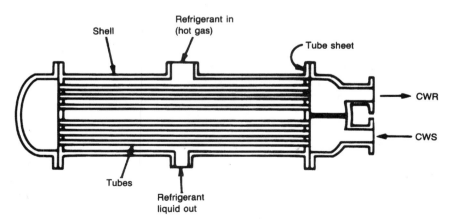

Figure 9.15 Shell and tubes of water-cooled condenser.

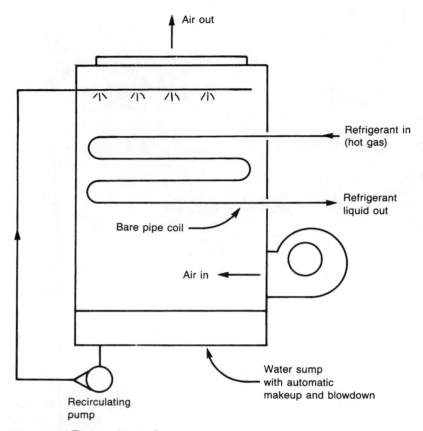

Figure 9.16 Evaporative condenser.

circulating pump. Water treatment is necessary to minimize corrosion and fouling. Bare tubes are used to minimize the effects of waterside fouling. This equipment utilizes the heat removed by evaporation of the water as water is sprayed over the coil. Capacity is a function of the ambient wet-bulb temperature, coil tube surface area, and airflow rate.

9.6   Cooling Towers

A cooling tower is a device for cooling a stream of water by evaporating a portion of the circulated stream. Such cooled water may be used for many purposes, but the principal concern in this book is its use as a heat sink for a refrigeration system condenser. An excellent discussion of cooling tower principles is found in Ref. 2.

The two main types of cooling towers are open-circuit and closed-circuit, described below. There are also two basic configurations: cross-flow and counterflow. In both arrangements, the water enters at the top of the tower and flows downward through it. In the counterflow arrangement, the air enters at the bottom and flows upward. In the cross-flow arrangement, the air enters at one side, flows across the tower, and flows out the other side.

Towers may be forced- or induced-draft, using fans (Fig. 9.17), or natural draft, utilizing convective chimney effects. Typical of this latter group are the large hyperbolic towers seen at many power plants (Fig. 9.18). In a forced-draft tower, the air is blown into and through the tower by the fans; in an induced-draft tower, the air is drawn through the tower. Forced-draft arrangements keep the fan out of the moist airstream. Induced-draft towers may obtain more uniform airflow patterns.

Towers are spray-filled, with the water distributed through spray nozzles, or splash-filled, where the water flows by gravity and splashes off the tower fill material. In either case, the objective is to maximize the evaporation effectiveness. The most important factors in this effort are (1) the effectiveness of spray or splash in atomizing the water, (2) the internal tower volume in which air and water come into contact, (3) the airflow rate through the tower, and (4) the water flow rate. Tower fill material used to be redwood. Now most fill material is made

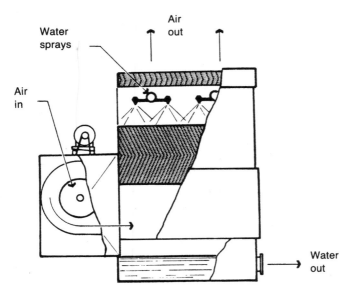

Figure 9.17 Forced-draft cooling tower. (*Courtesy of The Marley Cooling Tower Co.*)

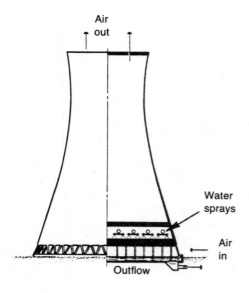

Air
out

Water
sprays

Air
in

Outflow

Figure 9.18 Natural-draft cool-ing tower. A flow is induced by chimney and stack effect. (*Courtesy of The Marley Cooling Tower Co.*)

of fiber-impregnated PVC or some similar plastic. Vitrified clay tile fill is used in some designs.

The two terms relating to tower performance are *range* and *approach*. The *range* is the difference between the entering and leaving cooling water temperatures. For HVAC practice, this is usually 10 to 20°F, although 8 to 10°F is common for vapor compression systems and 15 to 20°F is common for absorption systems. The *approach* is the difference between the leaving cooling water temperature and the design ambient wet-bulb (wb) temperature. This is usually between 6 and 12°F, with 7 to 10°F being typical.

9.6.1 Open-circuit cooling towers

In Fig. 9.19, there is only one water circuit, with a portion of the cooling water being evaporated to cool the remainder. Because the water is exposed to air, with all its contaminants, and absorbs oxygen, which is corrosive to most piping, the water must be carefully treated. To avoid increasing the concentration of solids as water is evaporated, blowdown must be provided: A portion of the water is wasted to the sewer, either continuously or intermittently. A blowdown rate equal to the evaporation rate is common, but the blowdown rate may be adjusted for the water quality. Ideally, treatment additives and the blowdown rate should be controlled automatically by a system which measures water quality and the solids concentration. Periodic sterilization of the water is also required to control algae and bacterial growth.

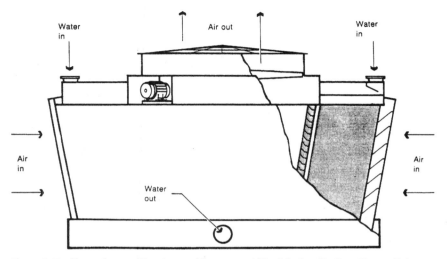

Figure 9.19   Cross-flow cooling tower. (*Courtesy of The Marley Cooling Tower Co.*)

9.6.2   Closed-circuit towers

The closed-circuit tower (Fig. 9.20) is designed to minimize corrosion and fouling in the cooling water circuit by making this a closed circuit. The cooling water flows through a bare tube coil in the tower, and coolant water in a separate circuit is sprayed over the coil and evaporated. This is essentially the same system as the evaporative condenser previously described. The coolant water circuit is open and needs treatment and blowdown. Because of the temperature differential through the tube wall, this system is slightly less efficient than the open-circuit tower, but the lower fouling effect improves the performance of, and decreases maintenance on, the condenser. This tower usually has a higher first cost than the open-circuit tower does.

9.7   Cooling Coils

A cooling coil is a finned-tube heat exchanger for use in an air-handling unit (AHU). Chilled water, brine, or refrigerant flows inside the tubes, and air is blown over the outside, across the fins and tubes. When used as an evaporator with liquid refrigerant, this coil is the evaporator in the refrigeration cycle and is called a *direct-expansion* (DX) *coil*.

9.7.1   Coil construction

The "standard" cooling coil has a galvanized-steel frame or casing and copper tubes, with aluminum fins bonded to the tubes. Other mate-

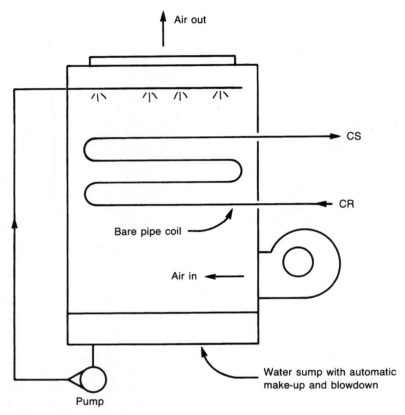

Figure 9.20 Closed-circuit cooling tower.

rials are available if required; e.g., copper fins are often used to avoid galvanic corrosion. Fins are usually the plate type, with the tubes expanded for a tight fit in holes in the fins. Some manufacturers use integral or spiral-wound fins. Spacing will vary from 6 to 14 fins per inch. Closer spacing increases the heat transfer and air pressure drop. Headers on water coils are cast-iron, steel pipe, or sometimes copper pipe, with flanged or threaded connections, depending on the size.

Coil dimension notation is shown in Fig. 9.21. From 1 to 12 rows is standard, but custom-made coils with any number of rows can be obtained. With an odd number of passes, the supply and return headers are on opposite ends. In a full-circuit coil (Fig. 9.22a), the number of passes (one pass means water flow from end to end of the coil) is equal to the number of rows fed. In a double-circuit coil (Fig. 9.22b), the tubes are arranged to provide only half as many passes as rows fed (e.g., a four-row coil with two passes), while in a half-circuit coil, the

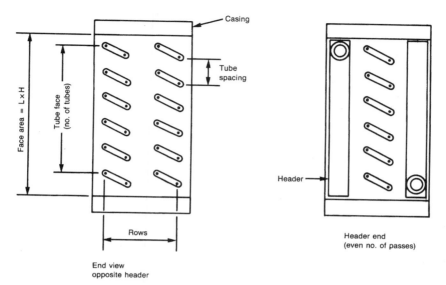

Figure 9.21 Dimensional notation for coils.

water flows from end to end twice as many times as there are rows fed — a four-row half-circuited coil has eight passes.

For direct-expansion coils, the same principles and rules apply, but instead of a supply header, a distributor is used, as shown in Fig. 9.23. This device is designed to distribute the refrigerant equally to each of the tube circuits. Note that all lines from the distributor are of equal length to provide equal pressure drops. The suction header is similar to the return header on a water coil, and is made of copper or brass for compatibility with the commonly used halocarbon refrigerants.

9.7.2 Coil thermodynamics

The thermodynamics and psychrometrics involved in the cooling-coil-to-air heat transfer process include (1) the psychrometric analysis of the cooling and dehumidifying process, (2) the effect on the heat transfer process of the air and water velocities, and (3) the effect of parallel flow and counterflow on coil heat transfer capacity.

Most air-cooling processes involve some dehumidification; this is the natural psychrometric response of the system if the air is cooled to near its saturation condition. The process is usually shown as a straight line on the psychrometric chart (Fig. 9.24) from some entering-air condition to the saturation curve (line *ABC*). The real process follows the curved line *AC*. The air cools sensibly to saturation, then subcools and dehumidifies down the saturation curve to point *C*. Point

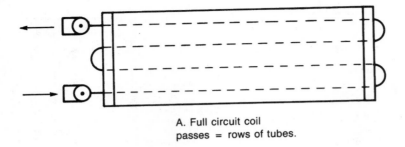

A. Full circuit coil
passes = rows of tubes.

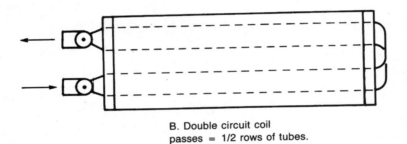

B. Double circuit coil
passes = 1/2 rows of tubes.

Figure 9.22   Coil circuit.

C is called the *apparatus dew point* (ADP) and is determined primarily by the temperature of the refrigerant or chilled water in the coil. Some of the air passes through the coil without being cooled, remaining at the condition of point *A*. The resulting mixture is at a condition such as point *B*. The reason for this is found in the geometry of the coil (Fig. 9.25). As air flows through the coil, some of the air impinges on the tubes and fins, but some air does not touch any part of the coil—it is bypassed. The percentage of air which remains at condition *A* is known as the *bypass factor*. Typically it is on the order of 5 to 10 percent for a four-row coil. This means that for design purposes there is a limited area in which point *B* can fall, which is defined by the system thermodynamics. The arbitrary selection of a point outside that area may look good to the designer, but the system won't actually work that way.

Another essential factor in the psychrometric analysis is the need for an apparatus dew point. It is impossible, thermodynamically, to have a cooling and dehumidifying process without an ADP (see line *AD* in Fig. 9.24).

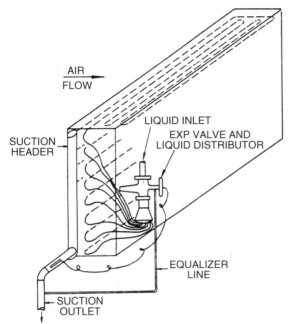

AIR
FLOW

LIQUID INLET

EXP VALVE AND
LIQUID DISTRIBUTOR

SUCTION
HEADER

EQUALIZER
LINE

SUCTION
OUTLET

NOTE: IF COMPRESSOR IS BELOW, THEN LOOP SUCTION TO
TOP OF COIL UNLESS PUMPDOWN CONTROL IS USED,
WHICH IS RECOMMENDED

Figure 9.23 Direct-expansion evaporation coil. (*Reprinted by permission from ASHRAE Handbook,* 1994 Refrigerant Systems and Applications, *Chap. 2.*)

Heat transfer from water or refrigerant to the air through the tube wall is influenced primarily by fouling and film factors (Fig. 9.26). The tube wall offers little resistance to heat flow. Dirt or oil on the fin or tube surface can be a detrimental insulator. The film factor—resistance to heat transfer due to a stagnant film of air or water along the tube wall surface (boundary-layer effect)—is also a detrimental insulator. The film resistance is greatest at low water and air velocities (low Reynolds numbers), hence the desire to keep airflows and water flows in the turbulent rather than laminar flow regimen.

Higher flow rates decrease the film resistance but lead to higher air and water pressure drops. Compromise is therefore necessary. One factor in the airflow rate selection is that the condensed water on the surface of the coil tends to carry over into the airstream at high velocities. It is considered good practice to select cooling-coil air velocities in the range of 400 to 600 ft/min and water velocities in the range of 3 to 6 ft/s.

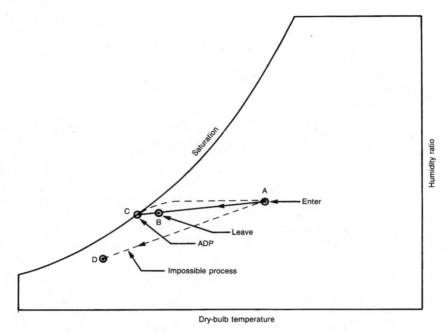

Figure 9.24 Cooling dehumidifying process on the psychrometric chart.

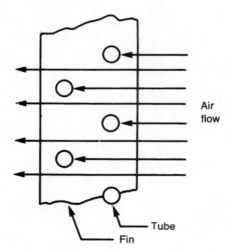

Figure 9.25 Partial section through a finned coil.

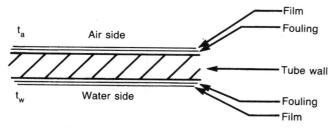

Figure 9.26 Heat transfer through a tube wall.

The use of *counterflow* or *parallel-flow* fluid paths has a considerable effect on coil performance. The two terms are illustrated in Fig. 9.27. In counterflow, the entering (coldest) water contacts the leaving (coldest) air, while leaving water contacts entering air. Thus the temperature difference between the two streams is fairly uniform throughout the coil. In parallel flow, the opposite effect occurs: The coldest water contacts the warmest air, and vice versa. Figure 9.28 shows the temperature relationships graphically. Counterflow is much more effective than parallel flow. The total heat transferred in the system is a function of the coil surface area, the heat transfer factor k, and the temperature difference. Because the temperature difference varies throughout the coil, it is described, for calculation purposes, by a single number called the *log mean temperature difference*, denoted by MED. So MED is determined by

$$\text{MED} = \frac{\text{GTD} - \text{LTD}}{\ln \ (\text{GTD}/\text{LTD})} \tag{9.2}$$

where GTD = greatest temperature difference, air to water
 LTD = least temperature difference, air to water
  ln = natural logarithm (base e)

(Sometimes MED is referred to as LMTD in the literature.) See Fig. 9.27a.

Referring again to Fig. 9.28, for the counterflow arrangement, GTD = 78 − 55 = 23°F, LTD = 58 − 45 = 13°F, and then MED = 17.5°F. For parallel flow, GTD = 78 − 45 = 33°F, LTD = 58 − 55 = 3°F, and MED = 12.6°F. This means that coil capacity would be reduced by about one-third if it were piped in parallel flow rather than counterflow, all else being unchanged. In some cases (e.g., when the leaving water temperature needs to be greater than the leaving air temperature), parallel flow will not work at all.

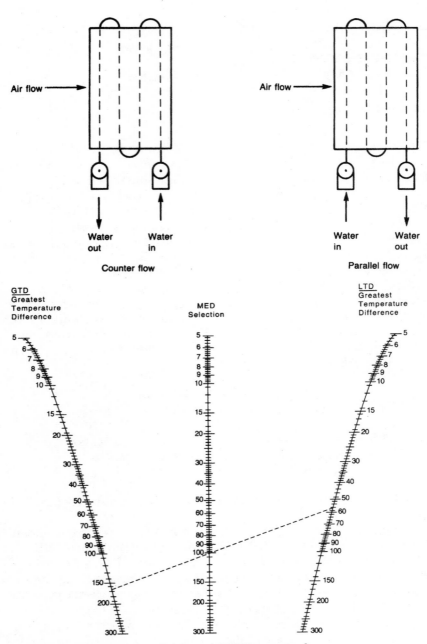

Figure 9.27 Counterflow, parallel flow, and MED nomograph.

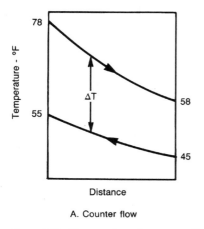
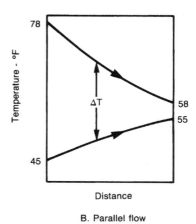

A. Counter flow

B. Parallel flow

Figure 9.28 Temperature for counterflow and parallel flow.

9.7.3 Coil selection

The criteria for coil selection include (1) geometry, (2) flow rate and design entering and leaving conditions for the air, (3) flow rate and design entering and leaving water temperatures, (4) coil characteristics, and (5) use of parallel flow or counterflow.

Geometry relates to the space available, including the space for piping connection and maintenance. Geometry also affects water velocities and therefore pressure losses. Fewer and longer tubes lead to higher velocities but double or half circuits can be used for adjustment. The airflow rate and conditions determine the total heat to be removed as well as the design ADP needed. Chilled water temperatures determine the water flow rate and indicate whether the desired ADP is possible. Coil characteristics include fin spacing, rows of tubes in the direction of airflow, and the k factor at design conditions. The k factor will vary with the amount of dehumidifying being done because a wet coil surface transfers more heat. The k factor for evaporating refrigerant in a DX coil is greater than that for a water coil, while the k factor for brine is less.

Most coil manufacturers can and will provide computerized coil selections, or software to be used in the design office to make selections based on the above criteria. It is also possible for the designer to make an accurate manual selection based on data in the manufacturer's catalog. One general equation which is often used is

$$\text{ROWS} = \frac{Q}{\text{MED} \times Ak} \tag{9.3}$$

where Q = total heat to be removed, Btu/h

A = coil face area normal to airflow, ft^2

k = heat transfer factor, Btu/h/ft^2 per row of tubes

ROWS = number of rows of tubes in direction of airflow

This requires that the manufacturer furnish tabular or graphic data for the value of k, similar to that in Fig. 9.29 or in some other form

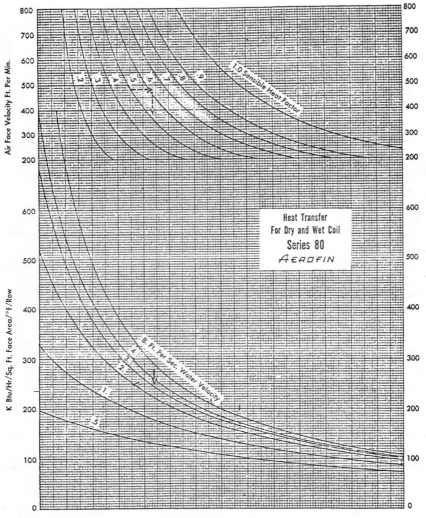

Figure 9.29 Coil selection graph (k factors). (*Courtesy of Aerifin Coil Co.*)

which allows for manual calculation. The chart shown is entered from coil face velocity, intersecting the sensible heat factor. The *sensible heat factor* is the sensible/total heat load ratio for the coil as determined from the calculations and the psychrometric chart. A projection downward will intersect the calculated water velocity curve. From this intersection, a horizontal projection will determine the *k* factor.

9.7.4 Effects of altitude

As previously noted, the density of air decreases with altitude. This has an effect on coil performance. At a higher elevation, for the same volume of air flowing, the mass flow rate of the air has decreased. The total cooling capacity of the air has decreased in similar proportion. In addition, there is a small change in the *k* factor due to the decrease in air density. Further, the air pressure loss decreases in proportion to the density change. The following manual selection procedure has proved satisfactory.

1. Determine the air density ratio at the design elevation (see Sec. 4.5 and Fig. 4.5). For example, the air density ratio at 5000 ft is 0.826.

2. Select the coil with a face area appropriate to the design (high-altitude) airflow rate, denoted by CFM, with an altitude face velocity in the recommended range of 400 to 600 ft/min.

3. Multiply the design airflow rate by the density ratio to obtain a standard-air CFM. Use this airflow rate and the equivalent face velocity with the manufacturer's data for *k* factors, etc., to determine the rows required and the pressure drop.

4. Multiply the standard air pressure drop by the density ratio to obtain the altitude design air pressure drop.

The manufacturer's computerized coil selection process should include altitude and temperature corrections.

9.8 Radiant Cooling

A radiant cooler is a ceiling or wall panel with a surface temperature less than the space temperature. It is particularly effective in achieving comfort in a location where the occupant is exposed to a warm radiant source, such as solar radiation through glass. Panels can be field-fabricated, by using water tubing built into the ceiling or wall construction. Factory-fabricated panels are available. It is essential that the panel surface temperature be kept higher than the dew point temperature of the air to prevent condensation. This limits panel applications to sensible cooling. Some other method must be used for latent cooling and dehumidification. Usually a secondary chilled water loop (Fig. 9.30) is required to provide water at the higher temperature needed to avoid surface condensation.

9.9   Evaporative Cooling

An evaporative cooling system utilizes the cooling effect of evaporating water as the heat sink. The use of this effect in condensing refrigerant and providing condensing water has been described. In many situations, it is possible to use evaporative cooling directly, at considerably lower cost than that for mechanical refrigeration. Historically this is the oldest form of air conditioning, and it has been used in the desert areas of the world for thousands of years, for both comfort cooling and food preservation. The canvas waterbags of an earlier generation are an example of applied evaporative cooling.

Figure 9.31 illustrates the basic evaporative cooling process. Water is mixed with the airstream in some manner—sprays are shown here. If the entering wet-bulb temperature t_{wb_1} is less than the entering dry-bulb temperature t_{db_1}, then some of the water will be evaporated. Because the process is essentially adiabatic, the heat required for evap-

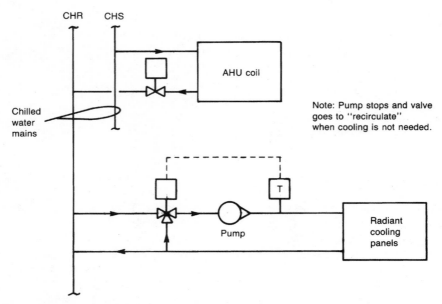

Figure 9.30 Secondary CHW loop for radiant panels.

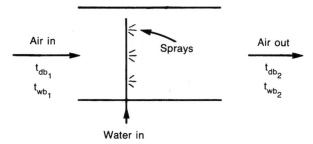

Figure 9.31 Evaporative cooling (sprays).

oration must come from the airstream, thereby lowering its dry-bulb (sensible) temperature. Then the leaving dry-bulb temperature t_{db_2} will be less than t_{db_1}. The wet-bulb temperature will remain constant, $t_{wb_2} = t_{wb_1}$, and the process can be shown on the psychrometric chart as a constant wet-bulb process (Fig. 9.32). If the system were 100 percent efficient, then t_{db_2}, would equal t_{wb_2} and point 2 would be on the saturation curve. No evaporative cooler is 100 percent effective. *Effectiveness* (eff) is defined as the ratio of the change of the dry-bulb temperature through the cooler to the initial difference between wet- and dry-bulb temperatures:

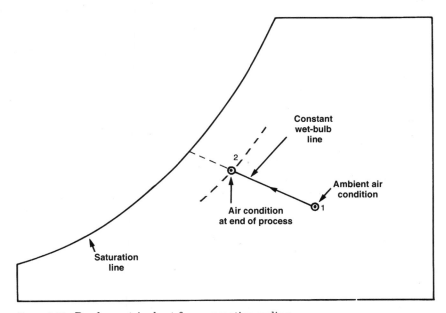

Figure 9.32 Psychrometric chart for evaporative-cooling.

$$\text{Eff} = \frac{t_{db_1} - t_{db_2}}{t_{db_1} - t_{wb_1}} \tag{9.4}$$

The enthalpy of the water has been added to that of the air, so the total heat of the airstream has actually increased. Nevertheless there is a sensible cooling effect, which can be very satisfactory in a dry climate. The resultant space humidity will be high compared to a mechanical refrigeration cycle. A very useful application of evaporative cooling is in providing relief to workers in hot-process situations, where larger airflows at moderate temperatures are most effective.

Evaporative coolers come in three types: *spray, slinger,* and *drip.* A spray-filled system, sometimes called an *air washer,* utilizes the atomization effect of the sprays to provide an air-water contact surface for evaporation. Baffles are then provided to eliminate any free moisture from the airstream. A recirculating pump, makeup water, and water treatment are needed. At the recommended air velocity of 500 ft/min, a 6-ft-long spray cooler will have an effectiveness of about 95 percent. A slinger is a rotating wheel, one edge of which is immersed in the water sump, so that as the wheel rotates, water is thrown into the air. The air passes through pads of fabric or shredded wood which are wetted by the water spray to further increase contact between water and air. The pads also act as free-moisture eliminators. Effectiveness is on the order of 40 to 60 percent depending on the air velocity. Drip-type coolers have pads which are wetted by water which is distributed by gravity from an overhead trough. Excess water may be wasted or recirculated by a small pump. This is the so-called swamp cooler which is very common in residences in dry climates. Effectiveness ranges from 30 to 50 percent. A *bleed* function is required for evaporative cooler basins just as for cooling towers, to avoid solids buildup on wetted surfaces. Chemical bacterial control is also recommended.

9.9.1   Two-stage evaporative cooling

A two-stage evaporative cooling system (Fig. 9.33) consists of a conventional cooler, as described above, preceded by a precooling section which gets its water supply from an evaporative water cooler. As shown on the psychrometric chart (Fig. 9.34), the precooling coil reduces both the wet- and dry-bulb temperatures of the air entering the second-stage evaporative cooler, thereby decreasing the final dry-bulb temperature and the resulting space humidity. While the system is more costly than single-stage evaporative cooling, it may be more cost-effective than mechanical refrigeration in some situations. This scheme is popular in the drier climates of the western United States.

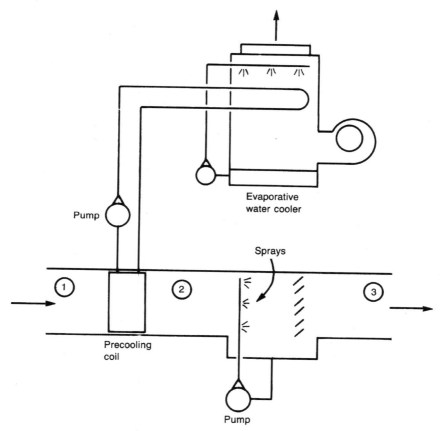

Figure 9.33   Two-stage evaporative cooling.

9.10   Refrigerants

Refrigerants for two-phase refrigeration cycles are selected on the basis of safety, suitability for the temperatures and pressures encountered, and effectiveness.

Ammonia is often used for low-temperature systems such as food and process cooling, ice rinks, etc. Propane finds some special applications. Until recently, when chlorinated hydrocarbons were determined to be harmful to the earth's ozone layer, R-11 (CCl_3F), R-12 (CCl_2F_2), and other similar compounds were in common use along with the less harmful refrigerant R-22 ($CHClF_2$). Consequent to international protocols which have set schedules for the elimination of damaging refrigerants from commerce, other replacements have been and are being developed. Part of the challenge is technical and part eco-

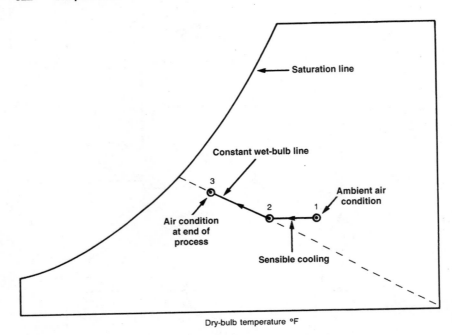

Figure 9.34 Psychrometric chart for two-stage evaporative cooling system.

nomic: first to find a fluid that has optimal characteristics and is safe; second to encourage manufacture in sufficient quantities to produce and distribute the fluid at an affordable price. R-123 ($CHCl_2CF_3$) has been developed as a near-equivalent replacement for R-11, with R-134a (CH_2FCF_3) replacing R-12. R-123 still comes under criticism for having some chlorine, although at a reduced level.

R-22 is used widely in residential and commercial air conditioning and even more in recent years in larger centrifugal, helical gear, and scroll compressor systems. It too will be phased out someday (scheduled for the period 2020–2030), but finding a suitable, widely accepted replacement has not come as quickly to solution.

For a full discussion of refrigerants and their properties, piping methods and industry practice, refer to the ASHRAE Handbooks.[3,4] Many manufacturers also publish manuals of recommended practice.

9.11   Summary

The HVAC designer has a wide variety of equipment available for providing cooling. This chapter has discussed some of that variety. Equipment manufacturers will provide detailed and helpful information to assist designers. However, designers must take care to ensure that they, and not the manufacturer, control the final system design.

References

1. F. Beseler, "Scroll Compressor Technology Comes of Age," *Heating / Piping / Air Conditioning,* July 1987, p. 67.
2. *Cooling-Tower Fundamentals,* The Marley Cooling Tower Co., 1985, Mission, KS.
3. ASHRAE Handbook, *1993 Fundamentals,* Chap. 16, "Refrigerants," and Chap. 17, "Refrigerant Properties."
4. ASHRAE Handbook, *1994 Refrigerant Systems and Applications.*

10

Equipment: Part 2

Heating

10.1 Introduction

Heating is the first word in the acronym HVAC. It is the most important part because without heating mankind would have difficulty in
surviving. Proper design of the heating system is even more critical
than that of ventilation or cooling. Human history began to develop
with the discovery and control of fire which increased people's ability
to survive in a harsh environment. In modern heating system design,
two primary concerns are proper system sizing, to achieve comfort,
and system reliability. Capital cost, operating cost, and pollution control are secondary in consideration. Pollution control is addressed by
code authorities. Energy conservation and operating costs go together
and have a considerable effect on life-cycle costs.

These concerns and many others are addressed in this chapter.

10.2 General

In a modern heating system, heating can be provided by

1. Fuel-fired boilers that produce steam, hot water, or thermal liquids
   for direct or indirect use

2. Furnaces, unit heaters, duct heaters, and outside-air heaters which
   provide hot air for direct circulation to the conditioned space

3. Waste heat furnaces and boilers which utilize the waste energy
   from some other source, such as an incinerator or refrigeration
   equipment

4. Solar energy collectors, both passive and active, which heat either water or air and, in some cases, solid materials

5. Heat pumps, either liquid or air

6. Direct-fired radiant heaters, either electric or natural gas

End users are provided heat by

1. *Direct air* — furnaces, duct heaters, outside-air heaters, reheat units, ducted heat pumps

2. *Indirect air* — coils and air-handling units, fan-coil units, unit ventilators

3. *Liquid* — radiators, convectors, liquid-filled radiant heaters

4. *Radiation* — direct radiation from panels, floors, or other radiators

10.3   Boiler Applications

Boilers can produce low-, medium-, or high-temperature water; low-, medium-, or high-pressure steam (including process steam); and thermal liquid.

10.3.1   Hot water boilers

Low-temperature water boilers (to 250°F) are the most widely used type for residential, apartment, and commercial construction. Medium-temperature water boilers (250 to 310°F) are generally used in industrial and campus-type facilities. High-temperature water (310 to 450°F) is used for extended campus-type facilities and industrial process facilities. Thermal liquid heaters are primarily found in industrial applications where both space heating and process heating are significant loads.

10.3.2   Steam boilers

Low-pressure boilers (up to 15 lb/in^2 gauge) are generally found in commercial, apartment house, and single-unit industrial facilities. They are used for space heating and domestic hot water, through end-use heat exchangers. Medium-pressure steam applications (15 to 150 lb/in^2 gauge) are generally found in campus-type facilities, hospitals, and industrial plants where there are significant process requirements. Power generation high-pressure steam boilers operate in the range of 150 to 900 lb/in^2 gauge or more with some degree of superheat in order to obtain good turbine efficiency. Waste heat from turbines is often used for space heating, domestic hot water, and process requirements.

10.4   Boiler Types

Boilers can be categorized in many different ways. For this book, the following categories are used.

10.4.1   Cast-iron sectional boilers

Cast-iron sectional boilers can produce hot water or steam at pressures up to 15 lb/in^2 gauge for steam and 30 lb/in^2 gauge for water. They are either atmospheric or power burner gas-fired, and they come in individual heat transfer sections which are modularized to obtain a range of capacities. The small sections allow for installation in spaces which are inaccessible to package boilers. They are easy to maintain and have the longest physical life of any type of boiler. Care must be taken to avoid the problem of thermal (cold) shock of these boilers for they fail if the castings crack.

10.4.2   Fire tube boilers

In fire tube boilers, the products of combustion are confined within a series of tubes surrounded by water. The most popular type is the *Scotch marine* boiler in which the combustion furnace is in the shape of a cylinder surrounded by water. Other types have steel firebox furnaces, brick-set firebox furnaces, and in some cases a combination of both. Capacities go up to about 1000 boiler horsepower (bhp) (1 bhp $\approx$ 33,480 Btu $\cdot$ h). Their popularity is due to their low first cost. Their useful life is less than that of either cast-iron or water tube boilers, and some fire-tube designs are susceptible to thermal shock under wide temperature differentials and sudden load shifts. The maximum operating pressure is usually 250 lb/in^2 gauge or less. Many older boilers had atmospheric-type burners. Current practice favors forced-draft burners. The Scotch marine boilers are all forced-draft design.

10.4.3   Water tube boilers

In water tube boilers, the water is inside the boiler tubes, and the products of combustion surround the tubes. There are a wide variety of configurations, including slant-tube (Fig. 10.1), bent tube (D type), C type, and express type. They range in size from small residential units to large utility boilers. They have an extended service life if proper water treatment and maintenance are provided. Water tube boilers may be factory-assembled and tested, package type, or field-assembled; the field-assembled boiler is more common in sizes above 200,000 lb/h capacity. Operating pressures of 150 to 900 lb/in^2 gauge or greater are used where process requirements are severe or where power generation is a consideration.

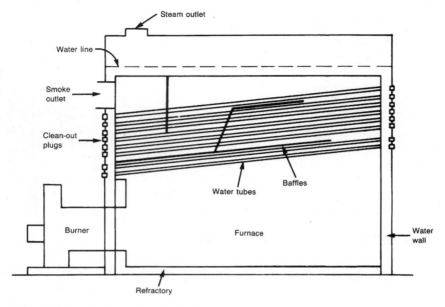

Figure 10.1 Slant-tube water tube boiler.

10.4.4 Thermal liquid boilers

Thermal liquid boilers are of the water tube type, but instead of water, a special thermal liquid is used. This liquid permits the generation of high temperatures—600 to 800°F—at low pressures. These units are often found in manufacturing facilities, with the thermal liquid used in processes. Steam is generated through a heat exchanger for use in space heating and other purposes. These boilers are prevalent in Europe but have seen limited application in the United States. Figure 10.2 shows a typical transport system for a package boiler, thermal liquid in this instance. Thermal liquids are often elusive in containment. Special consideration must be given to joint systems and device seals.

10.4.5 Steam quality

Heating and domestic hot water applications utilize saturated steam. Saturated steam is at a temperature and pressure that correspond to the saturation conditions discussed in Sec. 6.2 and is said to have 100 percent quality. Steam with some free moisture present has less than 100 percent quality (down to zero quality for condensed water). "Superheated" means that additional heat is applied to the steam to drive its temperature above the saturation temperature at the existing pres-

Figure 10.2 Transporting a package thermal liquid boiler. (*Courtesy of the International Boiler Works Co.*)

sure. In the boiler, this is accomplished in a special tube bank called a *superheater*. Superheat is required for many turbine and process applications, including cogeneration, but this steam must be "desuperheated" for use in normal heating and domestic water applications.

10.5   Combustion Processes and Fuels

The primary source of energy in a heating boiler is the combustion of a fossil fuel — coal, oil, or gas — or waste materials. The use of peat, garbage, sawdust, petroleum coke, and other waste products is increasing, but it is still a small fraction of the total fuel burned in this country.

Combustion is a process of burning — combining the fuel with oxygen and igniting the mixture. The result is heat release, absorbed through radiation, convection, and, to some degree, conduction.

10.5.1   The combustion process

The combustion process follows basic principles called the *three T's of combustion*. The first one is *time* — the time required for the air to properly mix with the fuel and for the combustion process to be completed. It is critical when waste materials are being combusted in conjunction with standard fuels. The second is *temperature*—the temperature at which the fuel will ignite, oxidation is accelerated, and the process of combustion begins. Ignition temperatures are well established for standard fossil fuels but must be carefully considered when waste or other organic-type materials are being burned. The third is *turbulence*—the process of thoroughly mixing the air and fuel so that each particle of fuel is in contact with the right amount of oxygen and combustion can continue to completion. The turbulence must be violent enough to ensure good contact between the fuel and the oxygen. Assuming there is enough combustion air to work with, inadequate turbulence is the most common cause of incomplete combustion. Inadequate turbulence can result in the generation of excessive amounts of carbon monoxide, and combustion may continue well beyond the furnace portion of the boiler.

10.5.2   The chemical reaction

In its simplest form, the combustion of natural gas (methane, CH_4) with air as a source of oxygen, the chemical reaction can be written

$$CH_4 + 2(4N_2 + O_2) \Rightarrow CO_2 + 2H_2O + 8N_2 + heat$$

This describes a perfect and complete or stoichiometric combustion process. In practice, the process is never perfect or complete. Some carbon monoxide is formed, and some contaminants, such as sulfur, are present and enter into the process. Sulfuric and nitric acids and nitrous oxides are often formed, along with other undesirable compounds.

10.5.3   Excess air

Because the combustion process is never perfect and perfect mixing of air and fuel is never achieved, every combustion process requires excess air. *Excess air* is the additional air that must be added to the theoretically perfect mixture to ensure as complete a combustion process as is practically possible. The larger the amount of excess air, the lower the combustion efficiency. Often overlooked is the possibility of condensation in the boiler or flue that has too much excess air. It can be reasoned that turbulence is a most important factor in the combustion process. Almost all of the newest boiler developments have been in burner design, in an attempt to improve the mixing of air and fuel to minimize excess air and to maximize combustion efficiency.

10.5.4 Combustion efficiency

The *combustion efficiency* is the ratio of fuel heat input minus the stack loss (through the chimney or vent), divided by the fuel heat input. Typical efficiencies for mechanically fired boilers range from 75 to 83 percent for new installations at full-load conditions. Firing at reduced capacity may reduce the combustion efficiency. Therefore it is desirable to match the boiler to the load as closely as possible or to use multiple boilers.

The *overall thermal efficiency* is the gross output in Btu/h divided by the fuel heat input in Btu/h. This rating takes into account the noncombustion losses from the boiler, such as radiation (see Fig. 10.3).

The *seasonal thermal efficiency* is the ratio of net delivered useful heat to gross fuel input and accommodates all system losses. Seasonal efficiencies for systems may range from 40 to 80 percent.

10.5.5 Fuels

Gaseous fuels include natural gas, manufactured gas, and liquefied petroleum gases (propane and butane).

Oil fuels include distillates or lightweight oils, especially no. 2, and residuals or heavy oils, nos. 4, 5, and 6. The residual oils have high viscosities and require preheating before they can be pumped and atomized. They are used only where proper handling and preheating can be provided, mostly in large boiler plants.

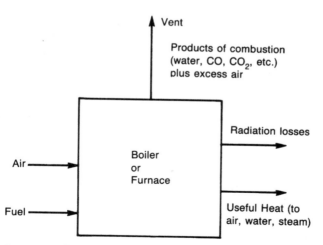

Figure 10.3 Boiler or furnace thermal input and output.

Solid fuels include bituminous and anthracite coals, coke, peat, and sawdust. A most critical factor in their utilization is the ash fusion temperature. Liquefied sodium compounds in the hot ash deposits may cool and scale up the convection banks of the boiler. Clinkers are an example of fused ash compounds.

Electricity can also be considered a fuel, and it is sometimes used to fire small steam and water boilers.

Waste materials are being used more and more in boilers, either in combination with gas, oil, or coal or as the primary fuel source. Many of these solid waste materials contain large amounts of impurities, such as chlorides, which can cause serious damage to boiler heat transfer surfaces. Under no circumstances should the utilization of solid waste be undertaken without the assistance of either the equipment manufacturer or a competent consultant, or both.

Wood became very popular for a time and has been in use for many years. A major problem with wood is its moisture content and resinous deposits. Supplies have become limited in many areas.

10.6 Fuel-Burning Equipment

Burners are devices for controlling the combustion process by mixing the fuel and air in the proper relationship and making the process efficient.

10.6.1 Coal burners

Coal and other solid fuels are fired automatically by means of stokers, pulverizers, or fluid-bed combustion systems. A *stoker* is a means of adding fuel on a metered basis to an existing fire. An underfeed stoker, normally applied to small boilers and furnaces, feeds fresh fuel from below the fire. The fuel is spread out on dump grates for the completion of combustion. A traveling-grate stoker (Fig. 10.4) has a moving or vibrating grate on which the fuel is deposited. The fuel burns as the grate moves so that at the end of the grate the ash is dumped to an ash pit from which it may be removed manually or mechanically. A spreader stoker feeds a traveling grate, or a dump grate, but the coal is deposited by throwing it onto the grate with a special feed device. A vibrating stoker (Fig. 10.5) is sloped so that the fuel moves down the grate by gravity from the feed end as lateral rods are moved back and forth. All these stokers include forced-draft and/or induced-draft fans to control the flow of combustion air. The balance between overfire air and underfire air is critical for complete combustion and reduction of particulate emissions.

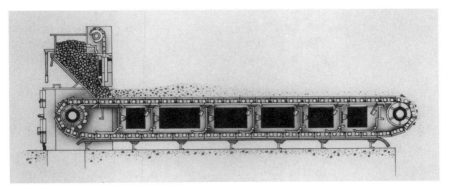

Figure 10.4   Coal-burning chain-grate stoker. (*Courtesy of Detroit Stoker Company.*)

Pulverized coal firing is found in larger (150 million Btu/h) boilers. Raw coal is fed through a mill which pulverizes the coal into coffee-ground to dust-size particles which are then introduced into the firebox through a burner tube similar to a gas burner. There is a violent mixing of coal particles and air to effect combustion.

An alternative concept for solid fuel and solid waste firing is the atmospheric fluidized-bed combustion system (Fig. 10.6). Although it has been used for many years in the sewage sludge combustion business, its application to power and heating boilers is relatively new. The solid fuel or waste is ground or crushed to a uniform size and then injected into a combustion bed, usually sand. Air is blown through beneath the bed, fluidizing the fuel and suspending it above

Figure 10.5   Coal-burning vibrating-grate stoker. (*Courtesy of Detroit Stoker Company.*)

Figure 10.6 Fluidized-bed combustion unit. (*Courtesy of the International Boiler Works Co.*)

the bed, where primary combustion takes place. Limestone or some similar material is mixed with or injected into the fuel, where it absorbs sulfur, thereby reducing the emission of sulfur compounds. A fabric filter (baghouse) is required to remove particulates from the process. These systems require a properly trained and experienced operating staff. The advantage of the system is the reduced emission of sulfur and nitrogen compounds. In the figure, note the overfeed stoker and underfire air supply.

10.6.2 Natural and Liquefied Petroleum (LP) gas burners

Gas burners are of the atmospheric, fan-assisted, or premixed type. The atmospheric burner (Fig. 10.7) is found in many residential applications and in commercial and industrial cast-iron boilers. It is also used in most direct-fired unit heaters. It depends on the inlet gas pressure and stack effect to provide combustion air and mixing. Primary combustion air is entrained by induction and mixes with the gas; the geometry of the burner is designed to provide an optimal fuel-air mixture. Secondary air is entrained over the fire to provide more complete combustion.

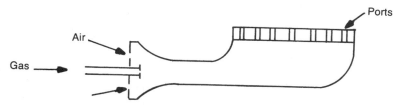

Figure 10.7 Atmospheric gas burner.

Forced-draft, fan-assisted or power burners (Fig. 10.8) use a fan to provide the combustion air, with a significant improvement in air-fuel mixing and efficiency compared with the atmospheric burner. Power burners are in wide use today in most heating boilers.

Premix burners mix the fuel and air in an internal mixing chamber so that optimum excess-air relationships are achieved. This burner has a very short, intense flame and is primarily found in applications where size or very high temperatures are significant.

Figure 10.8 Fan-assisted gas or oil burner. (*Courtesy of Gordon-Piatt Energy Group, Inc.*)

10.6.3 Oil burners

Atmospheric-type oil burners and rotary-cup burners were used in the past. Present-day burners are of the mechanical atomizing type (Fig. 10.9) that use an oil pump to develop an atomized oil spray through a nozzle. A fan provides air which is introduced in a swirling pattern at the nozzle to provide good mixing.

Oil burners in larger boilers may use a separate atomizing agent, such as compressed air or steam, for improving atomization (Fig. 10.10). This significantly improves the combustion process, but there is a cost penalty for the power used. In large boilers, the combustion improvement and control of the flame pattern are more important than the additional power cost.

About 20 years ago a new oil burner, the *low-excess-air* or *sleeve-type burner,* came to the market. It provides good combustion with as low as 2 or 3 percent excess air, compared with 15 percent or more for standard burners. It was developed by the British Admiralty for its warships and is now applied to industrial and utility applications. It should be considered in boilers of 100 million Btu/h and larger.

Figure 10.9 Oil burner with mechanical atomization. (*Courtesy of Gordon-Piatt Energy Group, Inc.*)

Figure 10.10 Oil burner with steam atomization. (*Courtesy of Gordon-Piatt Energy Group, Inc.*)

10.6.4 Ignition

Ignition is obtained by means of a standing gas pilot, an intermittent gas flame which is ignited by an electric spark, or direct electric ignition. The pilot or low fire must be proved by means of a thermocouple or photocell before the main gas or oil valve is allowed to open. Pilot burners are small, and it is not unusual to have several manifolded together. Most power-burner control systems have a purge and prepurge sequence to make sure there is not a combustible mixture in the boiler which might cause an explosion during start-up. Propane may be used to pilot oil burners where natural gas is not available.

10.6.5 Fuel-handling equipment

Gas burners require a fuel train connected to the utility gas distribution system downstream of the gas meter and pressure-reducing valve. Normally gas pressure is measured in inches of water, but pressures up to several pounds can be obtained if needed.

Oil fuels require a fuel-handling system which includes a storage tank or tanks, oil-heating systems at tanks and burners for heavy oils, oil filters, auxiliary atomizing equipment, if used, and pumps (see Fig. 10.11).

Coal handling has not changed basically in 100 years, except that it has become somewhat more mechanized. Coal is delivered from the mine to the silo or coal pile, from which it is moved to a day hopper by elevators or conveyors and is fed into the boilers. Ash is disposed of either mechanically or manually. In some large boilers, the fly ash is reinjected to use as much of the free carbon as possible. Entraining fly ash into overfire air also assists in developing turbulent conditions in the combustion chamber.

10.6.6 Controls

Controls for automatic fuel-burning equipment range from simple two-position on/off controls to full modulating systems which measure flue gas temperatures and constituents (O_2, CO, CO_2) and automatically adjust fuel-airflow ratios for maximum combustion efficiency (Figs. 10.12 and 10.13). There is sometimes a tendency to "oversophisticate" the control system. The controls should be as simple as possible, commensurate with the size and sophistication of the system and its operators. In all cases, control systems must include safety devices and procedures to prevent the development of hazardous conditions, including high pressure, high temperature, low water level, flame failure, and the like.

10.6.7 Environmental considerations

Natural and LP gas fuels require no pollution controls except for oxides of nitrogen in large boilers. Residual oils and solid fuels require

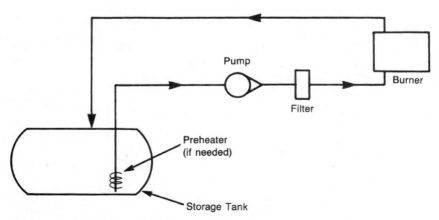

Figure 10.11 Elementary oil fuel-handling system.

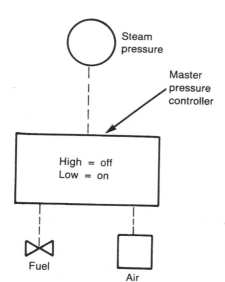

Figure 10.12 On/off burner control system. (*Courtesy of F. Govan.*)

tail-end control equipment to remove particulate matter and sulfur compounds. These are regulated by local, state, and federal codes. Such equipment includes fabric filters (baghouses), electrostatic precipitators, dry and wet scrubbers, and controls to maximize combustion efficiency. Some very exotic nitrogen and oxide control systems are being marketed, but their value is limited and costs are high. They should be evaluated as a developing technology.

10.7 Boiler Feedwater and Water Treatment Systems

Hot water boilers require very simple makeup systems. The hot water is used in a closed circuit, so water losses are minimal. Water softeners and small amounts of chemical treatment may be employed for oxygen scavenging and corrosion control. A simple *shot feeder* for adding chemicals is shown in Fig. 10.14.

Steam boilers may require elaborate makeup and feedwater systems. Condensate, steam trap, fitting, and blowdown losses may be as much as 15 to 20 percent of steam capacity. Some industrial plant systems are designed for 100 percent makeup.

In small boiler systems, condensate is returned by gravity to a receiver and then is pumped into the boiler as required. Makeup water is supplied to the condensate receiver along with water treatment chemicals. Periodic or continuous blowdown is required to remove the buildup of sediment and evaporated solids in the waterside of the

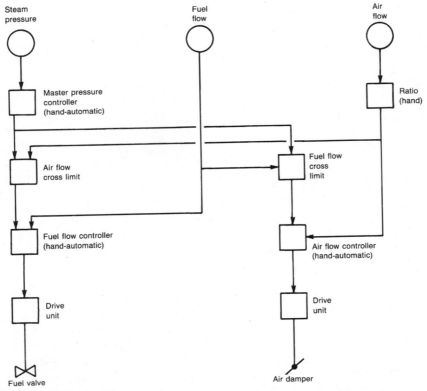

Figure 10.13 Full modulating burner control system. (*Courtesy of F. Govan.*)

boiler. Figure 10.15 shows a typical boiler feed system with a Hartford loop.

In larger steam boilers and all high-pressure boilers, the feedwater and treatment system can become complex (Fig. 10.16). Condensate is returned to a receiver from which it is pumped to a deaerating feedwater heater, which preheats the water and removes most of the dissolved oxygen; oxygen is very corrosive to the high-temperature waterside surfaces. Zeolite softening or charcoal filtering is frequently used for pretreatment of raw makeup water. Chemicals may be added. All this treatment must be automatically controlled, but as simply as possible. The water treatment program should be tailored to the specific conditions because all raw waters differ from each other. A water treatment consultant should be used. Most treatments include blowdown, pH control, and addition of chemicals to neutralize other contaminants. Great care must be taken when a new program is begun for an existing boiler system. A new program may loosen accumulated

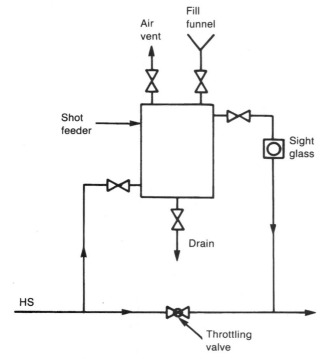

Figure 10.14 Shot feeder.

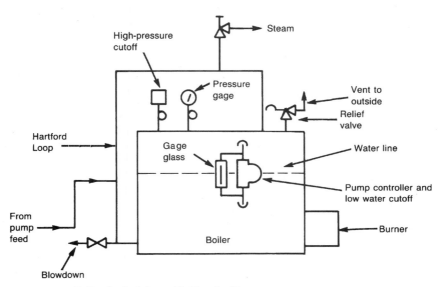

Figure 10.15 Boiler feed piping with Hartford loop.

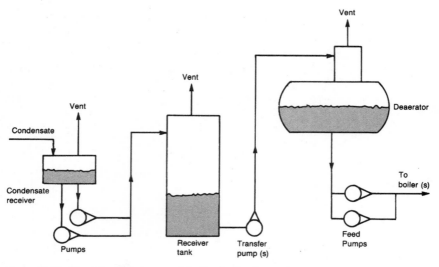

Figure 10.16 Condensate return and boiler feedwater system.

sludge and scale deposits, causing massive failures. Many boiler failures can be attributed to improper water treatment or overtreatment.

10.8 Boiler Codes and Standards

Boilers must be installed and operated in accordance with applicable codes and standards. The local code authorities will refer to one or more of the industry codes, especially the American Society of Mechanical Engineers (ASME) boiler code.[1] Other references will meet the standards of the American Gas Association (AGA), the Hydronics Institute (HYDI), the American Boiler Manufacturers' Association (ABMA), and insurance companies, such as Factory Mutual (FM) (particularly with respect to burner systems). Underwriters' Laboratories (UL) provides certification for some of the control and safety devices, such as relief valves. The proposed design of the boiler system should be submitted to the owner's insurance company to ensure compliance with requirements.

10.9 Boiler Design

There is little that a client or consulting engineer can do to influence the basic design of a boiler. However, clearly established performance criteria can be used to ensure a long-lived, efficient system. Some of these factors include

1. Heat release per square foot of flat projected radiant surface, a measure of the intensity of radiant heat transfer. A high value may contribute to a short boiler life.

2. Combustion volume—the physical volume in the furnace necessary for complete combustion. Too large or too small a volume may reduce efficiency.

3. Convection air unit tube spacing—critical when solid fuels are used. Small spaces can easily be blocked by scale or ash.

4. Combustion efficiency based on accepted test procedures.

5. Flue gas temperature—critical in the prevention of condensation and corrosion in the final sections of the boiler.

6. Physical size—especially important in existing structures where installation access is limited.

In the design and installation of larger boilers and high-pressure boilers, it is especially important that the design engineer have experience in the field, in order to properly evaluate the claims of competing suppliers.

10.10 Acceptance and Operational Testing

Residential and small commercial boilers are seldom tested individually for combustion efficiency. For larger boilers and high-pressure boilers, thermal testing in the field is usually required. The standard test is the ASME power test code, short form, input/output test method. It is expensive and time-consuming but provides accurate measurement of actual performance. There are other tests, but any simpler test is of questionable accuracy and will seldom yield consistent results when it is repeated.

For operational testing, the operator should have, at the least, a simple efficiency test kit, such as an "Orsat", a "Baccharach", or "Fyrite" gas absorption device. These devices or one of the newer automatic gas sampling devices can be used on a regular basis to measure efficiency and to indicate changes in performance. The instruments measure flue gas temperature as well as the percentage of oxygen, carbon monoxide, and carbon dioxide. By using a nomograph, the combustion efficiency can be calculated from these data. Significant changes— ±10 percent—must be investigated. Newer automatic gas sampling equipment which simultaneously reads flue gas, oxygen, and carbon monoxide is wonderful for regular testing of boiler performance.

10.11   Direct- and Indirect-Fired Heating Equipment

A *direct-fired* heater is one in which the fuel is converted to heat energy at the point of use. The usual fuel is either electricity or a fuel gas, either natural gas or liquefied petroleum gas (LP). Fuel oil is seldom used. Direct fuel firing in an occupied space requires background ventilation, enough to dilute the products of combustion.

Indirect-fired heaters utilize a heated fluid, e.g., steam or hot water, which is heated elsewhere and transported through a piping system to the point of use.

10.12   Heat Exchangers—Water Heating

Heat exchangers for steam to water or water to water are of the shell-and-tube type similar to those described in Chap. 9. A steam-to-water exchanger (Fig. 10.17) could also be called a *steam condenser*, because it is the latent heat of condensation which is being used to heat the water. The steam is in the shell, the water in the tubes. The system is controlled as shown in Fig. 10.18. For more accurate control at light loads, it is common practice to use two control valves in parallel, sequenced, with the smaller valve sized to handle one-third of the load. It is more difficult to control with higher steam pressures; for small heat exchangers it may be desirable to provide a steam-pressure-reducing station (see Fig. 6.1).

A water-to-water heat exchanger (Fig. 10.19) may be used in many ways. When it is used with high-temperature water (HTW), the HTW is always in the tubes. This provides an extra measure of safety since the tubes are more easily rated for higher pressures than is the shell.

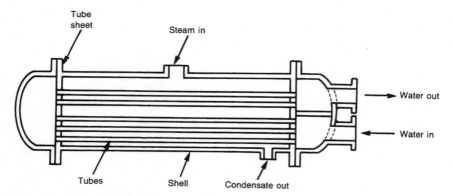

Figure 10.17   Steam-to-water heat exchange.

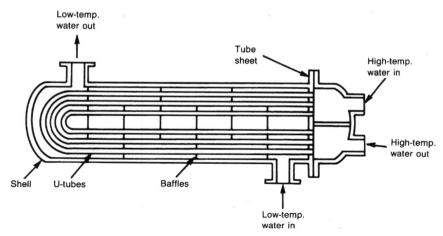

Figure 10.18 Control for steam to hot water heat exchanger.

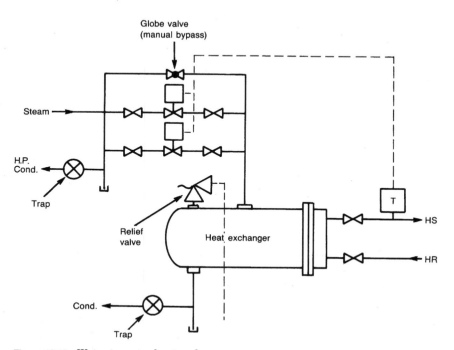

Figure 10.19 Water-to-water heat exchanger.

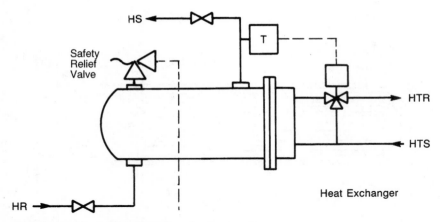

Figure 10.20 Control for water-to-water heat exchanger.

The control valve may be a two-way or a three-way type, as shown in Fig. 10.20. This may be part of a *cascade* arrangement, shown in Fig. 6.3. Two-way valves are used in variable-flow systems.

10.13 Heat Exchangers—Air Heating

A heat exchanger *coil* for air heating is of the finned-tube type, as described in Sec. 9.7. Steam-to-air coils are frequently made in a double-tube configuration (Fig. 10.21). The steam is supplied to the inner tube, which has a number of small metering orifices through which the steam passes to the outer tube. The result is a more or less uniform distribution of steam throughout the coil, providing uniform temperatures across the face of the coil and assisting in prevention of freeze-up of condensate in air preheating coils.

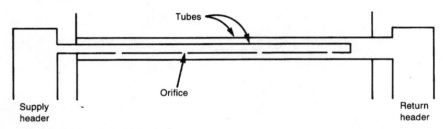

Figure 10.21 Double-tube steam coil.

10.13.1   Freeze protection for air preheating coils

One of the most difficult HVAC processes is the preheating of sub-freezing air. Many systems require large quantities of outside air, up to 100 percent. When outside air temperatures are below freezing, the steam condensate or water in the preheating coil may freeze instantly if proper precautions are not taken. Freezing usually results in rupturing of the coil tubes, requiring repair or replacement. This can also cause a shutdown of the air-handling unit (AHU) with consequent disruption of activities in the building areas served.

Figure 10.22 shows a preferred method of installing a steam preheat coil by using low-pressure steam (about 15 lb/in^2 gauge). A double-tube coil is used. The outside-air damper is interlocked to open when the AHU fan runs. The thermostat sensing bulb is in the airstream ahead of the coil. When the airstream temperature falls below 35°F, the two-position steam valve opens and remains fully open as long as freezing air is sensed. Because the condensate represents the real freezing problem, the steam trap must be oversized—for about 3 times the design flow rate—and a vacuum breaker must be provided. Note

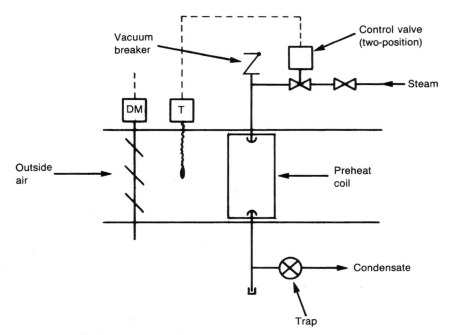

Figure 10.22   Control of steam coil to prevent freezing.

that condensing steam may create subatmospheric pressures in the coil such that vertical drop must be used to create pressure for condensate movement. A low-temperature safety device is provided to stop the supply fan if freezing air is sensed downstream of the coil.

The system just described is effective in preventing freeze-up but may produce excessively high temperatures in the air leaving the coil. To provide better control of downstream temperatures, the system shown in Fig. 10.23 is used. The coil is controlled as before, but face and bypass dampers have been added, modulated by a downstream thermostat. The bypass damper must be sized to have the same wide-open air pressure drop (100 percent flow through the bypass) as the coil-and-face damper combination at 100 percent flow through the coil.

For water coils, a different scheme must be used. Figure 10.24 shows the traditional method of preventing freeze-up in water coils. The three-way valve is modulated in response to a downstream thermostat. The circulating pump is sized to provide a minimum of 3 ft/s water velocity in the tubes. It has been determined empirically that this flow rate is sufficient to prevent freezing if adequate heating wa-

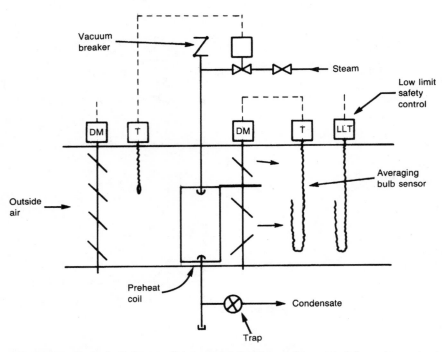

Figure 10.23 Control of steam coil to prevent freezing and to control downstream temperature.

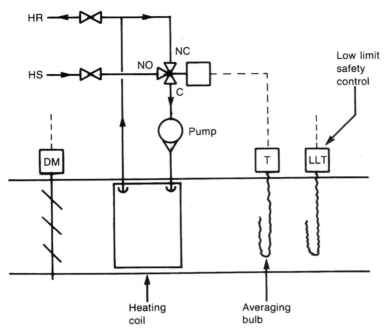

Figure 10.24 Freeze protection of hot water heating coil.

ter is available. The pump may be controlled to run only when the outside air temperature is below freezing.

The system of Fig. 10.25a can also be used. The same principle applies. Because the control valve is no longer in the pump circuit, somewhat less horsepower is needed.

One more common method of freeze protection in heating coils exposed to outside air is to create a glycol-filled *subsystem* including a heat exchanger, pump(s), and a heating coil. The glycol may be heated by steam or hot water. There must be a glycol fill mechanism and an expansion tank. The percentage of glycol used is related to how cold it gets. Fifty percent solutions are typical in subzero applications. The glycol solution must include corrosion inhibitors. See Fig. 10.25b for the freezing points of various concentrations of ethylene glycol and propylene glycol.

10.14 Unit Heaters and Duct Heaters

A *unit heater* is a package which includes a heating element and a circulating fan. It is designed for installation in or adjacent to the space to be heated. Units are made for horizontal discharge

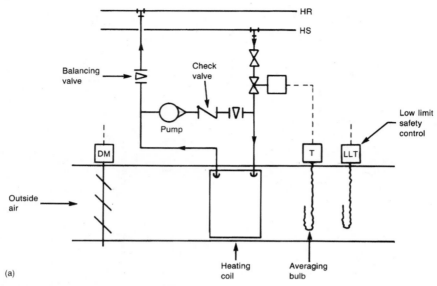

Figure 10.25a Freeze protection of hot water heating coil.

(Fig. 10.26) or vertical discharge (Fig. 10.27). Most unit heaters have propeller fans. Units with centrifugal fans may be used with ductwork to extend the area of coverage.

The heating element may be a steam or water coil or may be direct-fired by using fuel gas or electric resistance coils. Gas heaters require proper venting and safety controls. Unit heaters are normally controlled by a room thermostat which starts the fan and energizes the heating element simultaneously.

A duct heater (or duct furnace) is a unit heater without a fan and is installed in a duct or plenum. The duct heater depends on an AHU fan for air circulation. It may be the primary heating element—in the main duct or AHU plenum—or may be used for zone reheat control in branch ducts. Many package air-handling systems use duct heaters.

An outside air heater is a unit heater or duct heater used for pre-heating outside air, as required for exhaust makeup or combustion. To prevent freeze-up, gas or electric heating is used, with gas preferred on an energy cost basis. In some installations, codes allow the use of direct-fired unvented heaters—all the heat and products of combustion are in the airstream, but are so diluted as to pose no danger. This situation requires that all the supply air be exhausted.

Radiant unit heaters have no fans and utilize radiant heating rather than convective heating. For this purpose they are installed overhead and equipped with special high-temperature ceramic surfaces which

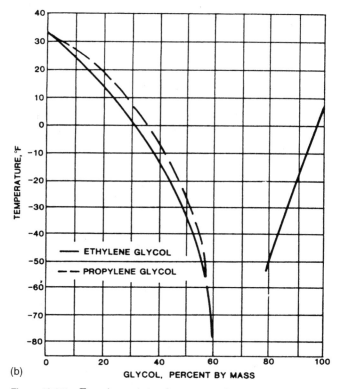

Figure 10.25b Freezing points of aqueous solutions of ethylene glycol and propylene glycol.

radiate primarily in the infrared spectrum. Radiant unit heaters are used mostly for "spot heating" at workstations in otherwise unheated or poorly heated buildings. They are also used for heating outdoor areas where people need to wait or stand in line, such as under theater marquees or in amusement parks. Radiant heating is a very efficient and economical method of achieving a level of comfort in an area

Figure 10.26 Horizontal unit heater. (*Courtesy of F. L. Reznor.*)

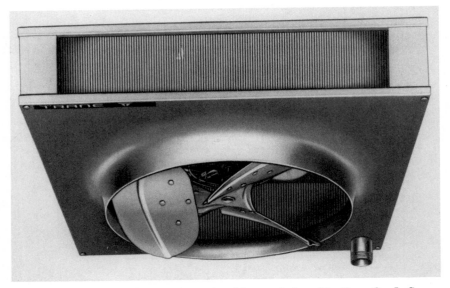

Figure 10.27   Vertical unit heater. (*Reproduced by permission of the Trane Co., LaCrosse, Wis.*)

which would be difficult or impossible to heat satisfactorily in any other way. Radiant heaters are sometimes used for snow melting at driveway entrances to parking garages.

10.15   Terminal Heating Equipment

Terminal heating equipment is equipment installed in or contiguous with the area served. In general, the heating source is remote — water or steam is used — but electric resistance heating is common. Duct heaters and some heat pumps can also be included in this category. In many cases, the terminal equipment is used for both heating and cooling (see Chap. 11).

10.15.1   Radiators and convectors

A *radiator* is a heating device which is installed in the space to be heated and transfers heat primarily by radiation. The most common example is the sectional cast-iron column radiator. There are many thousands of these in use throughout the world, although in new installations they have been largely supplanted by convector radiators or baseboard radiation. The heat source is hot water or low-pressure steam (5 lb/in^2 gauge or less). Small *electric radiators* include water and an electric immersion heater.

Radiators are rated in square feet of radiation or equivalent direct radiation (EDR). And 1 ft^2 EDR is equal to 240 Btu/h for steam at 1 lb/in^2 gauge and 180 Btu/h for water at 200°F. These ratings are no longer readily available but may be obtained from the Hydronics Institute,[2] formerly the Institute of Boiler and Radiator Manufacturers. Some representative data are available in the ASHRAE Handbook.[3]

Radiators are controlled in several ways:

1. Manually, by means of a globe valve.

2. Automatically, by means of a modulating or two-position valve. Self-contained valves, although they offer a rather coarse control, are very popular for this application.

3. In zones, by means of a zone control valve, sometimes with solar compensation. With zone control, orifices are used at radiator supply connections to ensure uniform distribution of steam throughout the zone.

4. In some small systems, by cycling the boiler or hot water pump.

5. By means of the vacuum system for steam heat. This requires a closed system in which the absolute pressure may be varied by a vacuum pump in the condensate return. The steam system may then be operated at subatmospheric pressures with a consequent reduction in steam temperature.

A *convector* is a heating device which depends primarily on gravity convective heat transfer. The heating element is a finned-tube coil or coils, mounted in an enclosure designed to increase the convective effect (Fig. 10.28). The enclosure (cabinet) is made in many different

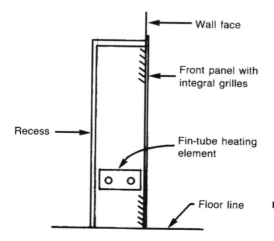

Wall face

Front panel with integral grilles

Recess

Fin-tube heating element

Floor line   **Figure 10.28** Convector.

configurations, including partially or fully recessed into the wall. The usual location is on an exterior wall at or near the floor. The capacity depends on the geometry—length, depth, height—and heating element design as well as hot water temperatures or steam pressure. Ratings are usually based on the test methods specified in Commercial Standard CS 140-47, *Testing and Rating Convectors.*[4] Refer to manufacturers' catalogs for specific data.

Baseboard radiation is designed for wall mounting in place of the usual baseboard. It is either a finned-tube system, similar to a convector but much smaller, or a cast-iron section, designed with convective heat channels to augment the radiant effect. Baseboard radiation is usually continuous along exterior walls. Blank covers may be used for appearance if the capacity is not needed.

Finned-tube, or finned-pipe, radiation uses larger tubing or pipe—1¼ to 2 in—with fins bonded to the pipe. The fins are typically 3½ to 4½ in square. The system is used mostly for perimeter heating, particularly of glass areas. Heat transfer is by convection, and a variety of enclosure types are available; some examples are shown in Fig. 10.29. Special enclosures are often made to suit an architectural decor.

All these heating elements may use either low-pressure steam or hot water as the heating source. Either one-pipe or two-pipe distribution systems are used, although two-pipe systems are more common in modern practice. Zoning by exposure, using solar compensated sensors, is a frequent practice. Electric baseboard radiation is also available. It is sometimes more economical, e.g., in an all-electric situation or where steam or hot water is not available.

10.15.2 Radiant panels

A *radiant panel* is a heating surface designed to transfer heat primarily by radiation. There may also be a convective component, and

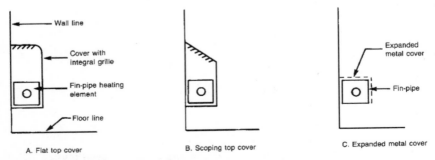

Figure 10.29 Typical fin-type enclosures.

in the case of floor panels, convective transfer may be predominant. Panels may be located in the floor, wall, or ceiling and may occupy part of or all the available area. Panel surface temperatures are limited by the physiological response of the building occupants. That is, too high a temperature may result in an uncomfortably warm feeling. Typical limitations are 80 to 85°F for floor panels, about 100°F for wall panels, and 120 to 130°F for ceiling panels. The heating source is hot water or electric resistance heating cable. Hot water supply temperatures should be consistent with the panel temperature limitations; for floor panels, e.g., the supply water temperature should be no more than 100°F.

Factory-assembled sidewall and ceiling panels and panel systems are available. Most panels are field-fabricated by using electric heating cable, copper tubing, or steel pipe embedded in the construction. For concrete floor panels, steel pipe was traditionally used (¾ or 1 in), because steel has an expansion coefficient similar to that of concrete. Corrosion at the concrete-pipe interface can be severe. In recent years, several plastic piping systems for radiant heating have come to market. Care must be taken in design, and quality control must be imposed on the installation. Electric heating cable may be used. Ceiling and wall panels use ½- to ¾-in copper tube or electric cable. Air venting is a serious problem, especially with floor panels.

Control systems are conventional, since radiant-heat-sensitive devices are not readily available. Floor panels are very difficult to control, since the relatively large mass provides a slow response. The supply water temperature is sometimes reset on the outside air temperature.

For a detailed discussion of panel heating, see the ASHRAE Handbook.[5]

10.15.3 Fan-coil units and unit ventilators

For a discussion of these devices see Chap. 11.

10.16 Heat Pumps

A *heat pump* is a mechanical refrigeration system arranged and controlled to utilize the condenser heat for some useful purpose, typically space heating. Systems may be package or built-up, air-to-air, water-to-air, or water-to-water. Earth-coupled systems are also used as a variation of the water-to-water concept.

10.16.1 Package heat pumps

A package heat pump is a factory-assembled system designed to provide either heating or cooling, as needed. The standard refrigeration cycle is modified as shown in Fig. 10.30. The key to the operation is the reversing valve. In the cooling position, refrigerant flow is directed first to the outdoor coil, which becomes the condenser. The liquid refrigerant then bypasses metering device 1 and flows through metering device 2 to the indoor coil. The metering device is a thermal expansion valve, throttling tube, or some other method of reducing the pressure. The indoor coil then becomes the evaporator, and cooling is provided. With the reversing valve in the heating position, refrigerant flow is reversed, the indoor coil becomes the condenser and provides heating; heat is extracted from the outdoor air. The changeover from heating to cooling may be automatic but is usually manual. Most package heat pumps are air-to-air. The heating capacity decreases as the outdoor air temperature decreases.

While most air-to-air heat pumps will operate satisfactorily down to 0°F outdoors, auxiliary heating will be needed except in very mild climates. Figures 10.31 and 10.32 illustrate the procedure for determining the auxiliary heat required. Figure 10.31 shows the method of calculating the net heating load as a function of temperature. For buildings with 24-h occupancy, solar heat effects should be ignored. Note that the net heat loss is less than the calculated heat loss because of internal heat gains due to people, lights, and other sources. In Fig. 10.32, this net heating load is plotted against the heat pump capacity from manufacturer's data as a function of temperature. The shaded area is the excess of load over capacity, requiring auxiliary heat. Almost any fuel can be used for auxiliary heat, but electric resistance is the most common.

In a water-to-air package heat pump, a water-to-air heat exchanger is substituted for the outdoor coil. A central source for heating or cooling the water can then, in effect, provide the auxiliary heat. Systems of this type are used in apartment houses and hotels to allow maximum control of the room environment by the occupant. The system arrangement is shown in Fig. 7.16. The water temperature is controlled at a range of values—perhaps 70 to 85°F—which is suitable as both a heat source and a heat sink for the heat pumps. In mild weather when some units are in heating mode while others are in cooling, the heat load on the system may balance, and the central boiler and cooling tower may be idle.

During the last thirty to forty years, several large office buildings and small college campuses have been constructed using water-to-water heat pumps, with capacities up to several hundred tons. These systems typically use well water. Two wells are used, one for supply and one for disposal. One possible arrangement is shown in Fig. 10.33. The supply and disposal wells are manually selected. Well water and return water are mixed, for both evaporator and condenser, on a tem-

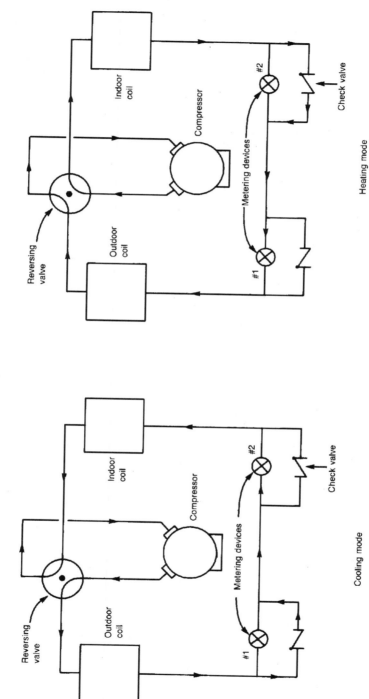

Figure 10.30 Package heat pump cycles.

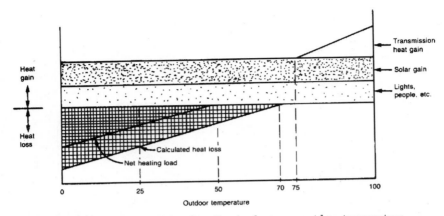

Figure 10.31 Estimating heating and cooling loads versus outdoor temperature.

perature basis. Under some conditions this system can become an internal source heat pump; i.e., when the exterior zone heating and interior zone cooling loads are in balance, or nearly so, little or no well water is needed. Internal source heat pumps without wells are used where there is sufficient internal cooling load to supply the net heating requirements under all conditions. Excess heat is disposed of through cooling towers. One problem with these systems relates to a high base electrical load for system pumping. Variable-flow piping schemes are helping to overcome this problem.

10.17   Heat Recovery and Reclaim

A *heat recovery system* intercepts and utilizes heat which would otherwise be "wasted" to a heat sink. The most common systems provide air-to-air heat interchange between supply and exhaust airstreams.

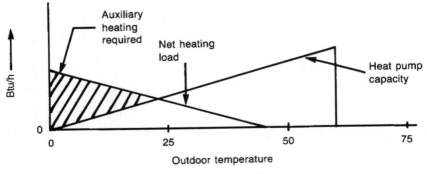

Figure 10.32 Heating load versus heat pump capacity.

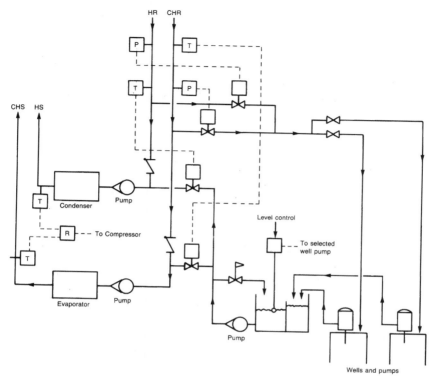

Figure 10.33 Large building heat pump, with well water source.

Use of refrigeration condenser heat for heating (as in a heat pump) or reheat is common. A less common application is heat recovery from a process, e.g., hot wastewater. Some heat recovery systems deal only with sensible heat; others include both sensible heat and latent heat and are called *total heat recovery systems.*

10.17.1 Coil loop heat recovery system

The coil loop or runaround heat recovery system consists of two heat exchange coils, a pump, and interconnecting piping (Fig. 10.34). The coils are usually placed in the outside air supply and exhaust airstreams because this will provide the greatest temperature gradient and therefore the maximum energy recovery. Sensible heat is transferred from the exhaust airstream to the outside airstream. Using eight-row coils, the system will recover up to 50 percent of the total available sensible heat energy. This is not entirely "free" energy because additional fan work is required to overcome the air pressure losses through the coils, and pump energy is required.

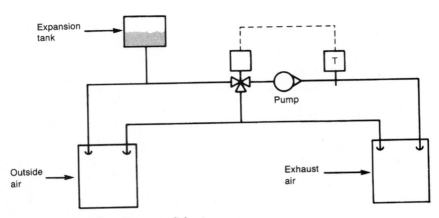

Figure 10.34 Coil loop (runaround) heat recovery.

The three-way control valve is used to keep the exhaust air coil water supply temperature above freezing to avoid frost formation. Where outside air temperatures are below freezing, a glycol solution is used.

These systems are always custom-designed, and have no geometric limitations; i.e., the coils may be widely separated or close together.

System calculations are iterative. The coil face areas are based on known air quantities, at about 500 ft/min. (The coils need not have the same face area.) A water quantity is assumed, and coil k factors are determined. The air temperatures for outside air and exhaust air entering the coils are part of the HVAC design. The water and/or brine temperatures are unknown, as are the air temperatures leaving the coils. By definition, the heat transfer quantities in the two coils must be equal to each other, and equal to the quantity transferred by the water. Four equations apply:

For air,

$$Q = \text{CFM} \times 1.08(t_{a_i} - t_{a_o}) \tag{10.1}$$

For the airside of the heat exchange coil,

$$Q = \text{MED} \times Ak \times \text{ROWS} \tag{10.2}$$

For the waterside of the coil,

$$Q = \text{GPM} \times (t_{w_i} - t_{w_o})(500) \tag{10.3}$$

For MED,

$$MED = \frac{GTD - LTD}{\ln(GTD/LTD)} \qquad (10.4)$$

where Q = heat transferred, Btu/h
$\quad$ MED = mean temperature difference
$\qquad A$ = coil face area, ft^2
$\quad$ CFM = air flow rate, ft^3/min
$\quad$ GPM = water flow rate, gal/min
$\quad a_i, a_o$ = air in, air out
$\quad w_i, w_o$ = water in, water out
$\quad$ GTD = greatest temperature difference, air to water
$\quad$ LTD = least temperature difference, air to water
$\qquad \ln$ = natural log (base e)

(See Sec. 9.7 for further discussion of these equations.)

Three or four iterations will usually suffice to obtain a satisfactory answer. If this indicates a possible frost condition on the exhaust coil, it will be necessary to recalculate, assuming that the three-way control valve will modulate so that water flows to the two coils will not be equal. The calculation can be more readily done on a computer or programmable calculator.

This system will also recover energy during the cooling cycle, but somewhat less than during the heating cycle, because the overall temperature difference will be less.

An interesting and different application of the coil loop system is shown in Fig. 10.35. The application was an air-handling system for electronics manufacturing. The problem was a clean room with a very low relative-humidity requirement, a high percentage of outside air to

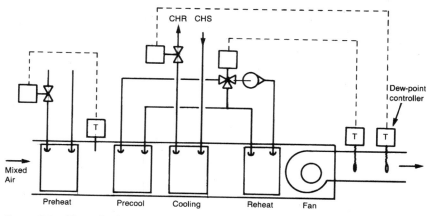

Figure 10.35 Precool-reheat internal coil loop heat recovery.

makeup exhaust, and a high airflow rate (cubic feet per minute per square foot) with a resulting small temperature difference between the supply air and room design. These criteria required a great deal of cooling to a low ADP and much reheat, as shown in the psychrometric chart of Fig. 10.36. The coil loop system was added to provide some reheat while also precooling the mixed air; it is effective above about 60°F outside air temperature. At cooling design conditions, as shown in Fig. 10.37, the coil loop system supplies about two-thirds of the needed reheat, while doing an equal amount of precooling. The three-way valve is controlled to avoid excessive reheating.

10.17.2 Rotary air-to-air exchanger

A rotary air-to-air heat exchanger, or heat wheel, is a revolving cylinder filled with heat transfer media through which the air passes (Fig. 10.38). Supply air flows through half of the wheel, exhaust air through the other half. A partition is arranged to separate the two airstreams, although some leakage and cross-contamination may occur. This may be minimized by use of a *purge* section between the two halves and by making the clean airstream "positive" with respect to the exhaust airstream. Heat transfer may be sensible only or sensible plus latent, depending on the type of heat transfer medium selected.

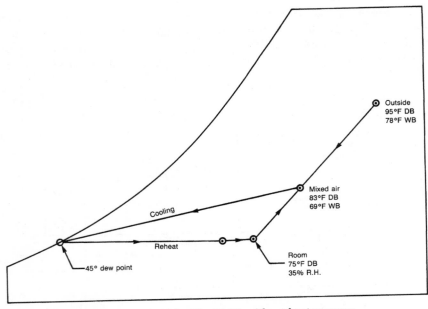

Figure 10.36 Psychrometric chart for Fig. 10.35, without heat recovery.

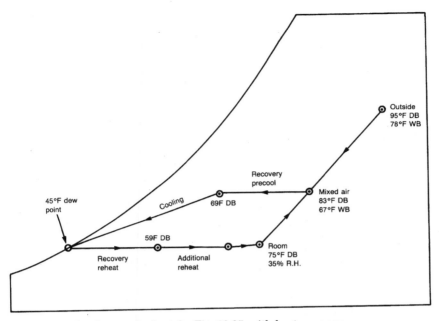

Figure 10.37 Psychrometric chart for Fig. 10.35, with heat recovery.

Even the sensible-heat-only design may involve some latent heat transfer if there is a buildup of hygroscopic dust on the medium. Capacity may be controlled by varying the rotational speed. This system requires that exhaust and supply airstreams be contiguous at the exchanger, which may pose problems. Remember that outside air intake and exhaust discharge louvers must have sufficient separation to avoid cross-contamination.

10.17.3 Heat pipe recovery system

A heat-pipe energy recovery system consists of a bank of closed tubes, each of which operates independently. The tube is lined with a capillary wick, partially filled with a suitable refrigerant and sealed (Fig. 10.39). One end of the tube is in the warm airstream, the other in the cold airstream; an external partition divides the two airstreams. The warm air vaporizes the refrigerant, and vapor migrates through the tube and is condensed in the cold end. The condensed liquid returns to the warm end through the wick. This is a passive system, being driven entirely by the temperature difference between the two airstreams. Tilting the tube to increase the liquid flow rate will increase the capacity; the capacity can be automatically controlled by varying the degree of tilt. The tube may be finned or bare. Several rows may

Figure 10.38 A rotary heat exchanger, for heat-wheel heat recovery. (*Courtesy of Cargocaire Engineering Corp., Amesbury, Mass.*)

be used, up to 12 or 14. Airflow rates usually range from 400 to 600 ft/min.

10.17.4 Twin-tower sorbent recovery system

This system consists of one or more towers handling supply air and one or more towers handling exhaust air (Fig. 10.40). A sorbent liquid is circulated between the supply and exhaust towers, usually being sprayed into the airstream to increase contact. Because the sorbent solution transfers latent as well as sensible heat, this is a total heat or enthalpy transfer system. It provides dehumidification of the outside airstream and is often used primarily for that purpose. The sorbent is an effective antifreeze, and the system will tolerate winter air temperatures as low as −40°F, but is limited on the high-temperature end to about 115°F. Additional fan and pump work is required. Recovery effectiveness in the 60 to 70 percent range is typical. This system had a historical use in the hospital environment, where large quantities of exhaust air were involved and humidification was required.

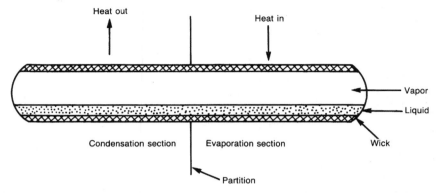

Figure 10.39 Heat-pipe energy recovery.

Makeup water for humidification is sometimes added in the system. Demineralized or deionized water is recommended.

10.17.5 Fixed plate air-to-air
heat exchanger

The fixed-plate air-to-air heat exchanger (Fig. 10.41) is defined by its name. The plates may be pure (without fins) or finned. The exchanger is arranged for cross-flow or counterflow. Efficiencies as high as 80 percent can be obtained with high velocities and the extended surface

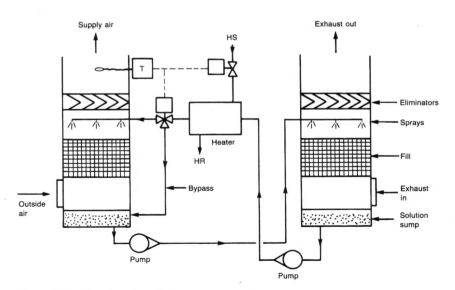

Figure 10.40 Twin-tower enthalpy recovery system.

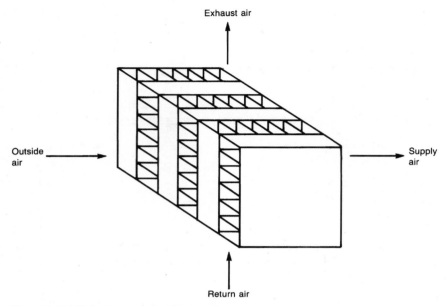

Figure 10.41   Plate fin cross-flow heat exchanger.

which results in high pressure drops. The most popular construction material is aluminum, but it can be any sheet material. The plates are formed from aluminum-alloy sheets with a surface coating of a lower-melting-point alloy. The plates are assembled in a jig and are dipped in a molten salt bath maintained at a temperature such that the surface alloy melts while the base alloy remains unaffected. The resulting structure is monolithic and easy to use. These exchangers are relatively more expensive than those described earlier. They are used extensively in aircraft because of light weight and suitability over a wide range of temperatures. As with rotary wheel and heat-pipe energy recovery devices, the supply and exhaust airstreams must be contiguous.

10.17.6   Thermosyphon recovery systems

A thermosyphon heat exchanger system utilizes the natural-convection circulation of a two-phase refrigerant to transfer energy between two airstreams.

The sealed-tube thermosyphon is similar to the heat pipe, but instead of a wick relies on gravity to return the liquid refrigerant. Thus the pipe must be installed vertically.

The coil loop thermosyphon is similar to the coil loop system but without a pump. The system may be bidirectional or unidirectional, as shown in Fig. 10.42. Individual loops have fairly low efficiencies—25 percent is typical. Loops may be installed in series to achieve higher recovery rates.

10.17.7 Condenser heat recovery

Reclamation of condenser heat is very common for reheat control in the cooling cycle, particularly where high airflow rates and/or humidity control is involved. Several methods are available.

The simplest method is to make the return (hot) condensing water available directly to the reheat coils (Fig. 10.43). If an open-circuit cooling tower is used, there will be problems with fouling in the coils. A closed-circuit tower should be used.

A double-bundle shell-and-tube condenser can be used, as shown in Fig. 10.44. The reheat loop functions continuously but may not use enough heat to satisfy the refrigeration cycle. The loop to the cooling tower is controlled to be used as required to make up the difference.

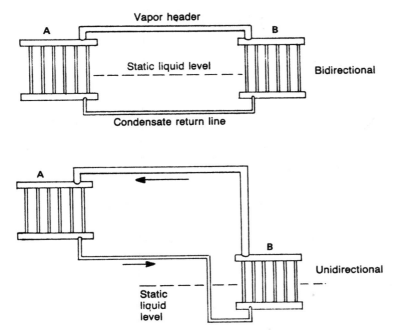

Figure 10.42 Coil loop thermosyphon heat recovery. (*Reprinted by permission from the ASHRAE Handbook,* 1996 HVAC Systems and Equipment, *Chap. 42.*)

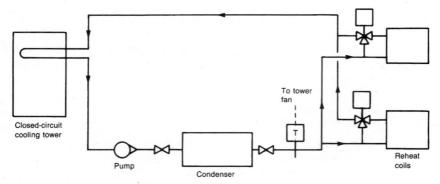

Figure 10.43 Condenser water is used for reheat.

10.18 Solar Heating

Solar energy may be utilized to provide part or all of a building's space-heating requirements, for domestic or process water heating, as the heat source in an absorption refrigeration cycle, and in many other useful ways. Technically, the most important problems have been solved. At this writing, active solar heating systems have such a high first cost that the economic recovery period is excessive in all but a few special cases. This will change only when conventional fuel costs increase to several times their present rate or when some way is found to decrease the first cost of solar heating systems. For a full discussion of solar heating design, see the chapter on solar energy in the ASHRAE Handbook.[6] Do not take this brief paragraph as understated

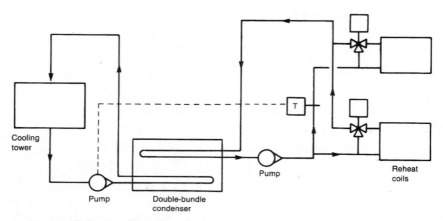

Figure 10.44 Reheating by using a double-bundle condenser.

regarding solar energy. In fact, many of the renewable energies—wind, tide, organic fuels, and hydroelectric and photovoltaic power—are solar-derived. There is great respect for the contribution of the sun to our communal well-being.

10.19   Humidification

Humidification is the process of increasing the moisture content of the air in a conditioned space. This operation is used wherever some low limit of humidity is required and is needed only when the drying effect of outside air (as in winter) cannot be compensated by humidity sources, such as people or functions, within the space. Moisture may be added directly in the space or in the air-handling unit serving the space.

Methods of adding humidity include sprays and atomizers, evaporators, and steam injectors.

10.19.1   Sprays and atomizers

Spray or atomizer humidifiers work by breaking the water into very small droplets so that it may readily be vaporized and diffused in the air. Spray humidifiers are usually mounted in the air ducts, where the unabsorbed water may be collected by an eliminator system and re-circulated by a pump (see the discussion of evaporative coolers in Sec. 9.9). Atomizers are designed to create a very fine mist and are often located within the conditioned space. Any spray or atomizing humidifier creates a sensible cooling effect because the heat of vaporization must come from the air. On the psychrometric chart, this is shown as a constant wet-bulb process (see Fig. 9.32). Control may be manual or automatic, by cycling the spray pump or atomizer slinger.

It is necessary to have adequate free or forced air circulation around the atomizer. Otherwise the air in its vicinity will become saturated, and the moisture will coalesce into droplets and fall as rain. Dissolved solids in the water may precipitate as a powder or scale on the humidifier. Using deionized spray water is helpful. Bacterial control at the device may be necessary.

10.19.2   Evaporators

The simplest form of evaporator is an open pan of water. Evaporation will take place, but very slowly even when air is blown across the water surface. Wicks of porous ceramic or similar materials may be added. When the base of the wick is immersed in water, water spreads through the wick by capillary action, increasing the area of contact between water and air. Although this is an improvement, it is still unsatisfactory for accurate humidity control.

The wetted-drum humidifier includes a fiber pad through which the air must pass, as through a filter. The pad is driven by a small motor, and the lower end of the pad passes through a water reservoir, so that the pad is continually wetted. This device is very effective. Control is accomplished by cycling the drive motor. The system acts as an evaporative cooler.

Evaporation rates may be increased and controlled by adding heaters, either immersion heaters in the pan (Fig. 10.45) or radiant heaters mounted above the pan (Fig. 10.46). The evaporation rate is controlled by cycling the heater. Steam heaters may be used, but electric heat is more common. Evaporation of untreated water leaves a residue of lime and salts which can cause severe maintenance problems. Periodic bleed or drip of the reservoir is often used. The use of distilled or deionized water will eliminate these problems. Package deionizers or deionizer services are available.

The amount of water to be evaporated and the amount of heating required can be calculated from a psychrometric analysis. Figure 10.47 is a psychrometric chart showing a typical heating and humidifying process. Design conditions have been assumed, for illustration, to be as follows:

- Room: 70°F, 40 percent RH; then $h = 23.7$ Btu/lb, $w = 0.0063$ lb$_w$/lb$_a$.

- Outside air: 32°F, 50 percent RH; then $h = 9.8$ Btu/lb, $w = 0.0020$ lb$_w$/lb$_a$.

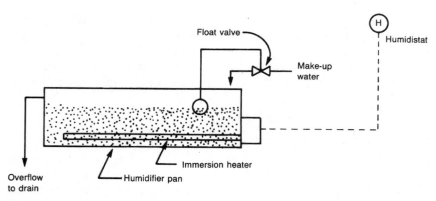

Figure 10.45 Heated-pan humidifier.

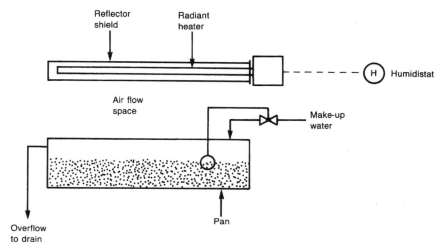

Figure 10.46 Humidifier with radiant heater.

- Airflow rate: 10,000 ft³/min with 20 percent outside air
- At design conditions, heat loss and internal humidity gains require an air supply condition of 85°F, with $w = 0.0060$.
- Mixed air: 62.4°F; then $w = 0.0054 \, \text{lb}_w/\text{lb}_a$.
- Latent heat of vaporization of water at 85°F = 1045 Btu/lb.

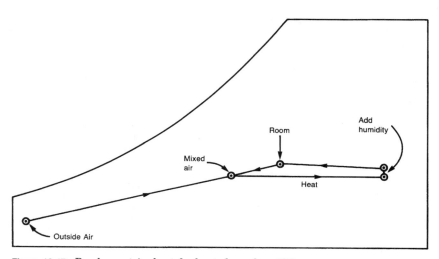

Figure 10.47 Psychrometric chart for heated-pan humidifier.

Then for air from 62.4 to 85°F,

Sensible heat = $(85 - 62.4)$ lb_w/lb_a

$$\times\ 10{,}000\ ft^3/min \times 1.08\ Btu/[h \cdot (ft^3/min) \cdot °F]$$

$$= 244{,}080\ Btu/h$$

Water required for humidification is

$$(0.0060 - 0.0054)\ lb_w/lb_a \times 10{,}000\ ft^3/min$$

$$\times\ 0.075\ lb_a/ft^3 \times 60\ mm/h = 270\ lb/h\ water$$

The heating required to evaporate the water is

$$270\ lb_w/h \times 1045\ Btu/lb = 282{,}150\ Btu/h$$

10.19.3 Steam humidifiers

Steam humidifiers are used extensively nowadays because of their simplicity and ease of control. The basic system consists of a modulating control valve and a perforated pipe, or grid, mounted in an air duct (Fig. 10.48). An enclosure around the grid is needed to remove condensate, especially at start-up, when the system is cold. The steam-jacketed grid is provided with an enclosure which is continuously heated so that only dry steam is delivered into the airstream. Control is provided by modulating the steam valve. A high-limit duct humidistat is recommended, as shown in Fig. 10.49, to avoid supersaturation conditions in the duct, with resulting condensation. Steam pressures should be 5 to 15 lb/in^2 gauge. Only clean steam should be used.

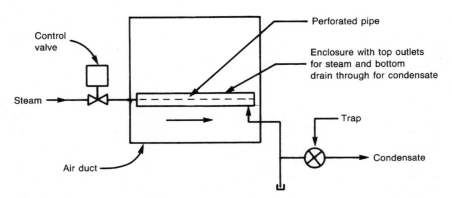

Figure 10.48 Enclosed-grid humidifier.

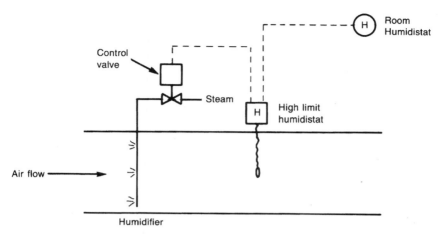

Figure 10.49 Control of steam humidifier.

The calculation for a steam humidifier is similar to that for a heated-pan humidifier. The only difference is that the steam is already in vapor form and is thus at a higher enthalpy. This results in some small increase in the dry-bulb (db) air temperature during the humidification process. For the conditions in the heated-pan humidifier example, steam at 5 lb/in² gauge has an enthalpy of 1156 Btu/lb. At 85°F the enthalpy of the water vapor (steam) is 1098 Btu/lb, for a difference of 58 Btu/lb. Then

$$58 \text{ Btu/lb} \times 270 \text{ lb/h} = 15,660 \text{ Btu/h}$$

and

$$\frac{15,660 \text{ Btu/h}}{(10,000 \text{ ft}^3/\text{min}) \times 1.08 \text{ Btu}/(\text{h} \cdot \text{ft}^3/\text{min} \cdot °\text{F})} = 1.45°\text{F}$$

which is the air temperature increase resulting from the humidification process.

10.20 Summary

This chapter has discussed methods of providing heating for buildings. It is by no means exhaustive. For example, residential stoves and furnaces have not been included, and solar heating has only been mentioned. Equipment selection and sizing depend mostly on data available from the manufacturers. Much additional information will be found in the ASHRAE Handbooks and in manufacturers' literature.

References

1. ASME, *Boiler and Pressure Vessel Code* (eleven sections), 1989.
2. Hydronics Institute (HYDI), 35 Russo Place, Berkeley Heights, N.J. 07922.
3. ASHRAE Handbook, *1996 HVAC Systems and Equipment,* Chap. 32, "Hydronic Radiators."
4. Commercial Standard CS 140-47, *Testing and Rating Convectors,* Department of Commerce, 1947. (This code has been withdrawn but not replaced.)
5. ASHRAE Handbook, *1996 HVAC Systems and Equipment,* Chap. 6, "Panel Heating and Cooling."
6. ASHRAE Handbook, *1995 HVAC Applications,* Chap. 30, "Solar Energy Utilization."

Equipment: Part 3

Air-Handling Systems

11.1 Introduction

By definition, air conditioning involves control of the air temperature, humidity, cleanliness, and distribution. It follows that an air-handling unit (AHU) of some kind is an essential part of an air conditioning system, though not necessarily of a heating-only system.

The function of the AHU is to provide air at a quantity, temperature, and humidity to offset the sensible and latent heat gains to the space (in the cooling mode) and the heat losses (in the heating mode), while maintaining the required temperature and humidity in the space. This can be most clearly shown on a psychrometric chart (Fig. 11.1). A typical cooling design room condition is 78°F dry-bulb (db) temperature and 50 percent RH. For illustration, a load of 120,000 Btu/h sensible and 30,000 Btu/h latent cooling is assumed. Then, for an assumed 20°F temperature difference between the room and supply air temperatures (58°F supply air), the design flow rate, designated CFM, will be

$$\text{CFM} = \frac{120,000}{20 \times 1.08} = 5555 \text{ ft}^3/\text{min} \tag{11.1}$$

The change in specific humidity Δw may be calculated as follows:

$$\Delta w = 30,000 \text{ Btu/h} \times \frac{1 \text{ min}}{5555 \text{ ft}^3} \times \frac{1 \text{ ft}^3}{0.075 \text{ lb}_a} \times \frac{1 \text{ h}}{60 \text{ min}} \times \frac{1 \text{ lb}_w}{1059 \text{ Btu}}$$

$$= 0.0011 \text{ lb}_w/\text{lb}_a \tag{11.2}$$

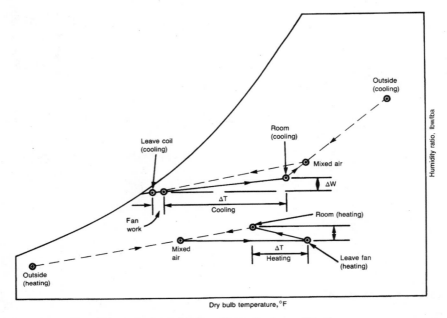

Figure 11.1 Psychrometric chart for draw-through air conditioning process.

The point defined by these two differential values can be plotted on the chart, as shown. The "validity" of this point must be verified, based on the cooling coil capability and the AHU arrangement, as discussed in Sec. 3.6. For a draw-through arrangement (i.e., with the supply fan downstream of the cooling coil), the supply air temperature will be greater than the coil leaving temperature because of heat added by fan work. For this example, if 5 hp is required, the temperature difference (TD) will be

$$\text{TD} = 5 \text{ hp} \times \frac{2545 \text{ Btu}}{1 \text{ hp} \cdot \text{h}} \times \frac{1}{5555 \text{ ft}^3/\text{min}} \qquad (11.3)$$

$$\times \frac{1 \text{ h} \cdot (\text{ft}^3/\text{min}) \cdot {}^\circ\text{F}}{1.08 \text{ Btu}} = 2.1^\circ\text{F}$$

Then a coil leaving condition of 55.9°F db and 55.5°F wb can be plotted, and this will probably be valid.

For a blow-through arrangement, the fan work causes an increase in the mixed-air temperature before the air goes through the cooling coil, and the process will be as shown in Fig. 11.2. In this case, it will be necessary to increase the supply air TD to 22°F to get a valid coil leaving condition. This will reduce the air quantity to 5050 ft³/min

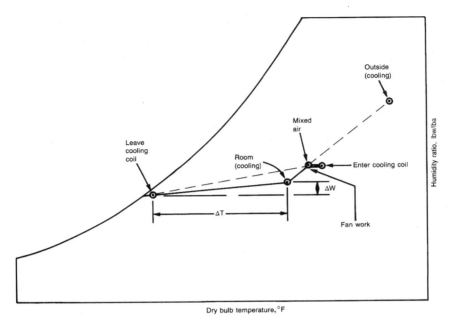

Figure 11.2 Psychrometric chart for blow-through air conditioning process.

and will require more care in the air distribution system to avoid cold air spillage and drafts. The Δw will be greater because of the reduced CFM.

Humidity control is not always required, but some upper limit will be inherent in any refrigeration-type cooling process—chilled water, brine, or direct expansion.

Supply air-handling equipment may be classified in several different ways:

1. *Type or arrangement.* The five basic arrangements are single-zone, multi-zone, double-duct, variable air volume (VAV), and induction.

2. *Package versus built-up.* Package equipment is factory-assembled, and when it is installed, it requires only connections for utilities and ductwork. The term *built-up* implies that most of or all the components are field-assembled.

3. *Self-contained.* A self-contained system includes internal thermal energy generation.

4. *Central station and terminal units.* Central station equipment is remote from and delivers air through ductwork to the conditioned

space. Terminal units are installed in or adjacent to the conditioned space. Terminal units are used in conjunction with central station equipment.

Exhaust systems may serve a single space or multiple spaces, and may include heat recovery, special filtration, and other special equipment.

11.2 AHU System Arrangements

Air conditioning practice includes only five basic AHU arrangements, although there are many variations on these basic concepts. Single-zone and VAV systems have similar, even identical, physical arrangements but use different control strategies. Multizone and double-duct systems are similar in arrangement and concept but are different enough to be considered separately. Induction systems are unique.

11.2.1 Single-zone AHU

A single-zone AHU is intended to serve only one room, or a group of rooms which are contiguous and which have similar load and exposure characteristics. The maximum area served by a single-zone AHU should not exceed 10,000 ft^2.

The typical single-zone AHU arrangement is shown in Fig. 11.3. This is a draw-through system, with the heating coil in the preheat position to protect the cooling coil from freezing air. The system is controlled as explained in Sec. 8.5.1. It is important to sequence the operation of the control valves to avoid simultaneous heating and cooling.

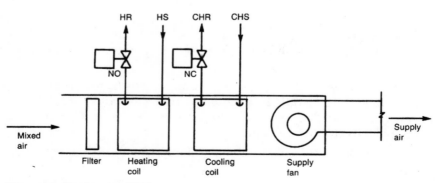

Figure 11.3 Single-zone AHU.

When one or more of the rooms served by a single-zone AHU has a load characteristic different from the other rooms, zone reheat must be provided by means of coils in the zone branch ducts (Fig. 11.4), by radiation, or by fan-coil units. Because reheat is potentially energy-wasteful, it may be preferable to use a different type of AHU, as described below.

A single-zone unit may be used to control humidity in the room. The unit is arranged as shown in Fig. 11.5. The cooling coil precedes the heating coil, which is therefore in the reheat position. Humidity control always requires additional energy—as reheat or in other ways. The cooling coil valve is controlled by either the space temperature or the space humidity, whichever creates the greater demand. If humidity controls, the temperature will tend to fall and the space thermostat will control the heating coil valve to provide reheat. The humidifier is used when required.

11.2.2 Multizone AHU

The typical multizone (MZ) AHU arrangement is shown in Fig. 11.6. Side-by-side hot and cold airstreams are provided. Each zone is provided with dampers to mix hot and cold air to satisfy the requirements of the zone. In this way, one zone may be heated while simultaneously another is cooled. The mixing dampers are located at the unit, with a separate duct run to each zone. Thus, economics and practicality limit

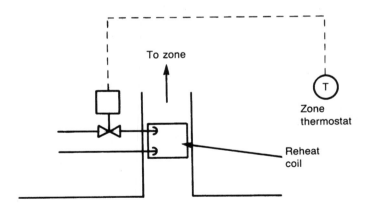

Supply duct main

Figure 11.4   Zone reheat coil.

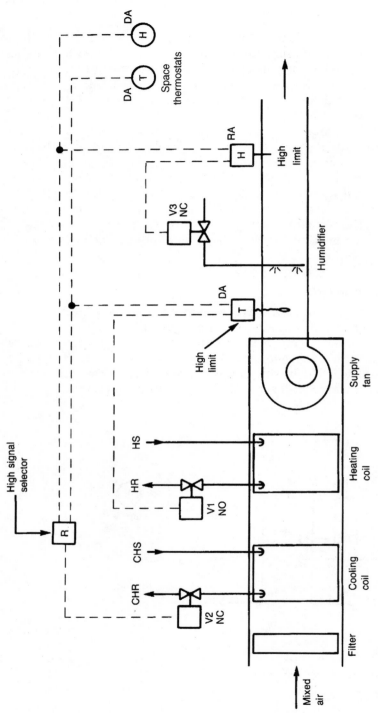

Figure 11.5 Single-zone AHU with humidity control.

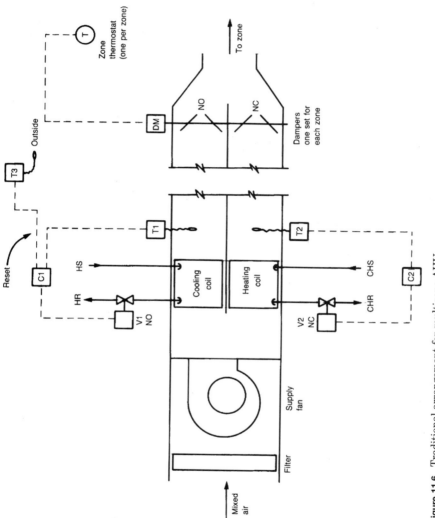

Figure 11.6 Traditional arrangement for multizone AHU.

the size of the typical MZ unit. The great majority of such units are the package type.

From an environmental control standpoint, the conventional MZ unit is less than ideal. Because the control is achieved by reheat, it is an energy waster. The three-duct MZ unit (Fig. 11.7) retains the control benefits while eliminating the energy waste, by adding a *bypass* duct (plenum). The sequence of control is described in Sec. 8.5.2.

11.2.3 Double-duct (dual-duct) AHU

The *double-duct* (DD) AHU uses the same principle of operation as the MZ unit. However, the hot and cold ducts are extended through the building, with a mixing box provided for each zone. Thus, the double-duct AHU can be as large or as small as desired. The conventional system (Fig. 11.8) has the same advantages and disadvantages as the multizone AHU. Many of the older systems installed in the 1950s and 1960s also included high-velocity/high-pressure duct systems to minimize the space occupied by the ducts. Electric energy was relatively inexpensive at that time, so the additional fan work was of no concern. Also 5 to 6 inches of total pressure across the fan was common, and 9 to 10 inches was not unusual. At today's energy prices, such a system may cost more for fan energy than for thermal energy on an annual basis.

Many of these older systems are being retrofitted to variable air volume by changing the heating coil to cooling, removing the mixing boxes, and using both heating and cooling ducts, in parallel, with new VAV boxes. In this way, the duct air velocity is reduced by about 50 percent with a significant saving in fan energy. Some reheat must be added for exterior zones.

The ideal dual-duct system is, perhaps, the two-fan system shown in Fig. 11.9 and described in detail in Sec. 8.5.3.

11.2.4 Variable-volume AHU

Unlike the AHU systems previously discussed, a VAV system supplies air at constant, or nearly constant, temperature and humidity. Capacity is controlled to match cooling load by varying the volume of air supplied to a zone. A *VAV box* is provided at each zone. The box includes a motorized damper (controlled by the zone thermostat) and usually some means of compensating for changes in static pressure in the supply duct. Such changes can affect the accuracy of control. The compensating device may be mechanical, e.g., a spring-loaded damper, or it may be a flow-sensing controller which is reset by the zone thermostat. The latter is given the anomalous description *constant vari-*

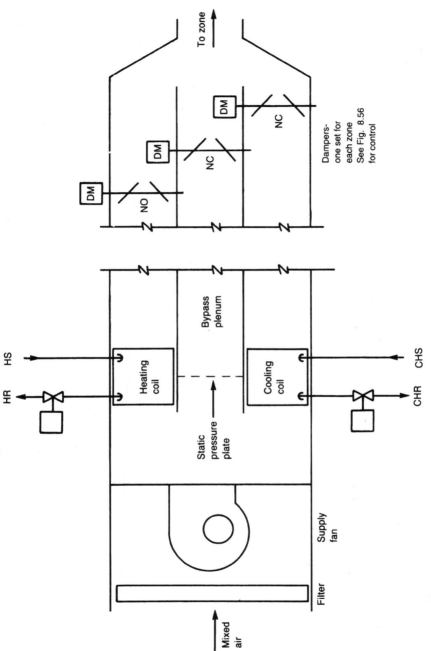

Figure 11.7 Three-duct arrangement for multizone AHU.

383

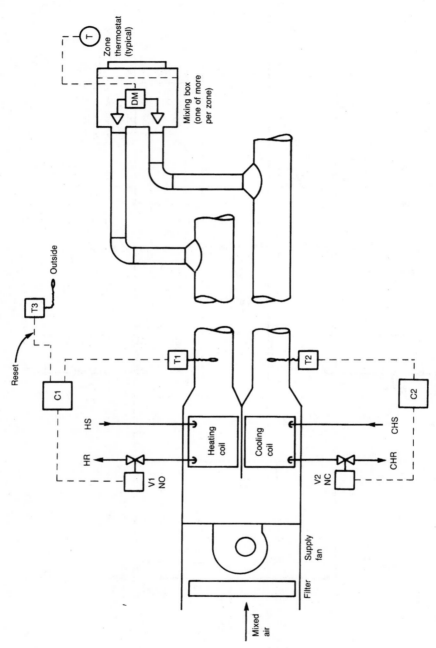

Figure 11.8 Traditional arrangement for double-duct AHU.

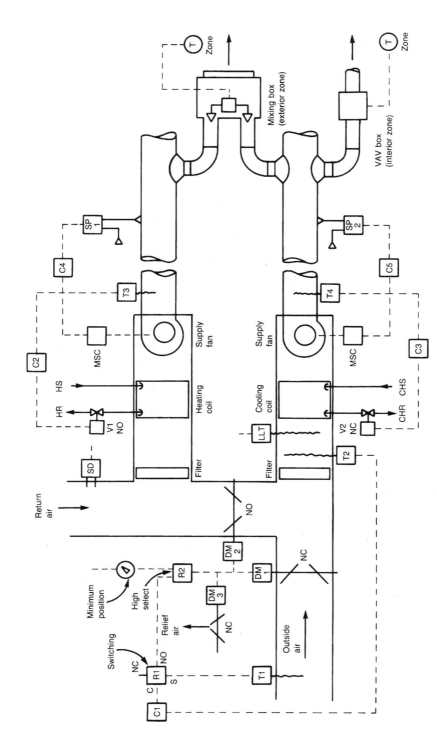

Figure 11.9 Two-fan system with double-duct AHU.

able-volume controller. While the zone supply volume could theoreti-cally go to zero, it is usual to provide a low limit of 35 to 40 percent of design airflow to maintain a minimum air distribution and venti-lation rate. Supplemental heating—reheat coils, radiation, fan-coil units—is required in zones with exterior exposure.

VAV systems were developed in response to the 1973 "energy crisis." The concept is based on the fan law which states that the fan horse-power (fan work energy) varies as the cube of the airflow, denoted by CFM. A reduction to 50 percent of the design CFM could result in a theoretical reduction to one-eighth of the design fan work. In practice, the method used to reduce the fan CFM determines the energy sav-ings, and the full theoretical savings is never realized, due to mini-mum system pressure requirements, to mechanical friction, and to air turbulence. Three methods are used to reduce fan CFM.

1. *Damper in duct, either upstream or downstream of the fan* (Fig. 11.10).   This forces the fan to "ride up the curve" (Fig. 11.11), i.e., to increase the fan pressure at the lower CFM. Little or no energy is saved.

2. *Inlet vane damper.*   The inlet vane damper alters the fan perform-ance, and a portion of the theoretical saving is realized. For actual savings, consult the fan manufacturer. See the discussion in Sec. 5.2.5.

3. *Fan speed control.*   Fan speed control allows most of the theoretical savings to be realized—except for mechanical and motor efficiency losses. Mechanical belt and variable-pitch pulley systems change

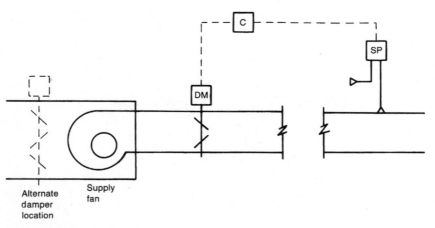

Figure 11.10 Volume damper for duct pressure control.

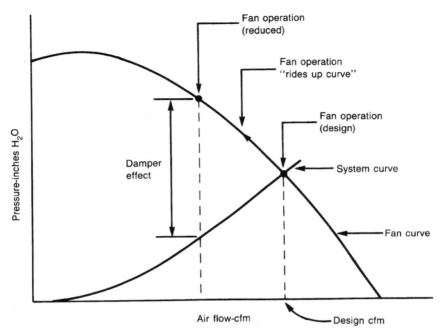

Figure 11.11 Fan and system curves for Fig. 11.10.

the fan speed while the motor speed remains constant. These systems are satisfactory only for small motors and are usually limited to residential and small commercial applications. Variable-speed clutch drives—hydraulic and magnetic types—allow constant motor speed. Some of these systems are satisfactory for large motors, but they have been largely superseded by variable-speed motor drives. Variable-speed motor drives of the variable-frequency type are the preferred method today (see the discussion in Sec. 8.3.3).

If no fan volume control device is used, the fan will nevertheless adjust its volume to match that of the combined VAV boxes by riding up the curve. In this case, the pressure in the duct system may increase beyond the compensation capacity of the boxes, resulting in poor control and noise. Recent technology allows direct digital control (DDC) devices at each box. These provide information to the fan volume controller to control the fan and system directly to required volume and temperature rather than indirectly to a duct static pressure at a fixed supply temperature.

The fan volume control devices described above maintain a constant static pressure at some point in the supply duct main, as shown in

Fig. 11.9. The sensor location is usually two-thirds to three-quarters of the distance from the fan to the most remote box, although the best location is near the inlet to the hydraulically remotest box.

Variable-volume supply may be obtained with constant fan volume by using a runaround bypass duct (Fig. 11.12). The bypass damper is controlled to maintain variable-volume supply at a constant static pressure in the supply duct, but without any change in fan volume.

When a return-air fan is used in a VAV system, controlling its volume to "track" that of the supply fan is difficult. In general, some fixed difference between the supply and return CFM is needed, to match the fixed exhaust CFM in the building. Because the two fans will always have different operating characteristics, it is not sufficient to simply track speed. Either flows or pressures must be measured. Various methods have been proposed for doing this, some involving complex and expensive control systems. The general rule is to avoid using return-air fans unless the return-air system has a high-pressure loss. Then a flow-sensing system such as shown that in Fig. 11.13 can be used. In this system, the return-air fan volume controller is reset by the supply airflow. Single-point flow sensing can be used, but greater accuracy is obtained with flow-measuring stations which measure velocities at several points across the duct. The building pressure with respect to the outdoors may be the best control signal, but it requires sensitivity to very small pressure changes, which in turn require a quality sensor or controller and a stable control system. Also, the outdoor pressure sensor is subject to variation of wind pressure and velocity. The indoor sensor is subject to stack effects.

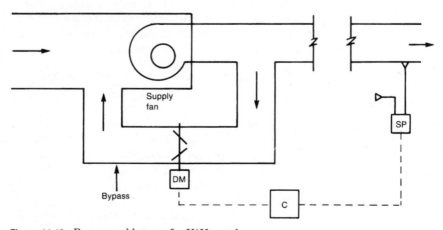

Figure 11.12 Runaround bypass for VAV supply.

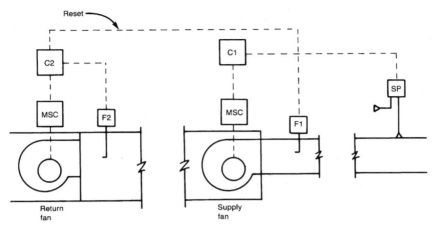

Figure 11.13 VAV return-air fan volume control.

11.2.5 Induction unit system

Induction unit systems are no longer common, but many were installed in the 1950s and 1960s. A central primary air system (single-
zone arrangement) supplies a constant volume of air at a constant
temperature (about 55 to 60°F for cooling) and a pressure of usually
6 to 8 in H_2O. The primary air temperature may be reset based on
outside conditions. The system handles up to 100 percent outside air;
the outside air volume must be sufficient to satisfy building exhaust
requirements and to provide some slight pressurization. At each zone
an induction unit is provided. This unit (Fig. 11.14) includes a large
face area, a low-pressure-drop coil used for additional cooling or reheat, a lint filter, and a supply grille for air delivery to the zone. Primary air is supplied to the unit through nozzles arranged to induce a
secondary airflow through the filter and coil. Dehumidification is accomplished at the primary air unit. Secondary chilled water to the
induction units is kept at a temperature high enough to avoid condensation, also avoiding the need for a drainage system. Induction
units may be floor-mounted, exposed, or ceiling-mounted, partially
concealed. The system may use less fan energy than a conventional
system, but piping and control systems become complex. The primary
air supply is constant, and the air supply to a zone may not be shut
off. Noise must be carefully attenuated. These systems found favor in
dormitory and hospital patient wing applications. They were used in
some instances as a perimeter system for office buildings.

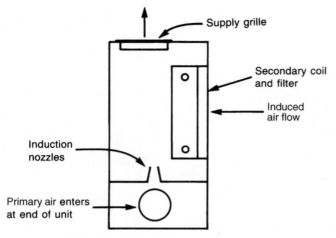

Figure 11.14 Induction unit.

11.3 Package Air-Handling Units

A package unit is factory-assembled, ready for installation either in total or in large segments. This class includes units for rooftop mounting, self-contained units, heat pumps, and split systems—units with an indoor section and an outdoor section. All these systems may or may not include factory-installed automatic controls and internal wiring and piping. Field installation may include connections for electrical, fuel, and water service; duct distribution systems; system controls; and room thermostats.

Package equipment is available in a wide range of capacities; some rooftop units will provide 100 tons (40,000 to 50,000 ft^3/min) or more of cooling. The advantage of the package unit is the saving in field labor. There are some disadvantages. Combinations of fan and heating/cooling elements may require some compromise for a specific application—one or more elements may be oversized. Efficiencies may be lower than optimum because most package equipment is made as small as possible for minimum clearances, etc. For the same reasons, maintenance may be more difficult. Typically, package equipment seems to be installed in less accessible places. Factory-set control strategies may or may not suit the designer's needs. Interface with building automation may be a challenge.

The designer should make sure that all listed capacities are based on tests of the package as built, rather than the individual components. The unit geometry can have an effect, usually detrimental, on performance (see Sec. 5.2). ASHRAE and the Air Conditioning and

Refrigeration Institute (ARI) publish a number of standards for testing and rating package equipment.

11.3.1 Rooftop AHU

The typical rooftop AHU is self-contained, although some are made for use with external sources of thermal energy. The self-contained system includes a direct-expansion cooling coil; a direct-fired heater, usually gas or electric; a refrigerant compressor with an air-cooled condenser and fan; a supply fan; air filter; and an economy-cycle outside-air control system with return, relief, and outside air dampers. Unit arrangement may be single-zone or multizone. If the dampers at the unit are removed, multizone units may be used as double-duct systems. Single-zone units may be used for VAV. A typical rooftop unit is shown in Fig. 11.15. The unit is mounted on a prefabricated curb, with all roof penetrations inside the curb. All controls are included, with only zone thermostats to be field-installed.

11.3.2 Split-system AHU

The split-system AHU consists of two packages. The outdoor section includes a refrigeration compressor, condenser (usually air-cooled),

Figure 11.15 Rooftop AHU. (*Courtesy of Mammoth, a Nortak Company.*)

and condenser fan. The indoor section includes the evaporator cooling coil, heating element, supply fan, and air filter. Control of outside air for ventilation is not usually included. The two sections are field-connected through refrigerant piping and electric wiring. Controls are included, but some field wiring is necessary. Heating may be obtained by means of any conventional fuel, usually gas, oil, or electricity. Split-system heat pumps are common. Split systems are usually small, ranging up to 15- or 20-ton capacity.

11.3.3 Package AHU with humidity controls

This is a special class of a package AHU, designed primarily for use in large mainframe computer rooms where a carefully controlled environment is required. The package is designed for installation within the computer room, as shown in Fig. 11.16. Supply air is discharged downward into the plenum space below the raised floor, from which it enters the computer equipment cabinets directly or is transferred into the room through floor supply registers. Return air enters the AHU directly from the room. The unit includes a supply fan, high-efficiency filters, humidifier, direct-expansion cooling coil, refrigeration compressor(s), and water-cooled condenser. Alternatively, a remote air-cooled condenser can be used, or a water coil with a remote source of chilled water. Electric reheat is typical, although hot water may be used. Units are available up to 20-ton capacity, and multiple-unit installa-

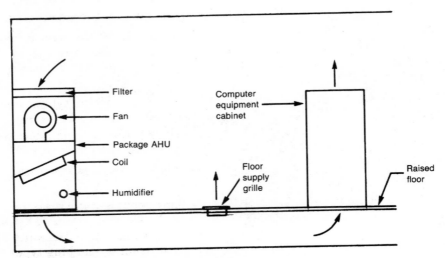

Figure 11.16 Computer room AHU installation.

tions are common. No ductwork is used in most down-flow applications. These units have found some application in low-grade cleanroom applications where humidity control is desired. Fan-assisted HEPA filters can be incorporated in a system.

11.4 Built-up (Field-Assembled) AHU

The built-up AHU is field-assembled from individual components selected by the designer. This allows the designer complete flexibility of size and arrangement. The built-up AHU tends to have a higher first cost than a package system but may be designed to be more efficient and easier to maintain. Criteria for making the choice between built-up and package equipment include the size of the system, budget, space available for equipment, and environmental requirements.

Built-up systems allow choices of styles and performance characteristics in every component.

Air intakes and discharges may be roof hoods or wall louvers or may be developed as an aspect of the building structure.

Dampers may be parallel-blade or opposed-blade, preferably with tight shutoff for outside air and relief air applications. They should have shaft bearings and sturdy linkages for long life.

Filters may be of any desired style and arrangement for the space allowed. Basic styles include flat replaceable fiberglass, washable media, pleated fabric media, bag type in various collection efficiencies, electronic precipitators, etc. Some filter styles can be automated for media renewal to reduce maintenance requirements.

Cooling may utilize any available economically viable service or combination of services, including evaporative cooling (direct or indirect), direct-expansion refrigeration, chilled water, cool irrigation water, well water, and the like.

Heating may utilize any available viable service including direct or indirect gas-fired devices, oil-fired devices, steam or hot water from a remote plant, warm water from chiller heat recovery (heat pump), geothermal water, solar heated water, or electric resistance coils.

Humidification may be by steam injection, air washers (including sprayed coils), or other wetted media.

Fans may be of any design which develops enough static pressure to handle the fan system component and distribution system pressure drops. Common fan types include centrifugal fans (forward-curved, backward-inclined, single-inlet, double-inlet, and plug type) and vane axial (fixed-pitch, variable-pitch, direct-drive, belt-driven).

There are special concerns with fans for noise and vibration control, airflow paths, maintenance access, and other site-specific conditions.

In all built-up systems, arranging for satisfactory operation and subsequent maintenance is an opportunity and a challenge. Since most cooling and humidification systems develop condensate and involve water, it is important to control and contain the water with membranes and floor drains. AHUs located above occupied spaces are of particular concern because vagrant moisture may damage high-value space beneath.

Sound and vibration control must be addressed early on and must be reviewed throughout the design of a project. Fan rooms should be located away from sensitive areas such as private offices and conference rooms. Acoustical consultants may be employed to ensure the success of the treatments. There seems to be no substitute for mass (masonry walls) in acoustical treatment. Sound traps may be needed at duct penetrations of the fan room envelope. Centrifugal fans develop major vibrations and noise in lower-frequency ranges while vane axial fans are more dominant in higher frequencies. Attention must be given to isolating the disturbing fan motion from the structure. Structure-borne noise and vibration may be expensive to attenuate in a postconstruction response, not to mention embarrassing to the designer.

Often in new construction, fan systems are erected while the structure is open, before the walls, roof, or ceiling is in place. Be sure to ask how the equipment can be replaced in case of component failure. The answer sometimes dictates multiple smaller components, oversized doors, knockout panels, removable roof sections, or corridors to the outside.

Note that many of the concerns for field-erected systems also apply to factory-built package equipment. The designer or specifier may reject some equipment sources for inadequately handling the design issues mentioned.

11.5 Terminal Units

A *terminal unit* is a part of a larger air-handling system—double-duct, VAV, or induction. The terminal unit is installed in or near to the zone which it serves and provides final control of the air temperature or volume in that zone. Included in this category are mixing boxes, VAV boxes, terminal reheat coils, and induction units.

11.5.1 Mixing boxes

Mixing boxes for double-duct systems are described in Sec. 8.5.3. The conventional box includes a constant-volume device to compensate for variations in static pressure in the hot and cold ducts.

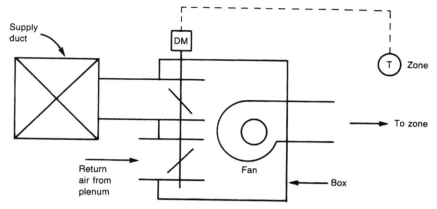

Figure 11.17 Fan-powered mixing box.

11.5.2 VAV boxes

The conventional VAV box is described in Sec. 11.2.4. Two other VAV boxes are used. One is the fan-powered mixing box (Fig. 11.17). The fan-powered box is a small fan-and-damper unit designed to circulate air at constant volume to the zone, while mixing return air with conditioned supply air from a central VAV unit. Unit capacities range from 100 to 1000 ft³/min. While the fan motors are individually small (all are fractional horsepower), in the aggregate a large amount of fan energy may be used. Fan noise must be carefully attenuated. The constant volume to the zones combined with central-station variable volume may be advantageous in some situations.

Another special VAV box is the *bypass box* (Fig. 11.18). This device includes a damper. Air is taken from a constant-volume supply system, and the quantity supplied to the zone is varied. The unused supply air is diverted directly to the return air plenum or duct. Because the primary system is a constant-volume one, there is no saving of fan energy as there is with the VAV systems previously described.

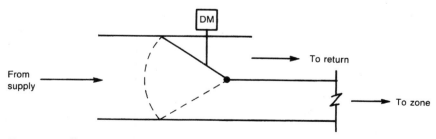

Figure 11.18 Bypass VAV box.

11.5.3 Induction units

These units are described in Sec. 11.2.5.

11.5.4 Terminal reheat

Reheat coils may be installed in conjunction with any of the terminal units described above or may be installed in any zone branch duct, similar to the arrangement shown in Fig. 11.4. When used with variable-volume, the reheat coil is controlled to provide heating only when zone airflow is at its minimum.

11.6 Individual Room AHUs

Individual room AHUs are used extensively in hotels, motels, and apartments. The principal advantages are economy—if the room is not occupied, the unit can be shut off—and convenience to the occupant, who can select heating, cooling, or nothing to suit her or his desires. Room AHUs include fan-coil units, unit ventilators, self-contained heat pumps, and water-to-air heat pumps.

11.6.1 Fan-coil units

A fan-coil unit includes a fan, a heating and/or cooling coil, a filter, and sometimes an outside air connection with a manual damper, all in a sheet-metal casing with supply and return connections (for concealed units) or grilles (for exposed units) (see Fig. 11.19). Ceiling- or floor-mounted arrangements are available. The coil may have one set of connections for both heating and cooling—two- and three-pipe systems—or may be split, with separate sections for heating and cooling. For a discussion of two-, three-, and four-pipe systems, see Sec. 7.8. A fan-coil unit depends on an outside source for its thermal energy. It ranges in size from 100 to 1000 ft^3/min.

11.6.2 Unit ventilators

A *unit ventilator* is a larger fan-coil unit—1000 to 2000 ft^3/min is typical—with an outside air connection which allows it to provide up to 100 percent outside air for ventilation and natural cooling (Fig. 11.20). The unit ventilator is used primarily in schoolrooms and similar spaces where high ventilation rates are needed. Controls are included with the unit. The typical control cycle provides for opening the outside air damper to minimum position when the fan is started, and increasing the outside air amount to 100 percent as the room temperature approaches the heating set point. Only minimal outside air is used when refrigerated cooling is required.

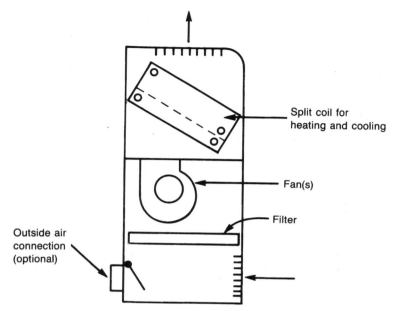

Figure 11.19 Fan-coil unit (floor-mounted).

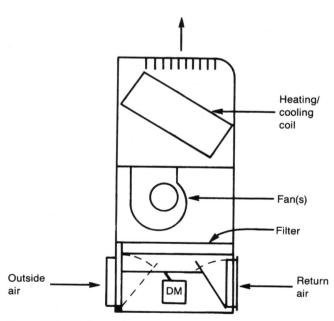

Figure 11.20 Unit ventilator.

11.6.3 Water-to-air heat pump

This system provides a small individual heat pump for each zone or room, with a central source of water at an intermediate temperature which allows each heat pump to provide either heating or cooling, as desired by the occupant. The central plant and distribution piping are shown in Fig. 7.18.

The initial cost for the water-to-air heat pump system compared to other systems of comparable duty varies from lower-cost if fewer larger zones are involved to higher-cost if many zones are involved. Cost is increased depending on the amount of outside air involved and the manner in which it is introduced. Operating costs for these systems may be higher than average. Depending on the occupancy schedule, the central circulating pump has to run whenever any of the heat pumps is in operation. Solenoid valves at each unit combined with a variable-speed pump can minimize this impact. The small reciprocating compressors wind up with a COP of 2 to 3, so at least one-third of the heating energy is electrically derived even though the other part may be "free." But recovered heat is also derived electrically, in part. The small compressors in each heat pump unit have an 8- to 12-year life expectancy. Typically located in the ceiling space, they can become a high-maintenance-cost item. Unit-generated noise is also a concern. These systems typically use a residential-type thermostat with heat-off-cool and fan-on-auto subbase, all of which results in a system which has the character of a residential system in its performance and comfort. Its best applications may be for elementary schools, which have relatively short operating hours, modest heating requirements, and limited summer use. Motel and hotel applications may make some sense where the space is unoccupied for more than half of most days.

11.6.4 Self-contained heat pump
or cooling unit with auxiliary heat

These are through-the-wall units, with an outside coil, an inside coil, and an auxiliary heating coil all in a single package. It has the lowest first cost of any individual AHU system. It is also the noisiest and typically least comfortable of all the systems discussed. These units are often found in hotel/motel applications.

11.7 Humidity Control

Any refrigeration-type cooling coil will provide some high-limit, possibly entirely acceptable control of humidity, intentionally or otherwise. Lower humidities may be obtained by subcooling with reheat or by adsorption-type dehumidifiers. Humidity may be added in several ways, as discussed in Sec. 10.19. Controlled dehumidification always requires the use of extra energy.

11.7.1 Dehumidification by refrigeration

This process requires a cooling coil utilizing chilled water, brine, or a refrigerant (Fig. 11.21). The air passing over the coil is cooled to a dew point corresponding to the desired relative humidity and space dry-bulb temperature. In Fig. 11.21, to maintain the space at 70°F db and 40 percent RH, a coil ADP of about 43°F would be required. To obtain this, it would be necessary to use brine at about 36°F entering. Chilled water cannot be safely produced at this low temperature. Direct expansion could also be used, with an appropriate suction temperature and superheat. Care must be taken to avoid coil icing. The system control and arrangement would be as shown in Fig. 11.5. Note the reheat requirement.

11.7.2 Chemical dehumidification

A chemical dehumidifier utilizes an adsorbent, such as silica gel, to remove unwanted moisture from the air. Figure 11.22 shows the continuous process. Trays of silica gel rotate through the conditioned

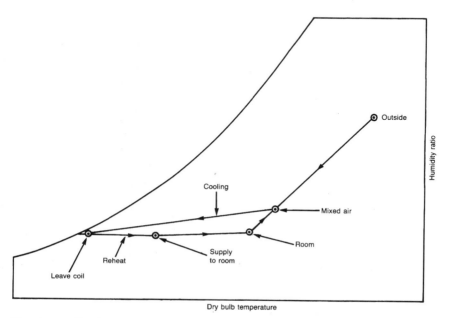

Figure 11.21 Psychrometric chart for dehumidification by refrigeration.

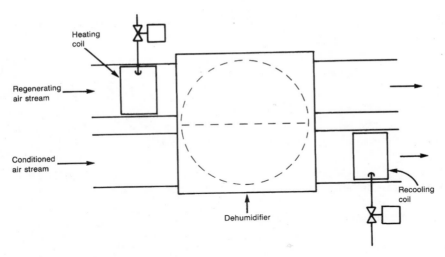

Figure 11.22 Chemical dehumidifier.

airstream—where moisture is adsorbed—and then through a hot regeneration airstream where the gel is dried. The adsorption process generates some heat, and additional heat is carried over from regeneration so that the process results in dryer but hotter air, as shown in Fig. 11.23. The air must then be recooled.

Another type of adsorption drying process uses two air-handling units, each with a lithium chloride-water spray system (see Sec. 10.17.5). The dry solution is used to adsorb moisture from the conditioned airstream and is then pumped to the wet unit, where it is regenerated by means of exhaust or outside air. This system also requires recooling. In both systems, the final humidity is controlled by controlling the regeneration air heater. Very low humidities—as low as 20 percent RH at 50°F—are possible with chemical adsorbents.

11.8 Control of Outside Air Quantity

Outside air economy-cycle control is described in Sec. 8.5.1. A *fixed* outside air quantity may be required to satisfy exhaust or ventilation requirements. This is provided as shown in Fig. 11.24 (less than 100 percent outside air) or Fig. 11.25 (100 percent outside air). With 100 percent outside air it is preferable to provide a time delay in the supply-fan motor starter circuit, so the damper will be at least 50 percent open before the fan starts. It is possible to control the outside and return air quantities to match a varying exhaust (Fig. 11.26). The space being exhausted is usually controlled at lower pressure than an adjacent space. The real problem here is accurate sensing of the very

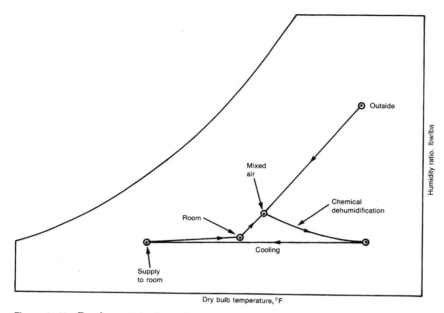

Figure 11.23 Psychrometric chart for chemical dehumidification.

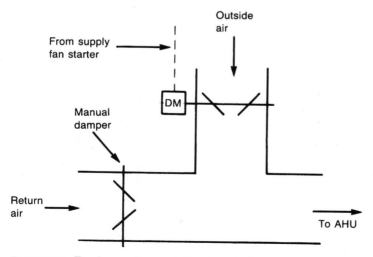

Figure 11.24 Fixed-percentage outside air control.

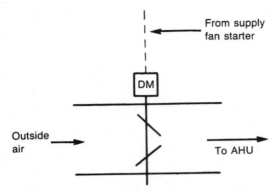

Figure 11.25   One hundred percent outside air control.

low-pressure differences used, while taking into account wind and stack effects. One of the better methods presently used provides an orifice in the wall between the two spaces where a velocity-measuring device (hot-wire anemometer or fluidic sensor) is located. Exhaust and makeup air quantities can then be adjusted to maintain a velocity corresponding to the desired pressure difference.

11.9   Effects of Altitude

The effects of higher than standard elevations and temperatures are discussed in Secs. 4.5, 5.3, and 9.7. Changes in the mass flow rate due

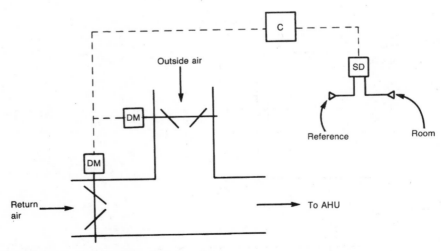

Figure 11.26   Outside air control pressure.

to altitude factors affect both heat transfer rates and the heat-carrying capacity of the airstream. At 4000- to 5000-ft elevation (western intermountain regions), the effect of altitude on air density varies from 10 to 15 percent. This is enough to create serious consequences if it is ignored.

11.10   Exhaust Systems

Power exhaust systems are required for many areas of a building. Code requirements govern exhaust from toilets, bathrooms, kitchens, and similar areas as well as many laboratory areas. Recirculation of air from hospital patient isolation rooms and soiled utility rooms is not generally allowed. Heat-, fume-, and moisture-generating processes should have collection hoods and exhaust systems. Laboratory and industrial fume hoods require large exhaust air quantities. Several standards cover the design of some types of exhaust systems.[1,2]

11.11   Smoke Control

The use of environmental air-handling systems for primary or supplementary smoke control is neither simple nor economical. There is a lack of good data on the design of adequate smoke control systems, and code requirements are often vague and conflicting. The best data available at this writing are contained in Refs. 4 and 5. Research in this area is ongoing. Automatic control systems can become quite complex and often include provision for manual override by firefighters. The designer is advised to study the latest research in this area.

Smoke detectors or fire detectors must be installed in most air-handling systems in accordance with local building and fire codes. Most of these codes are based on the recommendations of the National Fire Protection Association (NFPA),[6] particularly Standards 90A, 90B, and 91. If the AHU is not used for smoke control, then the smoke or fire detector must be used to stop the fan and to close the outside air damper.

11.12   Summary

While this chapter has dealt with air-handling units and systems, many of the system components are described in other chapters and frequent references have been necessary. The specialized areas of exhaust systems and smoke control have barely been touched upon, and detailed treatment is not possible within the scope of this book.

References

1. National Fire Protection Association, *Installation of Blower and Exhaust Systems for Dust, Stock, Vapor Removal or Conveying,* NFPA 91-1983 (ANSI-Z33.1-1982), 1979.
2. Fundamentals Governing the Design and Operation of Local Exhaust Systems, American Industrial Hygiene Association (AIHA) (ANSI-Z9.2-1995, Chap. 13).
3. R. W. Haines, "Air Flow Balance in a Laboratory," *Heating/Piping/Air Conditioning,* November 1981, pp. 159–160.
4. John H. Klote and James A. Milke, *Design of Smoke Management Systems for Buildings,* ASHRAE/Society of Fire Protection Engineers, Special Publication, 1992.
5. M. E. Dillon, "Recent Trends in Smoke Control," *ASHRAE Journal,* April 1987, pp. 27–31.
6. National fire codes, issued annually by National Fire Protection Association.

12

Electrical Features of HVAC Systems

12.1 Introduction

While most HVAC designers will have the support of a competent electrical design staff, it is important to understand certain fundamentals of electricity, power distribution, and utilization, because so many HVAC system devices are mechanically driven and controlled. This book cannot present electrical topics in great detail, but it can address several common topics and refer to more definitive works.

12.2 Fundamentals of Electric Power

Electricity is basically electrons in motion. Electromotive forces cause free or loosely bound electrons to move along or through a medium. Materials such as aluminum, copper, silver, and gold allow electrons to move freely and are called *conductors*. Materials such as porcelain, glass, rubber, plastics, and oils resist electron movement and are called *insulators*.

Forces that move electrons are magnetic. Moving a conductive wire in a way that cuts across a magnetic field induces a force or voltage in the wire. If there is a path for the electrons to follow, a flow will be established. The strength of the motive force is defined in volts, and the magnitude of the current is measured in amperes. The *resistance* to current flow is analogous to the friction loss of water flowing through a pipe. Voltage, current, and resistance are related to each other in the Ohm's law equation

$$E = IR \qquad (12.1)$$

where E = voltage, V

$\quad I$ = current, A

$\quad R$ = resistance, Ω

Power, defined as a force moving through a distance per unit time, is defined electrically by the equation

$$P = EI$$

where P = power in watts.

In direct-current (dc) systems, the voltage is applied in one direction only. In alternating-current (ac) systems, the voltage changes direction on a continuous basis; 60-Hz systems are common in the United States, while 50-Hz systems are common in Europe. (Hz = hertz, or cycles per second.)

12.3 Common Service Voltages

Many different voltages have been used over time for electrical service in and to buildings and complexes. Forty to fifty years ago, many—if not most—building distribution was single-phase at 120/240 V. A high-leg delta scheme was used to feed limited three-phase requirements. Current practice tends toward three-phase service in most locales. Smaller systems focus on 120/208 V, larger systems on 277/480 V. Control systems usually step down to 24 V.

Utility and complex distribution voltages are often found at 2300, 4160, 7200, and 12,470 V. Large motors are sometimes selected for 2300 or 4160 V if that works well with the distribution system. There is a sharp increase in the complexity and cost of electrical gear above 5 kV (5000 V) which precludes much use of the higher voltages. The HVAC designer may occasionally encounter 2300- or 4160-V motors on chillers. Competent help is needed in specifying electrical gear and protection for such applications.

Most motors and other user devices are rated to perform acceptably at nameplate voltage plus or minus 10 percent. Power companies generally commit to line voltage plus or minus 5 percent, with brownouts and outages allowed. This explains the common motor voltage versus system voltage relationships typically encountered. Table 12.1 illustrates these voltage rating–delivery relationships. It becomes apparent why industry has evolved away from the earlier 220/440-V motor ratings. The 240/480-V delivery systems were simply out of the motor service range much of the time.

12.4 Power Factor

In ac systems where the voltage is constantly changing from positive to negative and back again, current flow often lags the voltage. This

TABLE 12.1

Nominal line voltage, V	Probable service range, V	Nominal motor rating, V	Motor operating range, V
120	126– 114	115	126.5– 103.5
208	218– 197	200	270– 180
240	252– 228	230	253– 207
480	504– 456	460	506– 437
2300	2415–2185	2200	2420–1980
4160	4370–3950	4000	4400–3600

is particularly true of inductive loads such as motors, transformers, and magnetic fluorescent lighting ballasts (a type of transformer), all of which involve copper wire wound around a steel core.

As the voltage (electromotive force) propels electrons along the conductor, the electrons tend to momentarily gather or store themselves in the inductive body. It is as if the voltage has to tell the current to catch up. The net effect is that the *true* power (instantaneous voltage times amperage) is usually less than the *apparent* power (maximum voltage times maximum amperage). The *power factor,* denoted by PF, is then defined as the cosine of the phase angle between the voltage and the current. The power-defining equation for three-phase power evolves to

$$\text{Power} = EI\sqrt{3}(\text{PF}) \tag{12.2}$$

Since parasitic power losses in power distribution systems, as well as conductor capacity, are based on current flow

$$\text{Power loss} = (\text{current})^2(\text{resistance}) = I^2R$$

having a greater than necessary current flow out of phase with the voltage is detrimental to the overall electric system. Utility companies impose a cost penalty on consumers with poor power factors (usually less than 0.90). The biggest contributors to a poor power factor are inductive devices which are only partially loaded. The HVAC designer should avoid grossly oversized motors. The power factor is corrected by connecting capacitors to the line to offset the inductive effect. Capacitors have the opposite effect on current-voltage relationships from inductances. Sometimes motor specifications include capacitors. Or the capacitors may be installed in the motor control center. Less often, a bank of capacitors will be installed at a central point in the electric distribution system. Distributed power-factor correction is usually less expensive than central or consolidated correction. Central correction usually requires automated control of on-line capacitance, since the magnitude of on-line inductance varies with time.

12.5   Motors

Electric motors are devices which convert electric energy to kinetic energy, usually in the form of a rotating shaft which can be used to drive a fan, pump, compressor, etc. Single-phase motors are commonly used up to 3 hp, occasionally larger. Three-phase motors are preferred in electrical design for ¾-hp motors and larger, since they are self-balancing on the three-phase service. Motors come in various styles and with different efficiency ratings. The efficiency is typically related to the amount of iron and copper in the windings; i.e., the more iron for magnetic flux and the more copper for reduced resistance, generally the more efficient the motor. Open dripproof (ODP) motors are used in general applications. Totally enclosed fan-cooled (TEFC) motors are used in severe-duty environments. Explosion-proof motors may be needed in hazardous environments.

Motors are typically selected to operate at or below the motor nameplate rating, although ODP motors often have a service factor of 1.15, which implies that the motor will tolerate a slight overload, even on a continuous basis. Since motors are susceptible to failure when they are operated above the rated temperature, care must be taken in motor selection for hot environments such as downstream from a heating coil.

Motor windings are protected by overload devices which open the power circuit if more than the rated amperage passes for more than a predetermined time. This raises an interesting issue for a motor assigned to drive a fan that has a disproportionately high moment of rotational inertia. On start-up, a motor draws much more than the full-speed operating current. The time required to bring a fan up to speed may be too long if the motor doesn't have enough torque to both meet the load and accelerate the fan wheel. If the motor doesn't come up to speed within 10 to 15 s, it is likely that the motor protection will cut out based on the starting amperage. A motor sized tightly to a fan load may never get started. Therefore, it is important to size a motor for both load and fan wheel inertia. Fan vendors can help with this concern. This problem is particularly common on large boiler induced-draft fans where the dense-air, cold-start-up condition requires much more driver power than the hot operating condition.[1]

12.5.1   Motor rotation

In single-phase motors, the direction of motor rotation is determined by the factory-established internal wiring characteristics of the motor. Changing the connection of leads to the power source may have no effect on the direction of rotation. To make a change requires a change in an internal connection as directed by the manufacturer.

In polyphase motors, a lead sequence is established at the power plant. The motor presents three sets of lead wires which are connected to the three phases of the service. If a three-phase motor is found running backward, all that is needed to change the direction is to exchange any two leads.

12.6   Variable-Speed Drives

One of the most useful electrical developments in recent years has been the ac *variable-frequency drive* (VFD) for motor speed control. Electric speed control of motors is not a new concept—dc drives have been used for decades in the industrial environment—but low-cost ac drives suitable for the HVAC market are a relatively new product. These new drives typically use electronic circuitry to vary the output frequency which in turn varies the speed of the motor. Since the power required to drive a centrifugal fan or centrifugal pump is proportional to the cube of the fan or pump speed, large reductions in power consumption are obtained at reduced speed. These savings are used to pay for the added cost of the VFD on a life cycle cost basis. A quality VFD usually obtains greater energy savings than does a variable-pitch inlet vane or other mechanical flow volume control. In low-budget projects, the owner may forgo the higher-quality VFD service in favor of the lower-first-cost inlet vane damper for fans, or modulating-valve differential pressure control for pumps.

In applying a VFD to a duty, several factors need to be considered:

1. The VFD needs to be in a relatively clean, air conditioned environment. Since it is a sophisticated electronic device, particulates in the ambient air, wide swings in ambient air conditions, temperatures above 90°F, and humid condensing environments are all threatening to drive life expectancy.

2. The drive should be matched to the driven motor. Reduced motor speeds relate to reduced motor cooling while internal motor energy losses may be high in an inappropriately configured motor. High-efficiency motors are typically preferred for VFD service.

3. Drives and motors may be altitude-sensitive or may be affected by other local conditions. Drive and motor selection should be confirmed in every case by the drive vendor.

4. Some drives use a carrier frequency in the audible range, which may be emitted at the drive and/or at the motor. The noise may be objectionable. This is a difficult problem to abate in some applications. Some newer drives allow the carrier frequency to be set above the normal hearing range, which eliminates the noise problem, but may shorten motor life expectancy.

5. Some variable-speed drives impose "garbage" waveforms on the incoming utility lines or create *harmonic distortions* which affect the current flow in the neutral conductor of a three-phase power supply. The HVAC designer must work with the electrical design team to recognize and minimize this effect. Isolation transformers are not always effective in eliminating harmonic distortion back to the line.

6. If VFDs are applied to critical loads, it may be helpful to have bypass circuitry to run the motor at full speed in the event of a drive outage. This creates a concern for pressure control since the full-speed operation will develop a maximum pressure condition whether needed or not. Relief dampers may be considered. See Fig. 12.1 for a wiring schematic for a VFD installation.

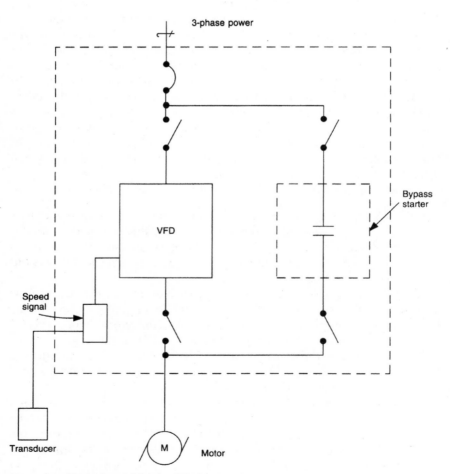

Figure 12.1 Typical variable speed drive controls.

7. Most VFDs can accept a remote input signal of 4 to 20 mA, or 0 to 10 V dc, derived from pressure transducers or flowmeters. The drives typically have a manual speed selection option if an occasional or seasonal speed change is all that is needed. The manual setting is also useful in a test-and-balance period.

12.7 HVAC–Electrical Interface

On a number of issues the HVAC designer must interface with the electrical designer, each sharing information and responding appropriately.

Motor loads: Motor sizes and locations derive from the HVAC equipment selections and equipment layouts.

Motor control features: HVAC control schemes determine many of the needed starter characteristics, e.g., hand-off-auto or start-stop, auxilary contact types and number, pilot light requirements, and control voltage transformer size if external devices needing control power are involved. Figure 12.2 is a form that can be used to communicate such information to the electrical designer. Be sure to coordinate the specification and control of two speed motors and motor starters.

Fire and smoke detection and alarm: The electrical designer is usually responsible for fire detection and alarm, if such is required. But building codes require smoke detectors in the airstream of recirculation fan systems larger than 2000 ft^3/min. If smoke is detected, fan systems are required to shut down. Similarly, if the building detection systems go into alarm, the fan systems must turn off. Further sophistication gets into smoke control in buildings, a separate topic by itself.

Lighting systems: The HVAC designer must fully understand the building lighting systems to be able to correctly respond to the cooling loads which develop. Any inordinately high lighting loads may stimulate discussion and evaluation of lighting fixture selection. Automated lighting control may be included as a feature of a building automation system.

Transformer vaults: Electric transformers typically lose 2 to 5 percent of the power load (winding losses) to the ambient air. Building transformers may wind up in underground vaults, in secure rooms, in janitor closets, or in ceiling spaces. Dissipation of the heat with ventilation is often a challenge. Note that even though the load may decrease, transformers seldom sleep; 24 h/day ventilation is required. Building HVAC systems which follow a time-clock schedule are inadequate for transformer rooms.

SCHEDULE OF MOTOR STARTERS AND CONTROLS

JOB NAME _____ SHEET_____ OF_____ JOB No. _____

_____ DATE_____ BY _____

MECH. SYMBOL	MECHANICAL EQUIPMENT SERVED	MOTOR DATA					STARTER DATA									REMOTE CONTROL				RELAYS			INTERLOCK FROM
		H.P.	Voltage	Phase	K.W. input	Windings	Type	NEMA size	No. speeds	Control voltage	Pushbutton	Selector switch	Pilot light	N.O. contact	N.C. contact	Control voltage	Pushbutton	Selector switch	Pilot light	P.E. relay No.	E.P. relay No.	Control relay No.	

Figure 12.2 Typical form that supplies information on motor control features to the electrical designer.

12.8   Uninterrupted Power Supply

Even the best of private and public power supplies are subject to variations in quality of delivered power and to occasional unplanned outages. At the same time, some types of electric loads cannot tolerate a power line disturbance or an interruption of power. Such loads may involve computer installations, communications and security installations, medical services, etc. Usually the power-consuming service supports a high value or critical function where the liability of interruption cannot be tolerated.

Uninterruptible power supply (UPS) systems provide continuity of power to a connected load, in and through power line disturbances, without a sign of the outage being seen by the load.

Earlier UPS systems had the character of electromechanical systems with a line voltage motor-driven generator which fed the load in parallel, with a backup battery installation which picked up the load when the generator faltered or dropped out.

Newer UPS systems use transistor-type technology, to convert the ac line voltage to direct current, in lieu of the motor-generator function. Batteries are still used as the storage medium to provide power when the primary service is interrupted.

UPS systems are of interest and concern to the HVAC systems designer in two ways:

1. UPS systems may be a useful, even required component of a critical HVAC service and will be included in the HVAC and electric system design.

2. UPS systems themselves create points of significant heat release which must be dealt with through ventilation, exhaust, or air conditioning.

The first interest is usually defined by the specific project and is a function of the supported service. The second is then dictated to the HVAC design and must be accommodated. If the UPS systems run 24 h/day, so must the related cooling, even if the general HVAC system is on time-of-day control.

Each transformation of power generally involves a release of 5 to 15, even 20 percent of the power handled. For example, a motor-generator set involves a 10 percent energy loss in the driving motor and another 10 percent energy loss in the driven generator, for an overall device loss of 20 percent or more of the power transformed. Electronic UPS systems may be more efficient with only 10 to 15 percent losses in the overall transformation process. This energy loss shows up as a heat rejection to the space. On a small scale, such as a UPS device for a single personal computer, the loss of 20 to 30 W out of 200 to 300 W total of unit capacity seems insignificant, but must be included in the capacity of the space air conditioning system. The factor is significant if many units are involved. Small UPS devices may be switched with the equipment served.

Where a larger UPS system is developed to serve a large load such as a mainframe computer, the heat rejection of the UPS system becomes a *spot load,* usually with a 24 h/day operating schedule.

Modern UPS assemblies with electronic voltage management technology usually require a stable environment between 60 and 85°F that is free of moisture condensation. The air will typically be filtered to reduce the potential of particulate collection on the circuit boards. Some device manufacturers can tolerate a wider range of temperature, but stable conditions free of rapid temperature swings seem to be a universal preference for maximum life of the equipment.

Many UPS units have self-contained fan-powered internal ventilation. The fans take in room air, blow it over the circuit boards and components and through the unit, and discharge it out an opening on the top or side (back, front), or toward the floor. The challenge is to capture the heat into the return air or exhaust air path while introducing supply air or makeup air into the room. UPS systems are now seldom treated with unconditioned outside air because summer temperatures in many locales rise above the range of desired operating conditions.

12.9   Standby Power Generation

While standby power generation is by definition an electrical service, the engines and related support issues are as much mechanical as electrical, perhaps more so. The mechanical HVAC designer needs to understand the unit function to participate effectively in the design.

A *standby generator,* sometimes called an *engine-driven emergency generator,* typically includes a reciprocating engine which may use natural gas, digester gas, propane, gasoline, or diesel fuel. The HVAC designer is typically responsible for an external fuel supply. Remote fuel storage tanks with local day tanks are often used. On some units the fuel tanks are mounted in the unit frame, which leaves the responsibility for design with the manufacturer. There are code limitations to the amount of fuel which can be kept inside a building.

The engine will have a heat rejector, usually either a unit-mounted or remote radiator. If it is remote, piping and concern for placement are involved. Engines are designed for pressures less than 15 lb/in^2, so radiator mountings more than 35 ft above the engine cannot be handled. Unit-mounted radiators typically draw cooling air from the room across the engine, expelling the air to the outside. This usually requires louvers and dampers to be open when the unit runs. The engine rejects approximately 10 to 15 percent of the heat value of the burned fuel to the room. The radiator cooling air usually picks up this heat. With a remote radiator, local room ventilation must pick up the engine radiant heat as well as the heat loss of the generator.

The engine shaft is usually directly connected to either an induction or a synchronous-type generator. The generator is similar to a motor working in reverse. Part of the engine shaft power which drives the generator is lost to the atmosphere in the transformation from kinetic to electric energy. These losses usually amount to 10 to 15 percent of the generator load.

The HVAC designer may be involved in the exhaust system design, which includes flexible connection of the exhaust manifold to exhaust stack, a silencer, extension of the exhaust to the atmosphere, insulation of the exhaust system, weather cap, and stack drain.

Reciprocating engines usually need a substantial reinforced-concrete base which is independent of the basic building structure. Special sound control features may need to be incorporated into the design. The point of exhaust discharge should be studied and kept remote from air intakes or sound-sensitive occupancies.

12.10   Electrical Room Ventilation

In rooms where electric devices consume electricity and give off heat, some sort of ventilation for cooling is required. Electronic installations may require mechanical refrigeration. Natural convection ventilation usually assumes a 10 to 20°F rise in the space which allows calculation of the probably required ventilating airflow quantities, assuming that the heat release can be estimated. The following estimating factors may be helpful.

- *Transformers:*   Assume that 3 to 5 percent of the active load will be dissipated in transformation. This may drop to 2 to 3 percent for more efficient units.

- *Elevator machine rooms:*   Figure all the elevator motor horsepower times a factor for the estimated percentage of time in use. Peak-use hours approach 100 percent. Consult the elevator vendor for temperature constraints and secondary losses from control panels, etc.

- *Motor control centers:*   These units generate some heat from control transformers and starter holding coils. This equipment does not hold up well in hot environments. Carry this observation over into plant design considerations.

12.11   Lighting Systems

Lighting design ranges from following a cookbook to a high-level art form. It is not the responsibility of the HVAC designer, but lighting imposes far-reaching consequences on the HVAC design. Nearly all lighting is derived from electricity. Only a fraction of the power is transformed to light, and virtually all the lighting-related energy is released to the space or ceiling plenum, where it must be addressed by the HVAC system. The HVAC designer should tell the lighting design team, and the other design team members, including the owner, about the possible impact of lighting layouts on the HVAC system.

In the first half of the twentieth century, most lighting was of the incandescent type. At that time, lighting levels were spartan, relative to system cost, operating cost, and availability of power. With the advent of fluorescent lighting in roughly the time period of World War II, it was perceived that productivity could be improved with increased levels of lighting in the workplace. The 1950s and 1960s then became a time of excess in lighting design, with high levels of illumination and consequent average imposed lighting loads of 4 to 6 W/ft^2, even more in some cases. The energy constrictions of the early mid-1970s called quick attention to the problem of conspicuous energy consumption for lighting. Public sensitivity combined with cost factors has helped reduce expectations and bring new lighting products to market. Common lighting designs for office space will now average 1.5 to 2.0 W/ft^2 of connected load. Multiple-level switching of lamps may reduce this even further for much of the time.

There is still the problem of high-intensity lighting using incandescent lamps for retail display and fine visual work. The incandescent lamp seems to offer a color spectrum that is closer to that of the sun than other lighting types. Where the lighting quality is truly important to the function of the space, incandescent fixtures should be questioned, but accepted. The consequence is the increased cooling capacity requirement and higher cost of power for lighting.

To quantify the impact of lighting power, 1 W/ft^2 extra will require approximately 0.15 $ft^3/(min \cdot ft^2)$ extra cooling air (15 to 20 percent more than average) and will require an additional 0.25 ton of cooling per 1000 ft^2 of building space. This is 25 to 30 tons of added cooling capacity to a 100,000-ft^2 building, related to only 1 W/ft^2 of lighting. Savings in HVAC equipment cost will often more than pay for improved lighting equipment.

12.12   National Electric Code

The bible of the electrical portion of the construction industry is the *National Electric Code* (NEC), also cataloged as NFPA 70.[2] This volume is given the weight of law in most parts of the United States. It defines in great detail what does and does not comprise an acceptable electrical installation. This volume is of interest to HVAC designers in several ways.

1. It defines required ventilation (natural or forced) for several types of areas housing electrical gear.
2. It prohibits spatial intrusions of unrelated ducts and piping into electrical rooms.
3. It defines acceptable electrical assemblies for many HVAC units which have electric components.

While the NEC is relatively coherent at face value, it sometimes becomes onerous in the local interpretation. Some jurisdictions prohibit any piping in an electrical room, yet fail to accommodate otherwise required roof drain lines or wet sprinkler fire protection. Heating and chilled water lines are generally prohibited even though they may be welded or have no joints in the room.

The NEC has stringent requirements when it comes to being able to shut off all electrical service in a motor control cabinet with a single disconnect. Yet some mechanical control circuits may interface with the electric power circuits and present a voltage of remote origin. The HVAC designer must work closely with the electrical designer to satisfy the needs of both disciplines.

12.13 Summary

In building construction, HVAC design is interwoven with the electrical design, and each discipline needs to be conversant with the other. Electrical-mechanical interfaces need to be fully communicated for complete designs to be achieved. The HVAC designer should have a working background in the fundamentals of electricity and electric control. Full presentation of the electrical needs of the HVAC system must be part of the HVAC design work, as must a complete understanding of the impacts of electrical heat releases on the building environment.

References

1. C. L. Wilson, "Fan Motor Time-Torque Relations," *Heating/Piping/Air Conditioning,* May 1971, pp. 75–77.
2. *National Electric Code, National Fire Protection Association,* Quincy, MA, 1990, revised every 3 years.

Design Documentation: Drawings and Specifications

13.1 Introduction

Design documents evolve from and include the designer's calculations, equipment selections, and sketches and are usually presented through formal drawings and specifications. These construction documents are the legal means by which the designer conveys the owner's expectations to the contractor. The importance of good documentation cannot be overemphasized.

An old adage says, "A picture is worth a thousand words." In construction, drawings are the picture, and specifications are the thousand words. But for projects of importance or great value where the work is accomplished by contract between owner and builder, it is important that there be good specifications in addition to the drawings to define the relationships between the parties. The purpose of this chapter is to review the nature of contracts and then to define drawing preparation and specification organization and writing well enough that the reader will have an understanding of and be prepared to practice the basic techniques of document preparation.

Since drawings and specifications become a part of a construction contract, they become legal documents. As such, they must define the work to be done clearly, completely, and unambiguously. Although this ideal is seldom achieved, designers must do their best to meet these criteria. Lawsuits involving millions of dollars have been filed based on the interpretation of a few sentences in the specifications or a lack of clear detail on the drawings.

At the same time as contract documents are identified as being of paramount importance to the relationship between owner and contractor, remember that document preparation has a cost which must be reasonable and a schedule which must be met. No project has an infinite amount of time or resources which can be allocated to the design effort. The result is that documents should be adequate, but not overdone, should be prepared deftly and in an organized fashion, all arranged to fully reflect the owner's hopes and the designer's intent.

13.2   The Nature of Contracts

In the United States, the law allows two or more individuals or companies or institutions to contract with each other for an exchange of goods or services. In HVAC work, a building owner, called the *owner*, will typically arrange with a vendor or installer, called the *contractor*, to furnish and install equipment and related material in a system. Often the HVAC work is performed in conjunction with the full complement of building construction. The agreement between the owner and the contractor contains the basic elements of any legal contract; i.e., there is a *work* or service or value committed to deliver, there is *compensation* for the work performed, and there is a *time* period of performance. All three components are required to establish a valid contract.

Most construction is undertaken by contract, where the specifications and drawings define the work to be done and the contract includes a description of compensations and a date of completion. There are often penalties for failure to perform in a timely manner and sometimes bonuses for early completion.

Persons signing the contract must be authorized to do so. This is self-evident in the case of a private individual, a proprietorship, or a partnership. In the case of a corporation, the board of directors must have given authority to the signator for his or her signature to be valid and binding on the company. Signatures are often witnessed or notarized.

13.3   Drawings

As already noted, contract drawings are legal documents and thus should avoid ambiguity. This consideration leads to several criteria which are typical of good drawings.

All these efforts take design time. The alternative—providing inadequate information, neglecting details, and careless checking—can take a great deal more time later on, can cost money for extras, and even can lose the confidence of the client in the designer's competence.

13.3.1 Drawing size and scale

Drawing size and scale should be appropriate for the work being described. Typical drawing sheet sizes are described both by letter and
by sheet dimension.

Size	Dimension
A	8½ × 11
B	11 × 17
C	18 × 24
D	24 × 36 (22 × 34)
E	30 × 42
F	36 × 48

Special sizes may be custom-ordered. Smaller sheets can be included
in a book of specifications. Larger sheets are almost always presented
as a set, except in the case of only one or two sheets, where they may
be neatly folded and placed in pockets in the specification book.

The drawing scale is determined by the amount of detail to be presented for the dominant aspects of the work. For building construction,
⅛ in = 1 ft is a common scale. A very large building may have a plan
view at ¹⁄₁₆ in = 1 ft for an overview and routing of major systems,
with sectionalized drawings at larger scale. Note that doubling a scale
uses 4 times the drawing area to make the presentation.

A common plan dimension for commercial and institutional construction is ⅛ in = 1 ft. For residential floor plans and for layouts
needing more detail and near-dimensional accuracy than can be expected of ⅛- and ¹⁄₁₆-in scales, ¼ in = 1 ft is used. Many mechanical
room layouts are presented at ¼-in scale. When much detailed presentation is needed, larger scales such as ½ in = 1 ft, ¾ in = 1 ft,
1 in = 1 ft, 1½ in = 1 ft, and 3 in = 1 ft are employed. For no apparent
reason, many civil engineers involved in construction use ³⁄₃₂-, ³⁄₁₆-,
and ⅜-in scales. No matter which scale is chosen, it is helpful and
important for all members of the design team to use common scale(s).
This helps to avoid errors and makes the overall drawing set easier
to read and interpret.

13.3.2 Drawing character

Line work, whether hand-drawn or computer- or plotter-generated,
should be clear, sharp, and accurate. Lettering should be neat, uniform, and legible. The appearance of the drawings can go far to establish the credibility—the acceptance or the lack of acceptance—of
the product, which is the design. Early in the development of *computer-aided design* (CAD) techniques, it was felt that computer-

generated drawings lacked the "character" of well-presented hand-drawn work. But over time, with improved software and increased CAD drafter skills, computer-generated drawings can have all the character of hand-drawn work plus a greater degree of consistency and accuracy.

13.3.3 Adequate information

Enough views, both plan and section, should be drawn to fully present the work. Details should be numerous and explicit. *Standard* details are often useful and save time, as long as the application is really standard. Failure to tailor standards to actual conditions can be an embarrassing, even costly experience.

13.3.4 Drawing legends

Symbols and abbreviations should be defined in a legend. There are many regional or office-specific legends, but no universally accepted industry standards, although ASHRAE, among others, has suggested a set of symbols in the ASHRAE Handbook *Fundamentals.*[1] Several government agencies have standards of their own which they require to be used on their projects.

13.3.5 Diagrammatic drawings

Schematic diagrams for system flow and control are very helpful to both the installer and the user. A complete and detailed schematic will answer most questions about concept and performance. Such diagrams are sometimes referred to as *flow diagrams, isometrics,* or *P&I* (process and instrumentation) *diagrams.*

13.3.6 Schedules

Equipment schedules with tabular equipment performance information should be on the drawings rather than in the specifications. Experience shows that the installer often doesn't have (or doesn't refer to) the specifications. After the job is completed, the drawings are almost always available while specifications have a way of disappearing. Equipment schedules then become a valuable resource for the owner. Some design offices include full equipment specifications on the drawings, but this seems to take more document preparation time and may make it more difficult to coordinate drawings and specifications if changes are needed.

13.3.7 Minimize sources of information

A good rule for presentation of quantitative information is to show it only once on the drawings and to not call out information on the drawings that is covered in the specifications, or vice versa. This reduces the potential for error as well as the amount of information which must be updated when changes are made.

13.3.8 Quality control—checking

Drawings should be carefully checked for errors and omissions, preferably by the system designer. Some offices use an independent checker to take a fresh look at the near-final product.

13.3.9 Use of computers in drafting

The use of computers in design, drafting, and specification writing has proliferated in recent years.[2] *Computer-aided design and drafting* (CADD) systems are available for lease or purchase or on a timesharing basis from computer service companies. As software has become more sophisticated, the systems have become easier to use. However, to use these systems requires a considerable investment in equipment and in training of personnel. A CADD system, properly designed and used, will almost always save time and money eventually, but time and costs may increase initially during the learning period. Many government agencies and private institutions are so committed to CADD that such capability is a requirement to obtain work.

13.4 Specifications

Most of the construction industry now uses a format developed by the Construction Specifications Institute (CSI) for specifications.[3] This is a standardized outline of contract and construction material with forms for the contract work and with section and subsection numbers for each technical topic. The system includes flexibility for adding information unique to a project. For details, contact the Construction Specifications Institute, 601 Madison St., Alexandria, Virginia 22314. See Fig. 13.1.

13.4.1 Contractual matters, or boilerplate

Preceding the technical specifications which with the drawing defines the actual work to be accomplished, there is usually additional documentation which establishes the contractual relationship between the

INTRODUCTION

MASTERFORMAT is a system of numbers and titles for organizing construction information into a regular, standard order or sequence. By establishing a master list of titles and numbers *MASTERFORMAT* promotes standardization and thereby facilitates the retrieval of information and improves construction communication.

Modern complex construction projects are designed and built by specialists using many different products and systems. Effective communication between the people involved in a project is essential for successful completion of the work. An important factor in effective communication is standardization. Information retrieval is nearly impossible without a standardized filing system familiar to the users. *MASTERFORMAT* serves the construction industry as a standard information filing and retrieval system and is an important tool for effective communication. It provides a uniform system for organizing information in project manuals, for organizing project cost data, and for filing product information and other technical data.

This is the 1988 edition of *MASTERFORMAT* and it replaces the 1983 edition. It is produced jointly by the Construction Specifications Institute (CSI) and Construction Specifications Canada (CSC).

HISTORY

Since its introduction in 1963, the 16 division format has been widely accepted as an industry standard in the United States and Canada. First published as part of the "CSI Format for Construction Specifications" it was later used as the basis for the "Uniform System for Construction Specification, Data Filing and Cost Accounting -- Title One Buildings" published in 1966. The "Uniform System" was developed and endorsed by the following organizations: American Institute of Architects, American Society of Landscape Architects, Associated General Contractors of America Inc., Associated Specialty Contractors, Construction Products Manufacturing Council, National Society of Professional Engineers, and Construction Specifications Institute. In 1966 a similar effort in Canada produced "The Building Construction Index" (BCI), based on the 16 division format which had been introduced by the Specification Writers Association of Canada, renamed Construction Specifications Canada in 1974.

The U.S. and Canadian formats were merged into a single format in 1972 and published as the "Uniform Construction Index" (UCI). The UCI was a comprehensive framework for organizing information contained in project manuals, as well as providing a basis for data filing and project cost classification.

In 1978, Construction Specifications Canada joined with the Construction Specifications Institute to produce the first edition of *MASTERFORMAT*, introduced by CSI as MP-2-1 and by CSC as Document 004E. It incorporated a complete organizational format for project manuals by including bidding requirements, contract forms, and conditions of the contract in addition to the 16 division format for specifications.

Figure 13.1

The first revised edition of *MASTERFORMAT* was published in 1983. It retained the basic principles of organization contained in the previous edition. However, revisions and additions were made to recognize the needs of the engineering disciplines. The concept of mediumscope titles and numbers was also introduced in the 1983 *MASTERFORMAT.*

THE 1988 EDITION

This 1988 edition is the latest step in the evolution of *MASTERFORMAT.* It includes revisions and additions needed to recognize new products and developments in the construction industry and is based on input from *MASTERFORMAT* users. During the next few years ongoing change will lead to further improvements and perhaps to new organizational concepts in future editions of *MASTERFORMAT.*

INDUSTRY ACCEPTANCE

Over the past ten years MASTERFORMAT has been used by more and more entities of the construction industry. Today the organization of specifications into 16 divisions is commonplace and *MASTERFORMAT* is the only system of section titles and numbers in widespread use in the United States and Canada.

In addition to its uses in the private sector in the United States it is approved for use by the Department of Defense. The McGraw-Hill Information Systems Company adopted *MASTERFORMAT* as a basis for their Sweet's Catalog Files of construction products in 1986. The R.S. Means Company adopted *MASTERFORMAT* as the basis of coding for their construction. cost data publications in 1987. *MASTERFORMAT* is the basis for numbering and titling Federal Construction Guide Specifications and is also used at state and municipal levels in the U.S. It is the framework for SPECTEXT, the master guide specification system developed by the Construction Science Research Foundation and marketed by CSI, and for other commercially available master guide specification systems.

Standardization of construction documents has long been promoted in Canada by CSC. Their effort has been supported by the Royal Architectural Institute of Canada, the Association of Consulting Engineers of Canada, the Canadian Construction Association, and the Housing and Urban Development Association of Canada. *MASTERFORMAT* is also used at provincial and municipal levels. In 1978 the Canadian federal government adopted *MASTERFORMAT* as the general basis of the titling and numbering system for their Government Master Specification (GMS). The GMS was later modified for use in the private sector as well, and is now marketed by CSC as the National Master Specification (NMS) .

The construction industry in both countries uses *MASTERFORMAT* for coding product literature, for organizing construction bids, and in commercially marketed cost estimating and cost accounting systems used by contractors.

Figure 13.1 *(Continued)*

contractor and the owner. Such documentation may include the following:

Invitation to bid (instruction to bidders):   A section which describes the nature of the project, establishes the time and place for submitting bids, and expresses any other qualification or action required of potential bidders.

Bid forms: A formal document for the contractor to fill out to propose a price for completing the work described by the contract documents. The proposal may call for a lump-sum price or may be written on a unit-price, hourly cost (time and materials), or cost-plus-fee basis.

Sometimes there is a *base bid* with additive or deductive *alternates.* The alternates give the owner a chance to adjust the cost of the project to meet available funding.

Bonds: Insurance policies. Since contractors sometimes fail to complete the work as contracted, some projects require a *payment bond* which guarantees to the owner that the contract will be completed, even by a third party if necessary.

General conditions: Often a lengthy document spelling out in great detail the behaviors required of the owner and the contractor in the conduct of the work. Standard industry forms of this document are often used [see the American Institute of Architects (AIA) form A-201 or Engineering Joint Documents Council (EJDC) form 1910-8].

In preparing elements of the contractual papers or boilerplate, it is helpful to use standard document forms which have a common origin and have been checked for legal consistency between the different forms. Such are available from the AIA and EJDC. Other entities such as federal and state institutions as well as corporations may have their own forms and formats. In this litigious U.S. society, it is imperative that these documents be clear and consistent.

13.4.2 Technical specifications

Following the information which describes the legal characteristics of the contract between the contractor (vendor and installer), the technical specifications describe in more or less detail which materials are to be used and how they are to be installed. The drawings, usually at reduced scale, indicate the form of the installation. Drawings and specifications complement each other. Drawings usually identify *quantity,* specifications cover *quality.*

There is no legally prescribed format for technical specifications. Any presentation of information which gets the contractor to do what the owner wants in a timely and cost effective manner can be considered a good specification. Decades ago, virtually every office had its own arrangement for specifications. Different components of construction could be found in almost any order, and the verbal expression could take almost any format. Roughly since World War II, there has been a gradual but discernible trend toward what is now a highly sophisticated and almost universally accepted arrangement for presenting technical specification information. The general format can probably be attributed to the federal government in its several agencies which oversee construction of thousands of projects.

As background for specification organization, recognize that many different trades may work on a single project, but for clarity and ease of bidding and quality control, it is helpful to group work for a given trade in one area or section of the specification. At the same time, recognize that a single contract between the owner and contractor for many elements of work creates a commitment for the contractor for all work, whether it is well arranged or not; but it may be difficult for the contractor to apportion work between trades, between subcontractors, if the specification is fragmented.

The format which has gained general construction industry acceptance has 16 divisions for major categories of work. These divisions are listed in Fig. 13.2. Each division is broken down into sections with

- Bidding requirements, contract forms, conditions of the contract
- Technical specifications

01000	General requirements
02000	Site work
03000	Concrete
04000	Masonry
05000	Metals
06000	Wood and plastics
07000	Thermal and moisture protection
08000	Doors and windows
09000	Finishes
10000	Specialties
11000	Equipment
12000	Furnishings
13000	Special construction
14000	Conveying systems
15000	Mechanical
16000	Electrical

Note: Other divisions are sometimes added in specialized construction such as for industrial plants. Not all divisions may be required for every project.

Figure 13.2 Technical specification format with 16 divisions.

as many sections and as much depth as needed to convey the criteria for the work (see Fig. 13.3).

Each section in turn has a format containing three parts, as indicated in Fig. 13.4.

As stated before, specifications can be written with more or less detail as desired. Specifications for residential housing are often abbreviated to 3 to 5 letter-size pages for the whole house, relying on the habit of the contractor for most detail. In contrast, specifications for large buildings contracted for through competitive bidding, which requires tight quality control, may be several hundred pages long.

Where many small items and work descriptions are involved, they may be lumped together in a single section. A major component which requires much detail to describe, such as a chiller or a boiler, may be given its own section.

The organization of specifications as described above leads to a file full of specifications for every imaginable product and construction method. To the extent that the writings are generic, they become *master specifications* and can be referenced over and over from one project to another in the design office. Such a file saves time and encourages consistency among the design staff.

There are pitfalls, however. Specification information becomes outdated and inaccurate as products and methods change. More challenging are the nuances between one job and the next which require careful editing of previous or master text to avoid major errors and costly mistakes. Keeping a master specification up to date requires a major commitment of time.

13.4.3 Specification language

While allowing every specification writer the latitude to be an individual in the drafting of specifications, there are a few suggestions which can be helpful to all:

1. Use good English, good grammar, and a good vocabulary. Don't use slang or colloquialisms (jargon).

2. Use as few words as possible without losing the meaning.

3. Use a direct rather than indirect form of expression. *Do it!* is more effective than *The contractor shall do it.*

   Many specifications have been written that use *shall* as the imperative. This is now seen as an awkward way of giving direction.

4. As with report writing, keep the sentence structure simple and clear. Avoid overly complex or long sentences and phrases.

5. Organize and present the writing in a consistent manner. Present similar information in the same ordered place in each section.

DIVISION I5 - MECHANICAL

Section Number	Title
15050	**BASIC MECHANICAL MATERIALS AND METHODS**
-060	Pipes and Pipe Fittings
	Aluminum and Aluminum Alloy Pipe and Fittings
	Concrete Pipe and Fittings
	Copper and Copper Alloy Pipe and Fittings
	Ferrous Pipe and Fittings
	Fiber Pipe and Fittings
	Glass Pipe and Fittings
	Hoses and Fittings
	Plastic Pipe and Fittings
	Pre-Insulated Pipe and Fittings
-100	Valves
	Manual Control Valves
	Self Actuated Valves
-120	Piping Specialties
-130	Gages
-140	Supports and Anchors
-150	Meters
-160	Pumps
-170	Motors
-175	Tanks
-190	Mechanical Identification
-240	Mechanical Sound, Vibration, and Seismic Control
15250	**MECHANICAL INSULATION**
-260	Piping Insulation
-280	Equipment Insulation
-290	Ductwork Insulation
15300	**FIRE PROTECTION**
-310	Fire Protection Piping
-320	Fire Pumps
-330	Wet Pipe Sprinkler Systems
-335	Dry Pipe Sprinkler Systems
-340	Pre-Action Sprinkler Systems
-345	Combination Dry Pipe and Pre-Action Sprinkler Systems
-350	Deluge Sprinkler Systems
-355	Foam Extinguishing Systems
-360	Carbon Dioxide Extinguishing Systems
-365	Halogen Agent Extinguishing Systems
-370	Dry Chemical Extinguishing Systems
-375	Standpipe and Hose Systems

Figure 13.3

DIVISION 15 — MECHANICAL *Continued*

Section Number	Title
15400	**PLUMBING**
-410	Plumbing Piping
-430	Plumbing Specialties
-440	Plumbing Fixtures
-450	Plumbing Equipment

      Domestic Water Heat Exchangers
      Drinking Water Cooling Systems
      Pumps
      Storage Tanks
      Water Conditioners
      Water Filtration Devices
      Water Heaters

-475	Pool and Fountain Equipment
-480	Special Systems

      Compressed Air Systems
      Deionized Water Systems
      Distilled Water Systems
      Fuel Oil Systems
      Gasoline Dispensing Systems
      Helium Gas Systems
      Liquified Petroleum Gas Systems
      Lubricating Oil Systems
      Natural Gas Systems
      Nitrous Oxide Gas Systems
      Oxygen Gas Systems
      Reverse Osmosis Systems
      Vacuum Systems

15500	**HEATING, VENTILATING, AND AIR CONDITIONING**
-510	Hydronic Piping
-515	Hydronic Specialties
-520	Steam and Steam Condensate Piping
-525	Steam and Steam Condensate Specialties
-530	Refrigerant Piping
-535	Refrigerant Specialties
-540	HVAC Pumps
-545	Chemical Water Treatment

15550	**HEAT GENERATION**
-555	Boilers
-570	Boiler Accessories
-575	Breechings, Chimneys, and Stacks
-580	Feedwater Equipment
-590	Fuel Handling Systems
-610	Furnaces
-620	Fuel Fired Heaters

      Duct Furnaces
      Gas Fired Unit Heaters
      Oil Fired Unit Heaters
      Radiant Heaters

Figure 13.3 (*Continued*)

DIVISION 15 — MECHANICAL *Continued*

Section Number	Title
15650	**REFRIGERATION**
-655	Refrigeration Compressors
-670	Condensing Units
-680	Water Chillers
	Absorption Water Chillers
	Centrifugal Water Chillers
	Reciprocating Water Chillers
	Rotary Water Chillers
-710	Cooling Towers
	Mechanical Draft Cooling Towers
	Natural Draft Cooling Towers
-730	Liquid Coolers
-740	Condensers
15750	**HEAT TRANSFER**
-755	Heat Exchangers
-760	Energy Storage Tanks
-770	Heat Pumps
	Air Source Heat Pumps
	Rooftop Heat Pumps
	Water Source Heat Pumps
-780	Packaged Air Conditioning Units
	Computer Room Air Conditioning Units
	Packaged Rooftop Air Conditioning Units
	Packaged Terminal Air Conditioning Units
	Unit Air Conditioners
-790	Air Coils
-810	Humidifiers
-820	Dehumidifiers
-830	Terminal Heat Transfer Units
	Convectors
	Fan Coil Units
	Finned Tube Radiation
	Induction Units
	Unit Heaters
	Unit Ventilators
-845	Energy Recovery Units
15850	**AIR HANDLING**
-855	Air Handling Units with Coils
-860	Centrifugal Fans
-865	Axial Fans
-870	Power Ventilators
-875	Air Curtain Units

Figure 13.3 *(Continued)*

This helps avoid the sin of omission and helps contractors and vendors become comfortable with the specification.

6. Don't duplicate information. If a change or correction is needed, make it in one location only.

DIVISION 15 — MECHANICAL *Continued*

Section   Title
Number

15880   **AIR DISTRIBUTION**

-885   Air Cleaning Devices
             Dust Collectors
             Filters
-890   Ductwork
             Metal Ductwork
             Nonmetal Ductwork
             Flexible Ductwork
             Ductwork Hangars and Supports
-910   Ductwork Accessories
             Dampers
             Duct Access Panels and Test Holes
             Duct Connection Systems
             Flexible Duct Connections
             Turning Vanes and Extractors
-920   Sound Attenuators
-930   Air Terminal Units
             Constant Volume
             Variable Volume
-940   Air Outlets and Inlets
             Diffusers
             Intake and Relief Ventilators
             Louvers
             Registers and Grilles

15950   **CONTROLS**

-955   Building Systems Control
-960   Energy Management and Conservation Systems
-970   Control Systems
             Electric Control Systems
             Electronic Control Systems
             Pneumatic Control Systems
             Self-Powered Control Systems
-980   Instrumentation
-985   Sequence of Operation

15990   **TESTING, ADJUSTING, AND BALANCING**

-991   Mechanical Equipment Testing, Adjusting, and
             Balancing
-992   Piping Systems Testing, Adjusting, and
             Balancing
-993   Air Systems Testing, Adjusting, and Balancing
-994   Demonstration of Mechanical Equipment
-995   Mechanical System Startup/Commissioning

Figure 13.3 *(Continued)*

7. Be knowledgeable about every aspect of the written specification. Don't refer to other works without having personal knowledge that the referenced material is germane to the discussion at hand.

8. One sure way to create ambiguity in a specification is to use terms such as *common industry practice, good workmanship, qualified*

Part I: *General Considerations*
           Relationship to other sections
           Scope of work covered by this section
           Quality assurance—the qualifications and criteria to be adhered to
           Codes and ordinances to be adhered to
           Submittals required of the contractor:
           ■ Fabrication drawings
           ■ Installation drawings
           ■ Drawings of record
           Warranties
           Field tests prior to order, fabrication, and installation
Part II: *The Product(s): Equipment and/or Material*
           Description of item(s)
           Methods of factory fabrication
           Factory tests
Part III: *Field Installation*
           Notes to the contractor to indicate the nature of the installation work
           Field testing after installation

Figure 13.4 Basic format for a specification section.

mechanics, and the like, which lack explicit definition. A good way to minimize both detail and ambiguity is to refer to the various codes and standards published by government agencies and trade associations.[1] Local building codes automatically apply, and specifications may exceed these but must avoid conflict with them. Industry standards apply only if the specification so states. Such references should be used only if the designer and the vendor or installer knows exactly what the code or standard says about the product or operation in question. Lack of specific information can lead to disaster.

13.4.4 Types of specifications

There are two kinds of specifications: the *performance* specification, which is based on performance criteria only, and the *or-equal* specification, which has a vendor-specific identification of what is wanted. A performance specification describes quality, materials, accuracy, and performance in specific terms, but without reference to a specific manufacturer or model number. A list of acceptable products may be included. An or-equal specification may also describe the product in more or less specific terms, but primarily states that the product is to be *equal* to a specific manufacturer's model number.

The or-equal specification is considered somewhat easier to write. However, when a substitute product is submitted, it may be difficult to evaluate because no two products are identical. The designer who rejects a submittal which differs from the specification must be prepared to defend her or his position and to demonstrate that the proposed substitute is, in fact, unsatisfactory. This kind of specification may lead to higher project costs, particularly if the selection is proprietary. On the other hand, if the market only has one vendor, or if there is broad variation in quality between vendors, or if the designer needs to match an existing piece of equipment, there may be no substitute for a tight, even proprietary selection.

Even though it is harder to write, a well-written performance specification may make it easier to evaluate subsequent submittals and can provide a clear basis for acceptance or rejection. *Well-written* means that the specification is explicit and as unambiguous as possible.

Government agencies often require that specifications be performance style, generic, nonproprietary, open to any product that can meet the criteria. This approach requires a very tight description to control quality, but must avoid listing proprietary singularities of given vendors to preserve the openness of the bidding. In writing generic specifications, the designer must be careful not to be an inventor, to specify something that nobody makes. This can happen when the designer lists the favorite aspects of several different sources of a similar product. This condition often results in disputations and higher costs.

Vendor-specific, or-equal specifications are easier to write because the product defines itself to some extent. If semiopen bidding is required, the specifications usually need to list three or more vendors of acceptable product.

The generic description with a listed vendor but wide-open bidding is used by some design offices to establish a known level of quality while leaving room for substitutions.

13.4.5 Automated word processing

Modern computer-based word processors with mass storage media and sophisticated software are a specification writer's dream come true. Previous work is immediately accessible for new work. New specification information is often delivered in disk format. Changes to the work can be made immediately, if not sooner, and the updated file can be printed within moments.

The qualifier in all this is that the specification writer is well off to have personal keyboarding capability. Changes can be made on the screen in the same time it takes to mark up a copy for a clerk to change, and the proofreading can be done at the same time. Large blocks of raw information are still easily delegated to the secretaries.

Some word processing programs have the ability to change paragraph numbering formats. This allows a specifier to readily adapt to formats of other offices. Computerized specifications are relatively simple to install and use. A personal computer with a good word processing program and large disk storage is required. Once the specification sections have been written and stored, they can be sorted, assembled, customized, and printed as required for each project. The techniques are now well known and readily available. The principal cost is for the initial purchase or writing and storing of the standard specification paragraphs and for the regular updating required to keep up with developing technology.

13.5 Summary

Valid contracts require a scope of work, a time of delivery, and a compensation for service. The best system designs will be unsatisfactory if they are not communicated properly to the contractor through the design documents, drawings, and specifications.

Document preparation is both an art and a science, but it can and should be mastered by HVAC system designers. The drawings and specifications complement each other, complete the design, and are part of the legally binding commitment of the contractor or vendor to the owner. Consistent drawing and specification formats and style help make a clear, concise directive to the contractor for required work.

References

1. ASHRAE Handbook, *1997 Fundamentals,* Chap. 34, "Abbreviations and Symbols."
2. ASHRAE Handbook, *1995 HVAC Applications,* Chap. 36, "Computer Applications."
3. Construction Specifications Institute, *MasterFormat.*

14

After Design: Through Construction to Operation

14.1 Introduction

While drawings and specifications in a sense are the designer's end product, this is not really the case. From the owner's point of view, the built and operating facility is the only real product. All other activity is only a prelude to the real thing. To truly succeed, the designer must follow the design through bidding and construction to start-up and eventual operation.[1]

Some parts of the in-construction work and project closeout work with transition of the project to the owner are called *system commissioning* and are assigned by specification to the contractor. Such work may include first-run inspections; preparation of operating and maintenance manuals; testing, adjusting, and balancing; and instruction of owner's operating personnel. But getting the project off on the right foot with the contractor, being available during construction, and helping the owner take over and get the project running well are also of serious concern to the project designer.

14.2 Participation during Construction

Designer participation during construction and beyond varies greatly with design office practice and with the owner's desires. Some design offices have one or more individuals dedicated to construction review in service to the owner. Some owners have their own people watch the job during construction, relieving the design office of any possible involvement, except to clarify a conflict in documents or a design error.

14.2.1 Bidding period

When the project design is completed, drawings and specifications are given to contractors with a request to prepare a proposal, to submit a bid. During the bidding period, the bidders may identify missing information and discrepancies or errors in the documents. The owner may identify additional services or installations which are wanted as part of the contract. In each case, the designer will prepare clarifying information which is added to the documentation by an addendum. Addendum information should have the same character and quality as the original documents.

When the contractors submit their bids, the bids may be opened publicly and read aloud, as in the case of most government institutions, or they may be opened privately, as is more common in industry and private enterprise. Public bid openings are often quite exciting if there are several bidders and if the bidding is quite close. Thousands, sometimes millions, of dollars of work for a contractor may hang on just a few dollars' difference in the bid offering.

As an aside, bidding is the marketing effort for the typical contractor and is an established mechanism for finding the right price for a scope of work. Preparation of a bid for a project is typically time-consuming and may involve considerable expense for a bidding contractor. Where several contractors with many subcontractors are involved in a take-off, yet only one team can succeed, it is clear that all bidders should be treated with respect for their effort.

14.2.2 Submittal (shop drawing) review

As a part of quality control for a project, the contractor is usually required to show her or his intentions for materials, equipment, and construction technique to the owner prior to ordering, delivery, and installation. The owner's representative, who may be the designer of the work or an associate or another appointed agent, then reviews the contractor's submittals for conformance to the specifications. The contractor is usually required to verify that he or she, too, has checked the submittal for quality and conformance. Even though contractually the submittal is checked only against the drawings and specifications, there are the background questions such as the correct size, coordination between trades (adequate structure, adequate access, correct electrical support). This joint review creates one more opportunity for checking the design in the sense of "measure twice, cut once."

Submittal information is usually stamped with a note of acceptance or rejection, with written clarification of observed deviations or omissions. See Fig. 14.1.

NO EXCEPTIONS TAKEN		AMEND-RESUBMIT	
MAKE CORRECTIONS NOTED		REJECTED-RESUBMIT	
REVIEWED BY:		DATE:	
RECOMMENDED BY:		DATE:	
CORRECTIONS OR COMMENTS MADE ON CONTRACTOR'S SHOP DRAWINGS DURING THIS REVIEW DO NOT RELIEVE THE CONTRACTOR FROM COMPLIANCE WITH CONTPACT DRAWINGS AND SPECIFICATIONS. THIS SHOP DRAWING HAS BEEN REVIEWED FOR CONFORMANCE WITH THE DESIGN CONCEPT AND GENERAL COMPLIANCE WITH THE CONTRACT DOCUMENTS ONLY. CONTRACTOR IS RESPONSIBLE FOR: CONFIRMING AND CORRELATING ALL QUANTITIES AND DIMENSIONS; FABRICATION PROCESSES AND TECHNIQUES; COORDINATING WORK WITH OTHER TRADES; AND SATISFACTORY AND SAFE PERFORMANCE OF THE WORK.			

Figure 14.1 Typical shop drawing review stamp.

Because construction is based on a contract between the owner and the contractor, the designer or submittal reviewer needs to be careful not to assume contractor responsibility in her or his review. In these days of increasing litigation, the language of review has evolved from *approved* to *no exceptions taken,* or some similar noncommittal expression, meaning that the reviewer finds no apparent fault, but still leaves the responsibility for conformance to the contractor. Such fancy footwork notwithstanding, material will occasionally show up on the job which is not really what was wanted or needed. Rejection at such a late date is awkward and sometimes embarrassing, but it is better handled sooner than later. The designer or reviewer needs to be acquainted with the design intent in this phase of the work.

14.2.3 Work oversight and field review

The designer can only *expect* what the designer *inspects!*

Periodically during construction the project designer or an assigned representative should visit the project to review the work accomplished and to answer questions that may arise. Sometimes a clarifying sketch or an explanation of design intent is all that is needed. Sometimes the designer will want to check material and equipment at the job site for conformance to documents and to submittal information. This is done to confirm to the owner that the project is being installed in a correct manner. Given the adversarial nature of fixed-cost construction (the owner wants the maximum for his or her dollars, the contractor wants to maximize the dollars left over), there is an element of fairness and integrity required of the designer in evaluating adequate contractor response to the document or contract requirements. Construction review is shaky ground for a neophyte. Experience at the hands of a good mentor is invaluable.

14.2.4 Change orders

In the process of construction, nearly always a condition will arise that is inadequately or incorrectly defined by the contract documents. Hopefully, the condition will be encountered before the constraints are cast in concrete or fabricated in steel. Upon identifying the problem, the construction team—designers and constructors—will seek a solution. Often an adjustment can be made which incurs no additional cost to the contractor, and the work proceeds. Sometimes correction of the problem creates additional cost and effort for the contractor, who then seeks added compensation. Such is granted by *change order* to the contract. A change order involves a documented scope of work, a price, and a time, and it becomes part of the contract when it has been agreed to by all parties. The pricing mechanism is sometimes awkward since the element of competitive bidding is gone. Even as some owners will try to obtain more service than the documents truly define, some contractors will seek compensation beyond the value or cost of the added work. In a field review, the designer must work hard to see that equity is maintained. When a design error is involved, the contractor is not interested in covering the cost, and some owners become an immediate designer's adversary. Design fees are typically inadequate to provide contingency funds, even for small items. Errors-and-omissions insurance protects against major lawsuits, but there is a cost range where designers must fend for themselves. Fortunate is the designer who works with an owner who realizes that no set of construction documents is perfect, that 2 to 3 percent of basic cost for added clarification is reasonable, and that openly working through problems is better than trying to hide or barter them away.

14.3 System Commissioning

The *commissioning* of an HVAC system is an old idea to some; yet generally it is a new concept, and those who now use the term have not yet arrived at a clear, concise definition. System commissioning certainly includes all the elements of *testing, adjusting,* and *balancing* (TAB) as well as training of operating and maintenance personnel. It may also cover continued observation of system operation as seasons change, with possible readjustment of controller gains as loads vary.

The basic idea is that the designer and contractor do not simply turn the system over to the owner and walk away without giving some assurance that things really do work as designed.

As the construction work comes to a conclusion, it is time to get all equipment and systems into operating condition and then make the transition of responsibility from contractor to owner. This usually involves a number of steps and many elements of the construction team. The designer acts in an advisory role to make sure that all aspects of the design intent are satisfied.

14.3.1   First-run inspection

There is an assumption that the basic work of each trade is complete, that power is available to run equipment, and that devices can be placed in operation without fear of damage. All equipment is verified as ready to run with shipping stops and protective wrap removed, protective covers in place, surfaces cleaned of construction spatter, rotating parts able to move freely, etc. Valves and dampers are placed in a normal position; fluid-handling systems have been flushed clean and pressure-tested. Insulation may or may not be finished, but work is at least in progress. Control installations are complete and may have a first-time setup.

With all components essentially there, the commissioning team goes from one device to the next, verifying basic function. Motors are jogged to verify rotation; fans and pumps are run for enough time to see that things are all right, that power is available where needed, and that automatic valves and dampers will stroke.

The first run does not require calibration or test of capacity or adjustment. It merely confirms that the systems are intact, ready for tuneup and test by the specialist.

14.3.2   Testing, adjusting, and balancing

With all systems in place and all devices capable of operation, the TAB team takes over to prepare the work for final delivery to the owner. This team is best managed by the general contractor, who has ultimate responsibility for the quality of the project and who can assign the work of all subcontracts. The work of the team then usually proceeds under the direction of an individual, or firm, independent of but working with the mechanical, sheet-metal, control, and electrical contractors, who can operate and adjust the systems without vested interest and with broader experience than any one of the specialized contractors might have. Such a team leader is often referred to as a TAB contractor but the need is broader than just TAB. What is needed is an operator who can place a system in operation and tune it up to optimal operating condition.

The system designer can be a help as well as a referee in the TAB work. The designer knows the intent of the design and the anticipated interrelationship between components. Some humility is required, however, for many—if not most—designers have little actual operating experience and may not know how to manipulate and operate systems as well as do some of the old hands.

It is helpful to have a member of the owner's eventual operating staff on hand for the start-up and TAB work. Early familiarity with and knowledge of the system minimizes questions later on. Some contractors, even designers, are reluctant to expose unfinished work to the owner's staff because the loose ends of a major work are sometimes embarrassing, but the owner's people are often a great help with corrective ideas.

The work of the TAB team proceeds from one system to the next, covering air handling, fluid handling, central plant, and controls. There are established procedures for TAB work for nearly every type of system. ASHRAE, the Associated Air Balance Council (AABC), the National Environmental Balancing Bureau (NEBB), and other agencies all have documentation for test and balance procedures. Note that the objective is not just to make the systems perform as the drawings say, but now to actually *do the job*. Design calculations are at best a prophetic estimate of what will be needed based on an owner's expressed intended use and an architect's designed structure. Real life may vary somewhat from the anticipated. The TAB team should be instructed and be willing to adjust to final needs. This does not mean willing to rebuild, unless defects are encountered, but to adjust within the capacity range of the systems.

One of the final products of the TAB work is a report which documents the work accomplished. The report usually includes reduced-scale drawings that show the nomenclature and location of all points of test and adjustment and shows in tabular form the quantities sought (design performance), the initial condition encountered, and the final condition attained, with analysis of any discrepancy and a report of corrective action. Part of the report may reflect work with the control contractor in verifying the calibration and proper function of each control device within the system context.

At the conclusion of the TAB work, the designer and owner might ask the TAB team to reread selected measurements to verify the integrity of the work. See Ref. 2 for further discussion of this topic.

14.3.3   Substantial completion

All the preceding work should bring the project to a point of *substantial completion*. This is usually defined as the time when the owner takes beneficial occupancy of the project with a "punch list" of items which the contractor has yet to complete. Substantial completion usually initiates the warranty period. Owners are recognizably reticent to award substantial completion; contractors are overly anxious to come to this point. If a project has a deadline with damages for late delivery or a bonus for early delivery, the potential consequences can be very important. The designer needs to be fair and helpful to both the owner and the contractor at this stage of work.

Many owners make a big mistake in failing to recognize that they should proactively take over the project, on the first day of beneficial use, for operation, maintenance, and improvement purposes. No project is perfect on inauguration day. Wear and tear begin the first time a device is turned on. For a system to still be well maintained and functional in 10 years' time it must be well maintained in the first year. Warranty ensures the quality of construction. It has nothing to do with operation and maintenance.

14.3.4 Operating and maintenance manuals and training

The preparation of these manuals and the training of operating and maintenance (O&M) personnel are usually, and rightly, the responsibility of the contractor. As part of the commissioning process, the designer and the owner's representative, with the operating personnel, should go through all the operating procedures described in the manual. Emergencies should be simulated to check emergency response procedures. The manuals should become a lifetime resource for the operators of the building and its systems.

14.3.5 Observation of operation

Many specifications require the HVAC contractor to certify in writing that the HVAC systems and controls are operating as designed and specified. This should not be sufficient to satisfy a competent, professional designer. There is no substitute for personal observation of system operation, not for just a few hours but for several times under varying load and climatic conditions. Among other things, this will expose any need to adjust controller gains and set points as system gains change with the load (see Chap. 8).

14.4 Summary

If the "proof of the pudding is in the eating," then the proof of a successful HVAC design is in the eventual operation and performance of the system in the hands of the owner. The designer, or someone he or she assigns, needs to follow the project from bidding and contract, through construction, through startup, to final takeover and operation by the owner. As elements of quality control, the designer needs to verify adherence to documents through submittal review and by in-construction review. And the designer needs to support the transition of the project from contractor to owner during start-up, test and balance, and instruction of the owner's personnel. With some inexperienced operating personnel, the designer may need to continue this effort into the operating period for a time. The bonus reward of all this effort is a functional project which serves as a credential in the search for the next design assignment.

References

1. ASHRAE Handbook, *1995 HVAC Applications,* Chaps. 32–39, regarding building operation and maintenance.
2. Ibid., Chap. 34, "Testing, Adjusting, and Balancing."

15

Technical Report Writing

15.1 Introduction

An HVAC designer is almost certain to be faced with the need to write reports. The ability to produce an organized, understandable, and succinct report will go far toward establishing credibility with both superiors and clients.

Writing a good report requires the same attributes needed for success in any other area: understanding of basic principles, planning and organization, and careful investigation.

Most reports are written in response to a problem, and describe the results of a study of that problem and its possible solutions. Thus, the problem-solving pattern outlined in Sec. 1.2 applies.

15.2 Organization of a Report

Figure 15.1 shows a typical report outline. There may be additional topics, but these are the essential items in any report.

Section 1 is sometimes called an *executive summary* because it is written for the executive who is not concerned with details but needs a synopsis.

The *authorization* describes the purchase order, letter, or other legal basis which led to the production of the report.

The *scope* is usually a part of the authorization and describes the objective and problems to be addressed. It is really a specification, and often provides the basis by which the report is judged for completeness and payment is authorized. Definition of the scope is the first step in the problem-solving process, and the value of the final report will relate strongly to the clarity of this definition.

SECTION	TOPIC	PAGE
I	SUMMARY AND RECOMMENDATIONS	

	A. Authorization	I-1
	B. Scope	I-1
	C. Acknowledgements	I-1
	D. Summary	I-1
	1. Existing Conditions	
	2. Alternatives	
	3. Cost Analysis	
	E. Recommendations	I-2
	Figures	
	Tables	

II	EXISTING CONDITIONS	
	A. General Description	II-1
	B. (Specifics)	
	C. (More specifics, etc.)	
	Figures	
	Tables	

III	ALTERNATIVES	
	A. General	III-1
	B. Alternative No. 1	III-2
	C. Alternative No. 2, etc.	
	Figures	
	Tables	

IV	COST ANALYSES	
	A. Basis	IV-1
	B. Cost Analyses	IV-1
	1. Alternate No. 1	
	2. Alternate No. 2, etc.	
	Figures	
	Tables	

| V | APPENDIX | |

Figure 15.1 Typical report outline and table of contents.

The *acknowledgments* provide recognition for those in the owner's organization who provided information for the report and assisted the report writer.

The *summary* is a synopsis of the material contained in the rest of the report and follows the same sequence.

The *recommendations* are based on the summary and represent the report writer's professional opinion. Ideally, the recommendations should be completely objective. In practice, they are always influenced to some degree by subjective considerations. The true professional will recognize and allow for some subjectivity.

The whole of Section 1 should not exceed two or three pages, one or two figures, and a summary table.

The rest of the report is composed of detail, as complete as possible but not verbose. Note that the summary follows the same sequence as the body of the report. Lengthy data summaries and calculations should be placed in an appendix. This material exists only to prove the conclusions of the report.

15.3 Writing with Clarity

Technical writing deals with factual data in objective, nonemotional ways for the purpose of presenting information on which to base decisions. Most technical reports are written not only for technical people but also for management and financial people, who may understand very little about the technical aspects. It is therefore necessary to use technical detail but to express it in simple terms, so that the audience can at least understand the principles involved.

The following suggestions should be helpful:

1. Technical terms should be defined or made clear in context.
2. Proceed stepwise, from the known to the unknown. The use of syllogisms, as employed in logical proof, is sometimes helpful. A syllogism consists of two *premises* or facts which, taken together, lead to a conclusion. When the steps become very detailed, complex, and lengthy, it may be desirable to simply say, "It can be shown." The proof can be included in the appendix, if needed.
3. Make sentences as simple and direct as possible. Avoid overly complex sentences with many phrases or clauses.
4. Writing should be factual. Any opinions expressed should be based on the writer's proven expertise and should be identified as opinions. Avoid subjective and emotive adjectives. These have no place in technical writing. Also avoid phrases like *it is clear that* or *it is apparent*. Do not use clichés.
5. Follow accepted grammatical rules concerning punctuation, syntax, agreement, and case. Grammatical errors reduce credibility.

15.4 Use of Tables and Figures

Figures and tables are useful in clarifying and summarizing a report. Often they can be used to capsule the report, with the text serving as an explanation of the data.

Figures can include any appropriate material, but usually consist of schematics, bar and line graphs, or pie charts. Tables can be used to summarize data that may take several pages of text to describe. Photographs are very useful in describing existing conditions.

These are the basic rules in preparing figures and tables:

1. Neatness counts. Good line work and lettering are required. The best written report may suffer from poor graphics.
2. Keep figures and tables as simple as possible. Avoid the use of data and detail which are not pertinent.
3. Organize tabular data in the same sequence as that used in the text. A good table can stand alone, through the use of clarifying notes, with little or no text required.
4. While it can be helpful, color is seldom used or required in engineering reports. Contrast can be achieved by shading, crosshatching, and variations in line width. The many stick-on halftone patterns used by commercial artists are helpful here.

15.5   Printing and Binding

The relatively recent developments in word processing capability, through the use of software programs run on a personal computer, have brought near publishing-house-quality document preparation to the engineering office, large or small. The quality of the final-copy reading material still depends on the ability of the typist and artist and on the nature of the software and the printer used, but the character of most material has been advanced by these modern techniques.

With recent word processing software, it is possible to incorporate graphics into the body of the text, to scan previously prepared material including photographs, and to print in high-quality formats with almost endless variations of type style and size. Text can be right- and/or left-justified or can be centered as desired.

Many offices use standard formats and forms for report presentation. Figure 15.2 illustrates a report form sheet which may be used to include text, tables, and figures. A design firm may be known to some extent by the quality of its report formatting and presentation.

Document quality is further defined by the printer used to make the camera-ready copy. Dot matrix printer technology has evolved to near typeset quality, but so-called laser printers go one step further and make a product almost indistinguishable from set type. Both dot matrix and laser printer manufacturers have been developing color-capable products, which may become the standard over time.

The reproduction method for office reports has evolved to photocopying or offset printing. Mimeograph and ink stencil work are in the past. The office copy machine has become so economical and conven-

ROGER W. HAINES   ★   CONSULTING ENGINEER   ★   LAGUNA HILLS, CALIFORNIA

Figure 15.2 Report page format.

ient that it dominates most small-volume copy work. Larger runs or high-profile reports are candidates for outside reproduction services.

Many clients and many engineering offices have standard report formats and "report form" sheets which include the company logo and borders for report pages. See Fig. 15.2.

There are many binding methods available, ranging from three-ring looseleaf styles to mechanical clip systems to standard bound-book techniques. Almost any binding system which delivers a neat and desirable end product may be satisfactory, with cost and client expectation being considerations. Many engineering offices use heavyweight covers, preprinted with the company logo and other data, with the report title overprinted in publication. The organization of the report in printing and binding should follow the outline of Sec. 15.2, as shown in Fig. 15.3. Figures and tables for each section may be embedded with the related paragraphs or grouped at the end of the section.

15.6 Letter Reports

Many preliminary and simple reports do not require the full formal report treatment. A letter report is like the summary section of the formal report. It should include all or most of the elements of the formal report and should be arranged in a similar fashion. It may even include one or two simple figures or tables. A letter report should be given the same careful study and planning as a formal report.

15.7 Summary

The writing of a good report is not a simple task. Like any other kind of writing, it takes practice, time, and effort. The neophyte writer may

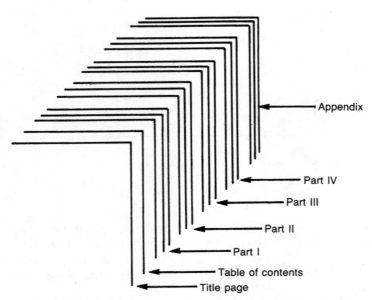

Figure 15.3 Report organization. Note that figures and tables for each part of the report go behind the text for that part.

read other reports, noting their strengths and weaknesses and thereby learning from them. Technical writing, like all other forms of communication, can serve to enhance or inhibit professional advancement and reputation. Recognize that good report writing is an art as well as a science. As art, there can be variations in format, in vocabulary, in artistic presentation. But every report should clearly identify the issues to be addressed and should clearly, succinctly, and accurately lead the reader from the problem to the solution. The writing style and presentation should enhance and give credibility to the report rather than detract from it.

Reference

1. R. W. Haines, *Roger Haines on Report Writing,* McGraw-Hill, New York, 1990.

Engineering Fundamentals: Part 1

Fluid Mechanics

16.1 Introduction

Fluid mechanics is a fundamental branch of civil, chemical, and mechanical engineering which deals with the behavior of liquids and gases, particularly while flowing. This chapter provides a brief review of the vocabulary and fundamental equations of fluid mechanics, and reminds the HVAC designer of the scientific principles underlying much of the day-to-day applied science calculations. See Ref. 1 for a fluid mechanics text for additional detail.

16.2 Terms in Fluid Mechanics

Many words are used in fluid mechanics which carry over into thermodynamics and heat transfer. A few of the fundamental terms are defined here for review.

Fluid: A liquid or a gas, a material without defined form which adapts to the shape of its container. Liquids are essentially incompressible fluids. Gases are compressible. Newtonian fluids are those which deform with a constant rate of shear. Water and air are newtonian fluids. Nonnewtonian fluids are those which deform at one rate of shear to a point and then deform at a different rate. Blood and catsup are nonnewtonian fluids.

Density ρ: Mass per unit volume, lbm/ft^3.

Viscosity μ: Resistance to shear, force · time/(length)^2.

Pressure P: Force per unit area.

Velocity V: Distance per unit time, ft/min, ft/s.

Laminar flow: Particles slide smoothly along lines parallel to the wall. Resistance to flow varies directly with the velocity.

Turbulent flow: There are random local disturbances in the fluid flow pattern about a mean or average fluid velocity. Resistance to flow is proportional to the square of the velocity.

Reynolds number Re: A dimensionless number relating fluid velocity V, distance as a pipe diameter D, and fluid viscosity μ:

$$\text{Re} = \frac{DV\rho}{\mu} \quad \text{for a pipe}$$

Reynolds numbers below 2100 generally identify laminar flow. Reynolds numbers above 3100 identify turbulent flow. Reynolds numbers between 2100 and 3100 are said to be in a transitional region where laminar or turbulent conditions are not always defined.

Turbulent flow is desirable in heat exchange applications, while laminar flow is desired in clean-room and low-pressure-drop applications.

Cavitation: When the local pressure on a fluid drops below the vaporization pressure of the fluid, there may be a spot flashing of liquid to vapor and back again. Such a condition can occur with hot water at the inlet to a pump. Such activity is called *cavitation*. It can be harmful to the pump through local erosion and interference with flow. Cavitation often sounds like entrained gravel or little explosions at the point of occurrence.

16.3   Law of Conservation of Mass

Fluid mechanics starts with the *law of the conservation of mass* (see Fig. 16.1), which states, "Matter can be neither created nor destroyed." This gives us a chance to set up an accounting system for all flows in a system and to know that our accounts of inflows, outflows, and storage must balance at every point in the system.

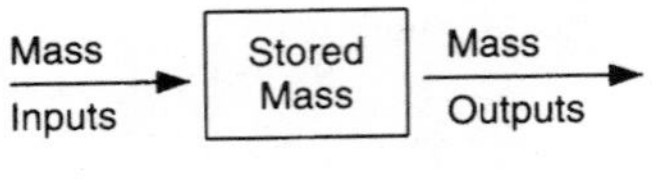

Mass ──→ Stored   Mass ──→
Inputs   Mass   Outputs

Figure 16.1   Conservation of mass.

16.4 The Bernoulli Equation (Law of Conservation of Energy)

Fluid mechanics studies focus on the Bernoulli equation (Navier-Stokes equations in more advanced mathematical analysis) which relates changes in energy in a flowing fluid (kinetic energy, potential energy, energy lost to friction, and energy introduced or removed) in terms of heat and work. If the study is observed over time, then all the terms are time-based and the work term is observed as power. The equation, similar to the conservation-of-mass equations, states that energy is conserved, that it cannot be destroyed, that it can be accounted for. See Fig. 16.2.

$$\frac{V_1^2 - V_2^2}{2} + g(h_2 - h_1) + \frac{1}{\rho}(P_2 - P_1) = \text{work}_{\text{in}} + Q_{\text{in}}$$

where V = velocity, g = gravitational constant, h = elevation, P = pressure, ρ = density, and Q is heat energy.

If this discussion seems somewhat theoretical, there are two brief equations derived from the above which are extremely useful in HVAC calculations. They are equations for estimating the theoretical horsepower of a fan or pump given the flow rate of water or air, the pressure drop to be overcome, and the nominal efficiency of the fluid-moving device.

For water:

$$\text{bhp} = \frac{\text{GPM} \times \text{head}}{3960 \times \text{eff}}$$

where GPM = water flow rate in gallons per minute, head = pressure rise across the pump in feet of water, eff = pump operating efficiency at calculation point, as a percentage, and the constant for water pumps is derived as follows:

$$\text{Constant} = \left(550 \frac{\text{ft} \cdot \text{lb}}{\text{s} \cdot \text{hp}}\right)(60 \text{ s/min})\left(\frac{1 \text{ gal}}{8.33 \text{ lb}}\right) = 3960 \left(\frac{\text{GPM} \cdot \text{ft}}{\text{hp}}\right)$$

For air:

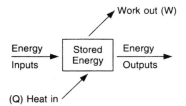

Figure 16.2 Conservation of energy.

$$\text{bhp} = \frac{\text{CFM} \times \text{SP}}{6356 \times \text{eff}}$$

where CFM = airflow rate in cubic feet per minute, SP = static pressure rise across the fan in inches of water, eff = fan operating efficiency at calculation point as a percentage, as for pumps, and the constant for fans is derived as follows:

$$\text{Constant} = \left(550\ \frac{\text{ft} \cdot \text{lb}}{\text{s/hp}}\right)(60\ \text{s/min})\left(\frac{1\ \text{ft}^3}{62.3\ \text{lb}}\right)(12\ \text{in/ft})$$

$$= 6356\left(\frac{\text{CFM} \cdot \text{in}}{\text{hp}}\right)$$

In each case, the derivation of the constant term is shown to illustrate how keeping track of units can help to solve problems if the constant is forgotten or if the information is given in other units. Note that the liquid pumping horsepower will increase with higher-density liquids and can be accommodated by multiplying the equation by the relative density of the fluid pumped compared to water. The same is true of the air equation. CFM is assumed to be for standard air (0.075 lb/ft^3 at 70°F). If heavier gases or hot thin air or air at altitude is being handled, the equation must be corrected by the relative density. The air formula is only valid for a near-atmospheric-pressure condition (14.7 lb/in^2 gauge $\pm$ 1 lb/in^2, say). More variance than that invokes principles of compressibility, which adds complexity to the calculation.

Fluid mechanics addresses friction loss in piping and duct systems. It requires attention to differences in elevation for pumping of "open" systems and teaches us to recognize static-pressure concerns in both closed and open systems.

Static pressure problems with standing columns of air or other gas nearly always are associated with buoyancy effects of warmer versus colder air, as in the induced *draft* of a chimney or the wintertime *stack effect* of a medium-rise or high-rise building.

16.5 Flow Volume Measurement

There are several different methods for measuring flow volume per unit time.

- *Direct liquid measurement:* This involves a mechanical measurement such as of the time required to fill a container of known volume or of observing the portion of a container filled in a given time.
- *Venturi meter:* A venturi is a smooth but constricted tube with a pressure tap at the wide point and the necked point. Since there are

no other effects, the change in static pressure from the wide to narrow sections can be used to determine velocity and flow volume.

- *Orifice plate meter:* An orifice plate is a plate with a carefully defined circular opening with a uniform edge characteristic. Laboratory measurement can identify a pressure drop across the plate for various flow rates. When the plate is installed between flanges with pressure taps, the field-measured pressure differential can be compared with the laboratory data to determine the flow rate (see Fig. 8.20).

- *Impact tube meter:* The total pressure in a flowing fluid is comprised of a velocity pressure component and a static or background pressure component:

$$P_t = P_{\text{vel}} + P_{\text{static}}$$

If a tube is directed into the flowing fluid in the opposite direction it will read total pressure P_t. If a second tube is inserted parallel to the flow so that it sees no velocity impact, it will read the local static pressure. The difference between the total pressure and the static pressure is the local velocity pressure, and it can be converted to velocity for any given fluid. In the turbulent region, the velocity pressure is proportional to the square of the velocity.

- *Equipment as a meter:* Almost any device set in a moving fluid stream can be used as a coarse flowmeter since the pressure drop across the element is proportional to the square of the velocity. Heat exchangers are often calibrated for the flow rate. Cooling coils can be read on both the airside and waterside.

16.6 Summary

Fluid mechanics issues show up in nearly every aspect of HVAC systems design. Pumps, fans, coils, heat exchangers, refrigeration systems, process systems, boilers, deaerators, water softeners and treatment systems, water supply and distribution, building plumbing and fire protection, etc., are all grounded in the physics of fluid mechanics. There is a direct analogy between electrical concepts and fluid flow concepts. Consider Ohm's law—E (voltage) = I (current) $\times R$ (resistance)—and compare voltage to pressure, current to fluid flow, and resistance to friction. Further recognize that storage of fluid in a tank is analogous to storage of electrons in a capacitor or inductive coil. An understanding of fluid mechanics leads to a rudimentary understanding of some aspects of electricity.

For further development of this topic, the reader is referred to the multitude of fluid mechanics textbooks or to the ASHRAE Handbook

Fundamentals, which has a significant summarized presentation of the topic. For a hands-on data and reference book focused on hydraulics, see Ref. 2.

References

1. ASHRAE Handbook, *1997 Fundamentals,* Chap. 2, "Fluid Flow."
2. Ingersoll-Rand, *Cameron Hydraulic Data,* 17th ed., Woodcliff Lake, N.J., 1988.

Engineering Fundamentals: Part 2

Thermodynamics

17.1 Introduction

Thermodynamics is an aspect of physics which deals with the energy characteristics of materials and with the behavior of systems undergoing changes in system energy levels. The field of thermodynamics is quite broad as well as deep, and can vary in presentation and in application from relatively simple to very complex. For the purposes of this book, a relatively simple presentation is adequate. The concepts of thermodynamics presented here are common to virtually all textbooks and reference books. For those who want greater detail, Chap. 1 of the ASHRAE Handbook *Fundamentals* is one presentation written at a college upper-division or graduate-student level.[1,2]

17.2 Thermodynamics Terms

One problem with understanding thermodynamics is that the basic terms *energy* and *entropy* are defined in relatives rather than absolutes.

Energy can be reduced to the concepts of heat and work and can be found in various forms: potential energy, kinetic energy, thermal or internal energy, chemical energy, and nuclear energy.

Potential energy is the energy of location or position of a mass in a force field. A body or a volume of water at the top of a hill has potential energy with respect to the bottom of the hill.

Kinetic energy is the energy of motion and is proportional to the square of the velocity as well as to the mass of the moving body.

Internal energy has to do with activity within the molecular structure of matter, and is typically observed with temperature measurements.

Chemical energy is related to the relationships between molecules in chemical compounds. When different molecules combine in chemical reaction, they may give off heat (exothermic reaction) or require heat (endothermic reaction).

Electric energy is related to electrons moving along a conductor.

Nuclear energy is the energy of atomic relationships between the fundamental particles of matter. Nuclear fission and fusion are reactions which release stored nuclear energy.

Heat is observed as energy in motion from one region to another resulting from temperature difference.

Work is an energy form which can be equated to the raising of a weight. This may be mechanical work, such as moving a mass in a force field, or it may be flow work, such as moving a liquid against a resisting force.

Enthalpy is a term used with energy units that combines internal energy with a pressure/volume or flow work term.

Property is a measurable characteristic of a system or a substance. Temperature, pressure, and density (the inverse is the specific volume) are all properties. The different kinds of energy, enthalpy, and entropy are all considered properties.

Temperature is a term used to quantify the difference between warm and cold or the level of internal energy of a substance. The original numerical designations were based on the difference between the freezing point and boiling point of water. The Celsius scale defined the difference in terms of 100 units with 0 as the freezing point and 100 as the boiling point. The Fahrenheit scale uses the freezing point of a salt solution as the 0 point with pure water freezing and boiling at 32 and 212°F, respectively. The lowest possible temperature, the condition at which molecular motion ceases, is called *absolute zero*. The absolute scale which uses the Celsius increment is called the *Kelvin scale*. It places absolute zero at $-273°C$, or the ice melting point of water at $+273K$. The absolute scale which uses the Fahrenheit increment is called the *Rankine scale*. It places absolute zero at $-460°F$, or the ice melting point of water at $+492°R$. There is no upper limit to a possible temperature.

17.3 First Law of Thermodynamics

The first law of thermodynamics sounds like the law of the conservation of mass, with different vocabulary. If the law of mass conservation asserts that *matter* can be neither created nor destroyed, then the first law of thermodynamics states that *energy* cannot be created or destroyed. This implies that the various forms of energy may be converted, one to another. It means that we can account for all energy conversions in a system with accuracy.

$$\text{Energy in} - \text{Energy out} = \text{Change in stored energy}$$

Entropy is used to define the unavailable energy in a system. In another sense, entropy defines the relative ability of one system to act on another. As things move toward a lower energy level, where one is less able to act upon the surroundings, the entropy level is said to increase. If we look at the universe as a whole, things are running down, so the entropy of the universe is said to be increasing.

17.4 Second Law of Thermodynamics

There are two classical statements of the second law of thermodynamics. The first was expressed by Kelvin and Planck: *No (heat) engine whose working fluid undergoes a cycle can absorb heat from a single reservoir, deliver an equivalent amount of work, and produce no other effect.* To understand this statement, recognize that for energy to be available at all, there must be a region of high energy level compared to a region of lower energy level. Useful work must be derived from the energy that would flow from the high potential region to the lower potential region. But 100 percent of the energy cannot be converted to work. If it were, the process would be dealing only with a single energy region, in violation of the Kelvin-Planck statement. The reality of this statement can be seen in our inability to extract energy from the environment unless there is a second, colder region to relate to.

The theoretical maximum efficiency η of a heat machine working between two energy regions is defined in terms of temperatures on an absolute scale as

$$\eta = \frac{T_H - T_L}{T_H} = 1 - \frac{T_L}{T_H}$$

where T_4 is the temperature of the high energy region, and T_L is the temperature of the low energy region.

As the temperatures approach equilibrium ($T_L = T_H$), the process efficiency tends toward zero.

The second statement of the second law is credited to Clausius, who said, "No machine whose working fluid undergoes a cycle can absorb heat from one system, reject heat to another system and produce no other effect."

Both statements of the second law place constraints on the first law by identifying that under natural conditions, things (including energy) run downhill. We must expend energy to make it happen otherwise. It takes energy to drive cold to hot, e.g., a refrigerator. It takes energy to raise a weight against gravity. A corollary is that there is no such thing as a perpetual-motion machine. We cannot get something for nothing. This means that in all the energy balances of the first law, we know that some things will not happen unless we expend or give up something to make them happen.

- *Example:* A heat resource cannot be fully converted to work. Work can be withdrawn from heat moving from a high-energy region to another, but not completely. This makes work (kinetic energy, electric energy, etc.) more valuable than conventional heat resources. Hence it is generally better to use fuels for heating and electricity for power, rather than to use power for resistance heating.

- *Example:* Refrigeration is a process of moving heat (thermal energy) from a cold region to a warmer region. Since this is counter to the nature of things, which is to run downhill, we must expend energy to make it happen (i.e., power to the compressor; steam or fuel to the absorption chiller).

17.5 Efficiency

Efficiency is simply defined for an energy conversion process as how much is obtained compared to how much was expended. For example, in a gas-fired hot water boiler, of every unit of fuel burned, approximately 80 to 90 percent is transferred to the circulated hot water while 10 to 20 percent goes up the chimney as products of combustion and uncaptured heat. The efficiency of the boiler is 80 to 90 percent.

Note that a heat transfer process in a heat exchanger does not utilize the word *efficiency*. All the heat taken from one side of the exchanger winds up on the other side. Even if the exchanger is fouled, there is an energy balance. However, a fouled exchanger will not transfer as much heat as a clean one. A comparison of fouled capacity to clean capacity is sometimes called *efficacy,* or effectiveness, but not *efficiency*.

17.6 Coefficient of Performance

The *coefficient of performance* (COP) is defined as

$$\text{COP} = \frac{\text{useful heat moved or obtained}}{\text{energy required to drive process}}$$

For a refrigeration cycle, the useful heat is the refrigeration effect

$$\text{COP}_{\text{cool}} = \frac{Q_{\text{evap}}}{Q_{\text{in}}}$$

In the classical sense, similar to the definition of efficiency, there is a theoretical limit to the COP defined in terms of the temperature (low) of the cold region T_L related to the temperature (high) of the warm region T_H

$$\text{COP}_{\text{max}} = \frac{T_L}{T_H - T_L} \qquad (17.1)$$

For a building chiller evaporating at 40°F (500°R) and condensing at 100°F (560°R), the maximum COP would be

$$\text{COP}_{\text{max}} = \frac{500}{560 - 500} = 8.3 \qquad (17.2)$$

The actual COP of these machines is on the order of 4 to 6, which reveals the inefficiencies of the real world. Studying the defining equation reveals that it takes more energy to refrigerate at lower temperatures, and it takes more energy to drive a process with a higher temperature differential between evaporator and condenser. We need to minimize the differential (often called *thermal lift)* as much as possible for the purposes of energy conservation.

For the heat pump cycle, the energy input becomes a benefit.

$$\text{COP}_{\text{heat}} = \frac{Q_{\text{evap}} + Q_{\text{in}}}{Q_{\text{in}}} = \text{COP}_{\text{cool}} + 1$$

In heat pump systems, the system designer should again try to work with the smallest possible thermal lift to get maximum beneficial effect for least input.

17.7 Specific Heat C_p

The *specific heat* of a substance is the amount of energy it takes to raise a unit mass of the substance by one degree in temperature. For HVAC work, the specific heat of water in the liquid state is 1 Btu/(lb · °F). Water in the solid state (ice) has a specific heat of 0.487 to 0.465 Btu/(lb · °F), which is easy to remember as 0.5 Btu/(lb · °F). Water as a vapor has a specific heat of 0.489 Btu/(lb · °F) which can also be rounded to 0.5 Btu/(lb · °F) as for ice. In a process where energy is added to or taken from a flowing stream of water, the specific heat term evolves to

$$[1 \text{ Btu/(lb} \cdot {}^{\circ}\text{F)}](8.3 \text{ lb/gal})(60 \text{ min/h} = 500 \text{ Btu/[h} \cdot (\text{gal/min}) \cdot {}^{\circ}\text{F]}$$

The heat-carrying capacity of a flowing stream of water is then

$$Q = \dot{M}C_p(T_2 - T_1) = \text{GPM}(500)(T_1 - T_2) \text{ Btu/h}$$

where $\dot{M}$ = mass flow rate
$\quad Q$ = energy transported, Btu/h
$\quad C_p$ = specific heat of water
$\quad T_1, T_2$ = temperature in, temperature out
$\quad$ GPM = fluid flow rate (water), gal/min

The same equation (a variation of the Bernoulli equation) can be adapted for any fluid (such as a glycol antifreeze solution) by substituting the specific heat and density of the substance for that of water.

The same derivation for air, which at standard conditions has a density of 0.075 lb/ft^3 and a specific heat of 0.24 Btu/(lb $\cdot$ °F), yields

$$[0.24 \text{ Btu/(lb} \cdot {}^{\circ}\text{F)}](0.075 \text{ lb/ft}^3)(60 \text{ min/h})$$

$$= 1.08 \text{ Btu/[h} \cdot (\text{ft}^3/\text{min}) \cdot {}^{\circ}\text{F]}$$

$$Q = \text{CFM}(1.08)(T_1 - T_2)\text{Btu/h}$$

17.8 Summary

Thermodynamics is an interesting and valuable study for the HVAC designer. Its principles define the concept of energy and identify which energy processes are possible and which are impossible unless forced. The first law of thermodynamics allows us to count and keep track of energy as if it were money in the bank. The second law of thermodynamics confirms that energy processes run downhill. This knowledge helps an HVAC designer to identify reality among the often overstated claims of overenthusiastic inventors and salespeople. The mathematical relationships of thermodynamics allow the designer to quickly and confidently calculate the energy flows in a process.

References

1. William J. Coad, "Fundamentals to Frontiers (Fundamentals of Thermodynamics)," *Heating/Piping/Air Conditioning*, February 1981.
2. William J. Coad, "Fundamentals to Frontiers (Unavailable Energy)," *Heating/Piping/Air Conditioning*, March 1982.

18

Engineering Fundamentals: Part 3

Heat Transfer

18.1 Introduction

This chapter presents a basic overview of heat transfer fundamentals, particularly as they apply to HVAC. For a detailed, rigorous treatment, the reader should refer to a good college-level text on heat transfer[1] or to the ASHRAE Handbook.[2]

18.2 Heat Transfer Modes

Heat is transferred between any two bodies by one or more of three modes: conduction, convection, and radiation. Thermal *conduction* refers to the direct transfer of energy between particles at the atomic level. Thermal *convection* may include some conduction but refers primarily to energy transfer by eddy mixing and diffusion, i.e., by fluids in motion. Thermal *radiation* describes a complex phenomenon which includes changes in energy form: from internal energy at the source to electromagnetic energy for transmission, then back to internal energy at the receiver. Radiation transfer requires no intervening material, and in fact works best in a perfect vacuum. In accordance with the second law of thermodynamics, net heat transfer occurs in the direction of decreasing temperature. In this text, the Fahrenheit ($°F$) scale is used, or for absolute temperatures the Rankine ($°R$) scale: $°R = °F + 460°$.

18.3   Thermal Conduction

For steady-state conduction in one direction through a homogeneous material, the Fourier equation applies:

$$q = -kA \, dt/dx \qquad (18.1)$$

where q = heat transfer rate, Btu/h
k = thermal conductivity, Btu/(h · ft · °F)
A = area normal to flow, ft^2
dt/dx = temperature gradient, °F/ft

The minus sign shows that heat flow takes place from a higher to a lower temperature.

In HVAC calculations, homogeneous barriers are never encountered—even when the solid barrier is homogeneous, there will be film resistance at its surface, as shown in Fig. 18.1. The heat transfer equation is then modified as follows:

$$q = UA(T_1 - T_2) \qquad (18.2)$$

where U is the overall coefficient of heat transfer per degree of temperature difference between the two fluids which are separated by the barrier. Usually, but not always, U is given in Btu per hour per square foot per degree Fahrenheit. The temperatures and the area A must be in units consistent with those of U.

Various building materials and combinations thereof have been tested to determine the *conductivity* k (Btu per hour per square foot per inch or foot of thickness per degree Fahrenheit) or *conductance C* (for a nonhomogeneous material such as a concrete block, in Btu per hour per square foot per degree Fahrenheit). The tests are made in a

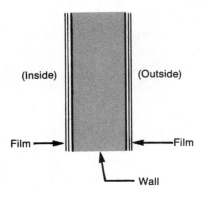

Figure 18.1 Conduction heat transfer through a wall.

"guarded hot box," designed so that heat transfer through the edges of the material is essentially eliminated. The results of these tests are tabulated and presented, with discussion, in the ASHRAE Handbook.[3]

The thermal conductivity k of any material is the reciprocal of its resistance R:

$$k = \frac{1}{R} \tag{18.3}$$

For barriers with material combinations which are not tabulated, the U factor may be calculated from the sum of the individual resistances. The general form of the equation is

$$\frac{1}{U} = R_1 + R_2 + R_3 + \cdots + R_n \tag{18.4}$$

Because resistance is the reciprocal of conductance or conductivity, a more specific form of the equation is

$$\frac{1}{U} = \frac{1}{f_o} + \frac{x_1}{k_1} + \cdots + \frac{x_n}{k_n} + \frac{1}{C_1} + \cdots + \frac{1}{C_n} + \frac{1}{f_i} \tag{18.5}$$

where f_o = outside film conductance
f_i = inside film conductance
x = thickness of homogeneous section with conductivity k

See Ref. 3 for a more detailed discussion. The incremental temperature drop through each element of the barrier is proportional to the resistance of the element. For example, in Fig. 18.1 if the wall is 6-in-thick perlite concrete with a k value of 0.93 per inch, and if the outside and inside film conductances are 4.00 and 1.46, respectively, then the overall U factor is

$$\frac{1}{U} = \frac{1}{4.00} + \frac{6}{0.93} + \frac{1}{1.46}$$

$$= 0.25 + 6.45 + 0.68 = 7.38$$

$$U = \frac{1}{7.38} = 0.136$$

If a temperature difference of 42°F is assumed, based on 72°F inside and 30°F outside, then the temperature gradient can be determined as shown in Table 18.1. This type of calculation is useful in determining the location where moisture condensation or freezing will take

TABLE 18.1   Temperature Gradient Information

Element	Resistance, $\dfrac{\text{h} \cdot \text{ft}^3 \cdot {}^{\circ}\text{F}}{\text{Btu}}$	Percentage of total R	ΔT	Temperature, °F
Outside air				30.0
Outside film	0.25	3.4	1.4	
Outside wall surface				31.4
Wall	6.45	87.4	36.7	
Inside wall surface				68.1
Inside film	0.68	9.2	3.9	
Inside air				72.0
Totals	7.38	100	42	

place, such as on inside window surfaces. To avoid problems, extra insulation, double glazing, or surface heating may then be used.

In HVAC practice, steady-state conduction seldom, if ever, takes place, because the outside air temperature and inside load conditions are constantly changing. The transient heat flow effects which result are functions of several variables, including the mass (storage effect) of the barrier. The sensible heat gain and cooling load factors discussed in Chap. 3 are approximations which allow the designer to compensate for these transients.

18.4   Thermal Convection

Thermal convection refers to heat transfer by eddy mixing and diffusion, as in a flowing airstream. In the typical airstream heating or cooling process, heat transfer takes place as a result of mixing with, and diffusion through, the air in the conditioned space. The final transfer is by conduction between air particles. Convection may be natural or free convection, due to differences in density, or it may be forced by mechanical means such as fans or pumps.

An HVAC process illustrating almost pure convective heat transfer is the mixing of two airstreams such as return air and outside air. If complete mixing takes place, the mixed airstream has a temperature (and humidity) resulting from a weighed average of the properties and masses of the two original air-streams. This is a result of convective eddy mixing and diffusion plus conductive heat transfer between particles.

A major HVAC application involving a combination of convection and conduction is heat exchange between two fluids such as refrigerant, water, steam, brine, and air, in many combinations. In general, the two fluids are separated by a barrier, usually the wall of a tube or pipe. Typical examples are the shell-and-tube heat exchanger (see Fig.

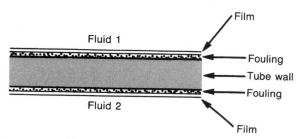

Figure 18.2 Heat transfer through a tube wall.

9.10) and the finned coil (see Fig. 9.21). In both cases, the barrier is a tube wall, as in Fig. 18.2. Heat transfer takes place within each fluid stream by convection, then by conduction through the wall and the contiguous films. The velocity of a fluid stream flowing uniformly in a conduit (tube or duct) is greatest at the center of the conduit and least near the edges (Fig. 18.3). This is due to friction of the fluid particles against the wall and against each other. The films of nearly motionless fluids on each side of the wall resist heat transfer, as noted above. Because the tubes in heat exchangers are usually copper, with its high conductivity factor, the films provide the major part of the resistance. Additional resistance is provided by the buildup of dirt, oil, or solids deposition on the tube surface. This is known as the *fouling factor*, and it is usually significant.

The film resistance is a function of the fluid velocity, being highest with laminar flow and lowest with turbulent flow. To estimate the degree of turbulence in a system, the *Reynolds number* Re is calculated:

$$\mathrm{Re} = \frac{DV\rho}{\mu} \tag{18.6}$$

where D = conduit diameter, ft
V = average fluid velocity, ft/s
μ = fluid viscosity, lb/(ft · s)
ρ = density, lb/ft³

The transition value of the Reynolds number is in the range of 2100

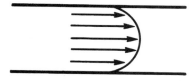

Figure 18.3 Velocity pattern for fluid flow in a conduit.

to 3100. Below 2100 flow is assumed to be laminar. Above 3100 turbulence is assumed. Between 2100 and 3100 it may be either, depending on various factors such as the roughness of the conduit. From the equation it is evident that laminar flow is equated with low velocities and high viscosities—i.e., other conditions being equal, oil will have a lower Reynolds number than water. The overall heat transfer rate increases abruptly as flow changes from laminar to turbulent.

For a finned-coil fluid-to-air heat exchanger, the general equation for heat transfer is

$$Q = kA \times \text{ROWS} \times \text{MED} \qquad (18.7)$$

where Q = total heat transfer, Btu/h
k = heat transfer coefficient per row per square foot
A = face area of coil normal to airflow direction
ROWS = number of rows of tubes in direction of airflow
MED = mean temperature difference

For a discussion and derivation of MED, see Sec. 10.17.1.

The value of k is greatly increased if steam or refrigerant is used in the coil; the effect of two-phase boiling or condensing is to increase the value of k over that obtained with single-phase flow.

For a shell-and-tube fluid-to-fluid heat exchanger, the equation for heat transfer is

$$Q = kA(\text{MED}) \qquad (18.8)$$

where k = heat transfer coefficient per square foot per degree Fahrenheit and A = total outside surface area of tubing in square feet. Again, two-phase flow increases the value of k. In Eqs. (18.7) and (18.8), the value of k must include the film and fouling factors. The heat exchanger manufacturer can provide values of k for a range of flow rates and fouling factors.

18.5 Thermal Radiation

Radiation heat transfer between two bodies takes place directly, by using electromagnetic energy across the intervening space. It is most efficient through a vacuum, because any intervening medium, even if transparent in the visible spectrum, will absorb some of the radiant energy.

Several mechanisms are at work in radiant energy transfer, such as surface emittance and absorptance, absolute temperature differences, and geometry.

18.5.1 Emittance, absorptance, reflectance, and transmittance

These terms describe properties of material surfaces as they relate to radiant heat transfer. A perfect *blackbody* absorbs all energy received and is said to have an absorptance of 1.0. This blackbody will also have an emittance of 1.0. (*Blackbody* is not a color description.) A perfect reflecting surface has a reflectance of 1.0 and absorbs none of the energy received. Not all the radiant energy received by a surface is reradiated, because some energy may be stored in the material, transmitted through it by conduction, or lost by convection.

Transmittance describes the ability of a body to allow some of (or all) the radiant energy impinging on it to pass through without being absorbed or reflected; this property is often called *transparency* or *translucence,* when it is related to the visible spectrum. An opaque body has zero transmittance.

Values of these properties for various materials and surface finishes are published in heat transfer textbooks and manufacturers' literature, in addition to the ASHRAE Handbook.[2,3,5] Painted surfaces in general have high emittances, regardless of color, ranging from 0.85 to 0.95, although metallic paints range from 0.40 to 0.60. Many surfaces that appear highly reflective have fairly high emittances. Window glass, which is highly transparent, can have a high absorptance and emittance. For all surfaces, the emittance varies with the angle at which energy impinges on the surface, being greatest when that angle is 90°.

18.5.2 Radiant energy transfer

Radiant energy is transferred between two surfaces which can "see" each other and are at different temperatures. One theory holds that radiant energy transfer takes place continually in both directions but that the two heat flows are equal at equal surface temperatures. One general equation for the net radiant heat transfer between two surfaces is

$$Q = \sigma F_e F_A A_1 (T_1^4 - T_2^4) \tag{18.9}$$

where Q = net heat transfer, Btu/h

σ = Stefan-Boltzmann constant = 0.173×10^{-8} [Btu/(h · ft² · °F)]

F_e = factor to correct for surface emittances not being equal to 1.0

F_A = factor to correct for geometric relationship of surfaces ($F_A = 1.0$ if surfaces face each other directly)

A_1 = face area of smaller surface, ft^2
T_1, T_2 = absolute (Rankine) temperatures of two surfaces

The absolute temperature difference is a major factor in radiant transfer.

Most of the radiant energy transfer takes place within the spectrum defined by wavelengths from 0.2 to 1.4 μm. Visible light ranges from 0.4- to 0.7-μm wavelength. The solar irradiation spectrum is shown in Fig. 18.4.

The areas of primary concern in HVAC use are solar energy effects, discussed in Sec. 10.18, and the *mean radiant temperature* (MRT). The MRT is a weighed average of all the surface temperatures surrounding and within a space, including people. A high value of MRT can cause discomfort even at a low air temperature. Very hot manufacturing processes are extreme examples.

18.6 Latent Heat and Moisture

Heat transfer due to condensation or evaporation may be considered a special form of convection. There is an important distinction. Heat transfer by conduction, convection, and radiation requires a temperature difference and is called *sensible heat transfer*. The processes of condensation and evaporation involve a change of state of a fluid at constant temperature, with heat added for evaporation or removed for condensation. This heat energy is called *latent heat*. In HVAC, the fluids used are water and refrigerants. Water is sometimes used as a refrigerant.

Moisture (water vapor) migration occurs through building construction materials in a manner similar to sensible heat transfer. The rate of moisture migration is a function of the permeability of the barrier

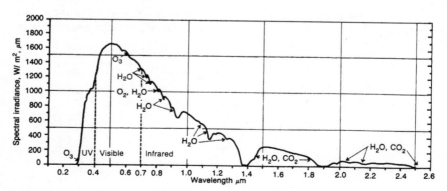

Figure 18.4 Special solar irradiation at sea level for air mass 1.0. (*Reprinted by permission from ASHRAE Handbook,* 1987 HVAC Systems and Applications.)

and the difference in vapor pressure across the barrier. The vapor pressure is related to the temperature and relative humidity (RH) of the air. Vapor pressures for saturated air (100 percent RH) are shown in Table 1 of Ref. 4, from which Table 19.1 of this book is abstracted. For example, the vapor pressure at 70°F, saturated, is 0.73966 inHg (inches of mercury). At 50 percent saturation it would be half of that amount. Note that vapor pressure is a function of temperature only, and at a given temperature and relative humidity, the vapor pressure is the same at any altitude or total atmospheric pressure.

Moisture sources of importance in HVAC are migration through building construction, moisture brought in by ventilation air, and moisture generated by people and processes within the building.

In general, vapor pressures will be higher at higher temperatures. To avoid problems, the moisture barrier should be on the warm side of the thermal barrier. When the air temperature on one side is below freezing, as in winter heating or a cold storage freezer, moisture migration into the barrier can result in condensation and freezing within the barrier, leading to damage or destruction of the barrier. Even small breaks, such as staple holes, in a moisture barrier can render it ineffective.

18.7 Summary

As noted in earlier chapters, the entire heating and air conditioning process is a series of fluid mechanics, thermodynamics, and heat transfer processes where energy is moved from one place to another. All three heat transfer mechanisms, conduction, convection, and radiation, are involved in nearly every system. All heat transfer is a function of a temperature difference between two points or entities. Conduction and convection are first-order functions of temperature $(T_1 - T_2)$ while radiation is a fourth-order function $(T_1^4 - T_2^4)$. Heat transfer by evaporation and condensation of moisture creates additional design issues which must be addressed by the HVAC system designer.

References

1. A. I. Brown and S. M. Marco, *Introduction to Heat Transfer,* 3d ed., McGraw-Hill, New York, 1958.
2. ASHRAE Handbook, *1997 Fundamentals,* Chap. 3, "Heat Transfer."
3. Ibid., Chap. 24, "Thermal and Water Vapor Transmission Data."
4. Ibid., Chap. 6, "Psychrometrics."
5. Ibid., Chap. 29, "Fenestration."

Engineering Fundamentals: Part 4

Psychrometrics

19.1 Introduction

Psychrometrics deals with the thermodynamic properties of moist air, which is the final heat transport medium in most air conditioning processes. The use of psychrometric tables and charts allows the designer to make a rational and graphic analysis of the desired air conditioning processes.

The general use of psychrometric data and charts began with the publications of Dr. Willis Carrier in the 1920s. In the 1940s, a research project conducted at the University of Pennsylvania by Goff and Gratch [sponsored by American Society of Heating and Ventilating Engineers (ASHVE)] resulted in new, more accurate data, which remained definitive until the results of further research were published in the 1980s.

This chapter deals with the subject rather briefly and simply, but in sufficient depth to provide an adequate background for HVAC design. For further study see Ref. 1.

19.2 Thermodynamic Properties of Moist Air

Moist air is a mixture of atmospheric air and water vapor. Dry air contains no water vapor. Saturated air contains all the water it can hold at a specified temperature and pressure. The properties of moist air can be evaluated by the perfect gas laws with only a small degree of error, which is not significant in most HVAC processes. The properties of interest in this discussion are the dry-bulb (db), wet-bulb (wb), and dew point temperatures; humidity ratio; degree of saturation; relative humidity (RH); and enthalpy and density.

19.2.1 Temperatures

The dry-bulb temperature T_{db} is the temperature of the moist air as read on an "ordinary" thermometer. When not otherwise defined, *temperature* means the dry-bulb temperature. In this text, the Fahrenheit scale is used.

The wet-bulb temperature T_{wb} is measured by a thermometer on which the bulb is covered with a wetted cloth wick. Air is blown across the wick, or the thermometer is moved rapidly through the air (as in the sling psychrometer), resulting in a cooling effect due to water evaporation. The amount of water which can be evaporated (and, therefore, the cooling effect) is limited by the humidity already present in the air. The temperature obtained in this manner is not the same as the thermodynamic wet-bulb temperature used in calculating psychrometric tables, but the error is small. The difference between the dry- and wet-bulb temperatures is sometimes called the *wet-bulb depression*.

The *dew point temperature* T_{dp} of moist air is defined by cooling the air until it is saturated and moisture begins to condense out of the mixture. For saturated air, all three of these temperatures are equal.

19.2.2 Humidity ratio

The *humidity ratio* w is the ratio of the mass of the water vapor to the mass of the dry air in a sample of moist air. The *specific humidity* is the ratio of the mass of the water vapor to the total mass of the moist air sample. Although the two terms are often used interchangeably, they are not identical.

19.2.3 Degree of saturation

The degree or percentage of saturation μ is the humidity ratio w of a moist air sample divided by the humidity ratio w_s of saturated air at the same temperature and pressure.

$$\mu = \frac{w}{w_s} \tag{19.1}$$

19.2.4   Relative humidity

The *relative humidity* ϕ is the ratio of the mole fraction of water vapor X_w in a moist air sample to the mole fraction of saturated air X_{ws} at the same temperature and pressure. The relative humidity is expressed as a percentage, from 0 percent (dry air) to 100 percent (saturated air). It can also be defined in terms of the partial pressures of the water vapor in the samples:

$$\phi = \frac{P_w}{P_{ws}} \tag{19.2}$$

Relative humidity values differ from percentage of humidity except at 0 and 100 percent.

19.2.5   Enthalpy

The *enthalpy h* is the total heat of a sample of material, in Btu per pound, including internal energy. However, in the ASHRAE tables and charts, the value of the enthalpy of dry air is arbitrarily set to zero at 0°F. This is satisfactory in terms of enthalpy differences, but enthalpy ratios may not be used. The enthalpy of a moist air sample is

$$h = h_a + w h_g \tag{19.3}$$

where h_a = enthalpy of dry air in sample
$\quad w$ = humidity ratio of sample
$\quad h_g$ = enthalpy of water vapor in sample (as a gas)
$\quad h$ = total enthalpy of sample (all at temperature of sample)

19.2.6   Volume and density

The *volume* of a moist air sample is expressed in terms of unit mass, in cubic feet per pound in this text. The *density* is the reciprocal of volume, in pounds per cubic foot.

19.3   Tables of Properties

The above-described properties and others are tabulated in Table 19.1, which is abstracted from an ASHRAE table. Table 19.1 is calculated for moist air at the standard atmospheric pressure of 14.696 lb/in^2 (29.921 inHg). At any other atmospheric pressure, these data will be different, because the partial pressure of water vapor is a function of temperature only, independent of pressure (see Sec. 19.7).

TABLE 19.1 Thermodynamic Properties of Moist Air

Temp. °F	Humidity ratio, lb_v/lb_a, W_s	Volume, ft³/lb dry air			Enthalpy, Btu/lb dry air			Entropy, Btu/(lb dry air·°F)			Condensed water			Temp. °F
		v_a	v_{as}	v_s	h_a	h_{as}	h_s	s_a	s_{as}	s_s	Enthalpy, Btu/lb h_w	Entropy, Btu/(lb·°F)	Vapor press. inHg p_s	
-8	0.0005139	11.377	0.009	11.386	-1.922	0.543	-1.378	-0.00422	0.00127	-0.00294	-162.63	-0.3325	0.024591	-8
-7	0.0005425	11.402	0.010	11.412	-1.681	0.574	-1.108	-0.00369	0.00134	-0.00234	-162.17	-0.3315	0.025959	-7
-6	0.0005726	11.427	0.010	11.438	-1.441	0.606	-0.835	-0.00316	0.00141	-0.00174	-161.70	-0.3305	0.027397	-6
-5	0.0006041	11.453	0.011	11.464	-1.201	0.640	-0.561	-0.00263	0.00149	-0.00114	-161.23	-0.3294	0.028907	-5
-4	0.0006373	11.478	0.012	11.490	-0.961	0.675	-0.286	-0.00210	0.00157	-0.00053	-160.77	-0.3284	0.030494	-4
-3	0.0006722	11.503	0.012	11.516	-0.721	0.712	-0.008	-0.00157	0.00165	0.00008	-160.30	-0.3274	0.032160	-3
-2	0.0007088	11.529	0.013	11.542	-0.480	0.751	0.271	-0.00105	0.00174	0.00069	-159.83	-0.3264	0.033909	-2
-1	0.0007472	11.554	0.014	11.568	-0.240	0.792	0.552	-0.00052	0.00183	0.00130	-159.36	-0.3253	0.035744	-1
0	0.0007875	11.579	0.015	11.594	0.0	0.835	0.835	0.00000	0.00192	0.00192	-158.89	-0.3243	0.037671	0
1	0.0008298	11.604	0.015	11.620	0.240	0.880	1.121	0.00052	0.00202	0.00254	-158.42	-0.3233	0.039694	1
2	0.0008742	11.630	0.016	11.646	0.480	0.928	1.408	0.00104	0.00212	0.00317	-157.95	-0.3223	0.041814	2
3	0.0009207	11.655	0.017	11.672	0.721	0.978	1.699	0.00156	0.00223	0.00380	-157.47	-0.3212	0.044037	3
4	0.0009695	11.680	0.018	11.699	0.961	1.030	1.991	0.00208	0.00235	0.00443	-157.00	-0.3202	0.046370	4
5	0.0010207	11.706	0.019	11.725	1.201	1.085	2.286	0.00260	0.00247	0.00506	-156.52	-0.3192	0.048814	5
6	0.0010743	11.731	0.020	11.751	1.441	1.143	2.584	0.00311	0.00259	0.00570	-156.05	-0.3182	0.051375	6
7	0.0011306	11.756	0.021	11.778	1.681	1.203	2.884	0.00363	0.00272	0.00635	-155.57	-0.3171	0.054060	7
8	0.0011895	11.782	0.022	11.804	1.922	1.266	3.188	0.00414	0.00286	0.00700	-155.09	-0.3161	0.056872	8
9	0.0012512	11.807	0.224	11.831	2.162	1.332	3.494	0.00466	0.00300	0.00766	-154.61	-0.3151	0.059819	9
10	0.0013158	11.832	0.025	11.857	2.402	1.402	3.804	0.00517	0.00315	0.00832	-154.13	-0.3141	0.062901	10
11	0.0013835	11.857	0.026	11.884	2.642	1.474	4.117	0.00568	0.00330	0.00898	-153.65	-0.3130	0.066131	11
12	0.0014544	11.883	0.028	11.910	2.882	1.550	4.433	0.00619	0.00347	0.00966	-153.17	-0.3120	0.069511	12
13	0.0015286	11.908	0.029	11.937	3.123	1.630	4.753	0.00670	0.00364	0.01033	-152.68	-0.3110	0.073049	13
14	0.0016062	11.933	0.031	11.964	3.363	1.714	5.077	0.00721	0.00381	0.01102	-152.20	-0.3100	0.076751	14
15	0.0016874	11.959	0.032	11.991	3.603	1.801	5.404	0.00771	0.00400	0.01171	-151.71	-0.3089	0.080623	15
16	0.0017724	11.984	0.034	12.018	3.843	1.892	5.736	0.00822	0.00419	0.01241	-151.22	-0.3079	0.084673	16
17	0.0018613	12.009	0.036	12.045	4.084	1.988	6.072	0.00872	0.00439	0.01312	-150.74	-0.3069	0.088907	17
18	0.0019543	12.035	0.038	12.072	4.324	2.088	6.412	0.00923	0.00460	0.01383	-150.25	-0.3059	0.093334	18
19	0.0020515	12.060	0.040	12.099	4.564	2.193	6.757	0.00973	0.00482	0.01455	-149.76	-0.3049	0.097962	19
20	0.0021531	12.085	0.042	12.127	4.804	2.303	7.107	0.01023	0.00505	0.01528	-149.27	-0.3038	0.102798	20
21	0.0022592	12.110	0.044	12.154	5.044	2.417	7.462	0.01073	0.00529	0.01602	-148.78	-0.3028	0.107849	21
22	0.0023703	12.136	0.046	12.182	5.285	2.537	7.822	0.01123	0.00554	0.01677	-148.28	-0.3018	0.113130	22
23	0.0024863	12.161	0.048	12.209	5.525	2.662	8.187	0.01173	0.00580	0.01753	-147.79	-0.3008	0.118645	23
24	0.0026073	12.186	0.051	12.237	5.765	2.793	8.558	0.01223	0.00607	0.01830	-147.30	-0.2997	0.124396	24
25	0.0027339	12.212	0.054	12.265	6.005	2.930	8.935	0.01272	0.00636	0.01908	-146.80	-0.2987	0.130413	25
26	0.0028660	12.237	0.056	12.293	6.246	3.073	9.318	0.01322	0.00665	0.01987	-146.30	-0.2977	0.136684	26
27	0.0030039	12.262	0.059	12.321	6.486	3.222	9.708	0.01371	0.00696	0.02067	-145.81	-0.2967	0.143233	27
28	0.0031480	12.287	0.062	12.349	6.726	3.378	10.104	0.01420	0.00728	0.02148	-145.31	-0.2956	0.150066	28
29	0.0032984	12.313	0.065	12.378	6.966	3.541	10.507	0.01470	0.00761	0.02231	-144.81	-0.2946	0.157198	29
30	0.0034552	12.338	0.068	12.406	7.206	3.711	10.917	0.01519	0.00796	0.02315	-144.31	-0.2936	0.164631	30
31	0.0036190	12.363	0.072	12.435	7.447	3.888	11.335	0.01568	0.00832	0.02400	-143.80	-0.2926	0.172390	31
32	0.0037895	12.389	0.075	12.464	7.687	4.073	11.760	0.01617	0.00870	0.02487	-143.30	-0.2915	0.180479	32
32*	0.003790	12.389	0.075	12.464	7.687	4.073	11.760	0.01617	0.00870	0.02487	0.02	0.0000	0.18050	32*
33	0.003947	12.414	0.079	12.492	7.927	4.243	12.170	0.01665	0.00905	0.02570	1.03	0.0020	0.18791	33
34	0.004109	12.439	0.082	12.521	8.167	4.420	12.587	0.01714	0.00940	0.02655	2.04	0.0041	0.19559	34
35	0.004277	12.464	0.085	12.550	8.408	4.603	13.010	0.01763	0.00977	0.02740	3.05	0.0061	0.20356	35
36	0.004452	12.490	0.089	12.579	8.648	4.793	13.441	0.01811	0.01016	0.02827	4.05	0.0081	0.21181	36
37	0.004633	12.515	0.093	12.608	8.888	4.990	13.878	0.01860	0.01055	0.02915	5.06	0.0102	0.22035	37
38	0.004820	12.540	0.097	12.637	9.128	5.194	14.322	0.01908	0.01096	0.03004	6.06	0.0122	0.22920	38
39	0.005014	12.566	0.101	12.667	9.369	5.405	14.773	0.01956	0.01139	0.03095	7.07	0.0142	0.23835	39

478

n			x											n
40	0.24784	0.0162	8.07	0.03187	0.01183	0.02004	15.233	5.624	9.609	12.696	0.105	12.591	0.005216	40
41	0.25765	0.0182	9.08	0.03281	0.01228	0.02052	15.700	5.851	9.849	12.726	0.110	12.616	0.005424	41
42	0.26781	0.0202	10.08	0.03375	0.01275	0.02100	16.175	6.086	10.089	12.756	0.114	12.641	0.005640	42
43	0.27831	0.0222	11.09	0.03472	0.01324	0.02148	16.660	6.330	10.330	12.786	0.119	12.667	0.005863	43
44	0.28918	0.0242	12.09	0.03570	0.01374	0.02196	17.155	6.582	10.570	12.816	0.124	12.692	0.006094	44
45	0.30042	0.0262	13.09	0.03669	0.01426	0.02244	17.653	6.843	10.810	12.846	0.129	12.717	0.006334	45
46	0.31206	0.0282	14.10	0.03770	0.01479	0.02291	18.164	7.114	11.050	12.877	0.134	12.743	0.006581	46
47	0.32408	0.0302	15.10	0.03873	0.01534	0.02339	18.685	7.394	11.291	12.908	0.140	12.768	0.006838	47
48	0.33651	0.0321	16.10	0.03978	0.01592	0.02386	19.215	7.684	11.531	12.939	0.146	12.793	0.007103	48
49	0.34937	0.0341	17.10	0.04084	0.01651	0.02433	19.756	7.984	11.771	12.970	0.152	12.818	0.007378	49
50	0.36264	0.0361	18.11	0.04192	0.01712	0.02480	20.306	8.295	12.012	13.001	0.158	12.844	0.007661	50
51	0.37636	0.0381	19.11	0.04302	0.01775	0.02528	20.868	8.616	12.252	13.033	0.164	12.869	0.007955	51
52	0.39054	0.0400	20.11	0.04415	0.01840	0.02575	21.441	8.949	12.492	13.065	0.171	12.894	0.008259	52
53	0.40518	0.0420	21.11	0.04529	0.01907	0.02622	22.025	9.293	12.732	13.097	0.178	12.920	0.008573	53
54	0.42030	0.0439	22.11	0.04645	0.01976	0.02668	22.621	9.648	12.973	13.129	0.185	12.945	0.008897	54
55	0.43592	0.0459	23.11	0.04763	0.02048	0.02715	23.229	10.016	13.213	13.162	0.192	12.970	0.009233	55
56	0.45205	0.0478	24.11	0.04884	0.02122	0.02762	23.850	10.397	13.453	13.195	0.200	12.995	0.009580	56
57	0.46870	0.0497	25.11	0.05006	0.02198	0.02808	24.484	10.790	13.694	13.228	0.207	13.021	0.009938	57
58	0.48589	0.0517	26.11	0.05132	0.02277	0.02855	25.131	11.197	13.934	13.262	0.216	13.046	0.010309	58
59	0.50363	0.0536	27.11	0.05259	0.02358	0.02901	25.792	11.618	14.174	13.295	0.224	13.071	0.010692	59
60	0.52193	0.0555	28.11	0.05389	0.02442	0.02947	26.467	12.052	14.415	13.329	0.233	13.096	0.011087	60
61	0.54082	0.0575	29.12	0.05522	0.02528	0.02994	27.157	12.502	14.655	13.364	0.242	13.122	0.011496	61
62	0.56032	0.0594	30.11	0.05657	0.02617	0.03040	27.862	12.966	14.895	13.398	0.251	13.147	0.011919	62
63	0.58041	0.0613	31.11	0.05795	0.02709	0.03086	28.582	13.446	15.135	13.433	0.261	13.172	0.012355	63
64	0.60113	0.0632	32.11	0.05936	0.02804	0.03132	29.318	13.942	15.376	13.468	0.271	13.198	0.012805	64
65	0.62252	0.0651	33.11	0.06080	0.02902	0.03178	30.071	14.454	15.616	13.504	0.281	13.223	0.013270	65
66	0.64454	0.0670	34.11	0.06226	0.03003	0.03223	30.840	14.983	15.856	13.540	0.292	13.248	0.013750	66
67	0.66725	0.0689	35.11	0.06376	0.03107	0.03269	31.626	15.530	16.097	13.577	0.303	13.273	0.014246	67
68	0.69065	0.0708	36.11	0.06529	0.03214	0.03315	32.431	16.094	16.337	13.613	0.315	13.299	0.014758	68
69	0.71479	0.0727	37.11	0.06685	0.03325	0.03360	33.254	16.677	16.577	13.650	0.326	13.324	0.015286	69
70	0.73966	0.0746	38.11	0.06844	0.03438	0.03406	34.097	17.279	16.818	13.688	0.339	13.349	0.015832	70
71	0.76528	0.0765	39.11	0.07007	0.03556	0.03451	34.959	17.901	17.058	13.726	0.351	13.375	0.016395	71
72	0.79167	0.0783	40.10	0.07173	0.03677	0.03496	35.841	18.543	17.299	13.764	0.365	13.400	0.016976	72
73	0.81882	0.0802	41.11	0.07343	0.03801	0.03541	36.743	19.204	17.539	13.803	0.378	13.425	0.017575	73
74	0.84684	0.0821	42.11	0.07516	0.03930	0.03586	37.668	19.889	17.779	13.843	0.392	13.450	0.018194	74
75	0.87567	0.0840	43.11	0.07694	0.04062	0.03631	38.615	20.595	18.020	13.882	0.407	13.476	0.018833	75
76	0.90533	0.0858	44.10	0.07875	0.04199	0.03676	39.583	21.323	18.260	13.923	0.422	13.501	0.019491	76
77	0.93589	0.0877	45.10	0.08060	0.04339	0.03721	40.576	22.075	18.500	13.963	0.437	13.526	0.020170	77
78	0.96733	0.0896	46.10	0.08250	0.04484	0.03766	41.592	22.851	18.741	14.005	0.453	13.551	0.020871	78
79	0.99970	0.0914	47.10	0.08444	0.04633	0.03811	42.633	23.652	18.981	14.046	0.470	13.577	0.021594	79
80	1.03302	0.0933	48.10	0.08642	0.04787	0.03855	43.701	24.479	19.222	14.089	0.487	13.602	0.022340	80
81	1.06728	0.0951	49.10	0.08844	0.04945	0.03900	44.794	25.332	19.462	14.132	0.505	13.627	0.023109	81
82	1.10252	0.0970	50.10	0.09052	0.05108	0.03944	45.913	26.211	19.702	14.175	0.523	13.653	0.023902	82
83	1.13882	0.0988	51.09	0.09264	0.05276	0.03988	47.062	27.120	19.943	14.220	0.542	13.678	0.024720	83
84	1.17608	0.1006	52.09	0.09481	0.05448	0.04033	48.238	28.055	20.183	14.264	0.561	13.703	0.025563	84
85	1.21445	0.1025	53.09	0.09703	0.05626	0.04077	49.445	29.021	20.424	14.310	0.581	13.728	0.026433	85
86	1.25388	0.1043	54.09	0.09930	0.05809	0.04121	50.681	30.017	20.664	14.356	0.602	13.754	0.027329	86
87	1.29443	0.1061	55.09	0.10163	0.05998	0.04165	51.949	31.045	20.905	14.403	0.624	13.779	0.028254	87
88	1.33613	0.1080	56.09	0.10401	0.06192	0.04209	53.250	32.105	21.145	14.450	0.646	13.804	0.029208	88
89	1.37893	0.1098	57.09	0.10645	0.06392	0.04253	54.582	33.197	21.385	14.498	0.669	13.829	0.030189	89
90	1.42298	0.1116	58.08	0.10895	0.06598	0.04297	55.951	34.325	21.626	14.547	0.692	13.855	0.031203	90
91	1.46824	0.1134	59.08	0.11150	0.06810	0.04340	57.355	35.489	21.866	14.597	0.717	13.880	0.032247	91
92	1.51471	0.1152	60.08	0.11412	0.07028	0.04384	58.794	36.687	22.107	14.647	0.742	13.905	0.033323	92
93	1.56248	0.1170	61.08	0.11680	0.07253	0.04427	60.271	37.924	22.347	14.699	0.768	13.930	0.034433	93
94	1.61154	0.1188	62.08	0.11955	0.07484	0.04471	61.787	39.199	22.588	14.751	0.795	13.956	0.035577	94
95	1.66196	0.1206	63.08	0.12237	0.07722	0.04514	63.343	40.515	22.828	14.804	0.823	13.981	0.036757	95

TABLE 19.1 *(Continued)*

Temp. °F	Humidity ratio, lb_w/lb_a, W_s	Volume, ft³/lb dry air v_a	v_{as}	v_s	Enthalpy, Btu/lb dry air h_a	h_{as}	h_s	Entropy, Btu/lb dry air·°F s_a	s_{as}	s_s	Condensed water Enthalpy, Btu/lb h_w	Entropy, Btu/lb·°F	Vapor press., inHg p_s	Temp. °F
96	0.037972	14.006	0.852	14.858	23.069	41.871	64.940	0.04558	0.07968	0.12525	64.07	0.1224	1.71372	96
97	0.039225	14.032	0.881	14.913	23.309	43.269	66.578	0.04601	0.08220	0.12821	65.07	0.1242	1.76685	97
98	0.040516	14.057	0.912	14.969	23.550	44.711	68.260	0.04644	0.08480	0.13124	66.07	0.1260	1.82141	98
99	0.041848	14.082	0.944	15.026	23.790	46.198	69.988	0.04687	0.08747	0.13434	67.07	0.1278	1.87745	99
100	0.043219	14.107	0.976	15.084	24.031	47.730	71.761	0.04730	0.09022	0.13752	68.07	0.1296	1.93492	100
101	0.044634	14.133	1.010	15.143	24.271	49.312	73.583	0.04773	0.09306	0.14079	69.07	0.1314	1.99396	101
102	0.046090	14.158	1.045	15.203	24.512	50.940	75.452	0.04816	0.09597	0.14413	70.06	0.1332	2.05447	102
103	0.047592	14.183	1.081	15.264	24.752	52.621	77.373	0.04859	0.09897	0.14756	71.06	0.1349	2.11661	103
104	0.049140	14.208	1.118	15.326	24.993	54.354	79.346	0.04901	0.10206	0.15108	72.06	0.1367	2.18037	104
105	0.050737	14.234	1.156	15.390	25.233	56.142	81.375	0.04944	0.10525	0.15469	73.06	0.1385	2.24581	105
106	0.052383	14.259	1.196	15.455	25.474	57.986	83.460	0.04987	0.10852	0.15839	74.06	0.1402	2.31297	106
107	0.054077	14.284	1.236	15.521	25.714	59.884	85.599	0.05029	0.11189	0.16218	75.06	0.1420	2.38173	107
108	0.055826	14.309	1.279	15.588	25.955	61.844	87.799	0.05071	0.11537	0.16608	76.05	0.1438	2.45232	108
109	0.057628	14.335	1.322	15.657	26.195	63.866	90.061	0.05114	0.11894	0.17008	77.05	0.1455	2.52473	109
110	0.059486	14.360	1.367	15.727	26.436	65.950	92.386	0.05156	0.12262	0.17418	78.05	0.1473	2.59891	110
111	0.061401	14.385	1.414	15.799	26.677	68.099	94.776	0.05198	0.12641	0.17839	79.05	0.1490	2.67500	111
112	0.063378	14.411	1.462	15.872	26.917	70.319	97.237	0.05240	0.13032	0.18272	80.05	0.1508	2.75310	112
113	0.065411	14.436	1.511	15.947	27.158	72.603	99.760	0.05282	0.13433	0.18716	81.05	0.1525	2.83291	113
114	0.067512	14.461	1.562	16.023	27.398	74.964	102.362	0.05324	0.13847	0.19172	82.04	0.1543	2.91491	114
115	0.069676	14.486	1.615	16.101	27.639	77.396	105.035	0.05366	0.14274	0.19640	83.04	0.1560	2.99883	115
116	0.071908	14.512	1.670	16.181	27.879	79.906	107.786	0.05408	0.14713	0.20121	84.04	0.1577	3.08488	116
117	0.074211	14.537	1.726	16.263	28.120	82.497	110.617	0.05450	0.15165	0.20615	85.04	0.1595	3.17305	117
118	0.076586	14.562	1.784	16.346	28.361	85.169	113.530	0.05492	0.15631	0.21122	86.04	0.1612	3.26335	118
119	0.079036	14.587	1.844	16.432	28.601	87.927	116.528	0.05533	0.16111	0.21644	87.04	0.1629	3.35586	119
120	0.081560	14.613	1.906	16.519	28.842	90.770	119.612	0.05575	0.16605	0.22180	88.04	0.1647	3.45052	120
121	0.084169	14.638	1.971	16.609	29.083	93.709	122.792	0.05616	0.17115	0.22731	89.04	0.1664	3.54764	121
122	0.086860	14.663	2.037	16.700	29.323	96.742	126.065	0.05658	0.17640	0.23298	90.03	0.1681	3.64704	122
123	0.089633	14.688	2.106	16.794	29.564	99.868	129.432	0.05699	0.18181	0.23880	91.03	0.1698	3.74871	123
124	0.092500	14.714	2.176	16.890	29.805	103.102	132.907	0.05740	0.18739	0.24480	92.03	0.1715	3.85298	124
125	0.095456	14.739	2.250	16.989	30.045	106.437	136.482	0.05781	0.19314	0.25096	93.03	0.1732	3.95961	125
126	0.098504	14.764	2.325	17.090	30.286	109.877	140.163	0.05823	0.19907	0.25729	94.03	0.1749	4.06863	126
127	0.101657	14.789	2.404	17.193	30.527	113.438	143.965	0.05864	0.20519	0.26382	95.03	0.1766	4.18046	127
128	0.104910	14.815	2.485	17.299	30.767	117.111	147.878	0.05905	0.21149	0.27054	96.03	0.1783	4.29477	128
129	0.108270	14.840	2.569	17.409	31.008	120.908	151.916	0.05946	0.21800	0.27745	97.03	0.1800	4.41181	129
130	0.111738	14.865	2.655	17.520	31.249	124.828	156.076	0.05986	0.22470	0.28457	98.03	0.1817	4.53148	130
131	0.115322	14.891	2.745	17.635	31.489	128.880	160.370	0.06027	0.23162	0.29190	99.02	0.1834	4.65397	131

*Extrapolated to represent metastable equilibrium with undercooled liquid.

SOURCE: Abstracted by permission from ASHRAE Handbook, *1997 Fundamentals*.

It is possible to calculate new values for a table similar to Table 19.1 at a different atmospheric pressure, by starting from the standard values in the table.[2] More accurately, new tables should be calculated by using the basic psychrometric equations.

19.4 Psychrometric Charts

The psychrometric chart is a graphical representation of psychrometric properties. There are many charts available from various equipment manufacturers and other sources. In this text, the ASHRAE chart in Fig. 19.1 is used.

The basic coordinate grid lines of the ASHRAE chart are the enthalpy, which slopes up to the left, and the humidity ratio, which is horizontal. The slope of the enthalpy lines is carefully calculated to provide the best possible intersections of property lines. Dry-bulb lines are uniformly spaced and approximately vertical; the slope of the lines changes across the chart. Wet-bulb lines slope similarly to enthalpy lines, but the slope increases as the temperature increases and no wet-bulb line is parallel to an enthalpy line. This is because of the heat added to the mixture by the moisture as it changes from dry to saturated air. Spacing between wet-bulb lines increases with temperature. The enthalpy lines (except every fifth line) are shown only at the edges of the chart to avoid confusion. A straightedge is needed to determine a value of enthalpy within the chart. Volume lines are uniformly spaced and parallel.

Relative-humidity lines are curved, with the 100 percent line (saturation) defining the upper boundary of the chart. These lines are not uniformly spaced. (Percentage of saturation lines would be uniformly spaced but are not used in HVAC design.)

When any two properties of a moist air sample are known, a state point may be plotted on the chart (Fig. 19.2) that identifies the values of all the other properties. Typically, the known properties are those most easily measured, i.e., dry- and wet-bulb temperatures and relative humidity or dew point temperature.

19.5 HVAC Processes on the Psychrometric Chart

Any HVAC process may be plotted on the chart if the end state points are known and sometimes if only the beginning state point is known.

19.5.1 Mixing of two airstreams

A very common HVAC process is the adiabatic mixing of two airstreams, e.g., return air and outside air, or hot and cold streams in a

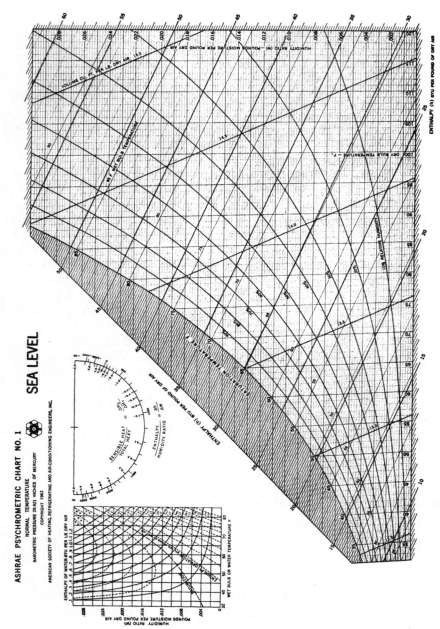

Figure 19.1 The ASHRAE psychrometric chart. (*Reprinted by permission.*)

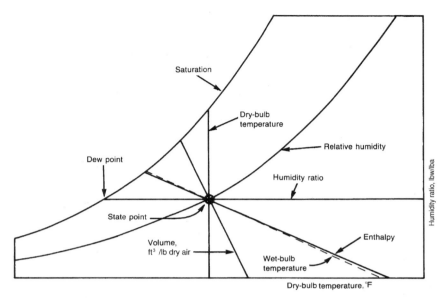

Figure 19.2 A state point on the psychrometric chart.

dual-duct or multizone system. The ASHRAE chart is a Mollier-type chart. On a Mollier chart, a mixing process may be shown as a straight line connecting two initial state points (Fig. 19.3, points *A* and *B*). The mixture state point *C* will be on the line located such that it divides the line into two segments with lengths proportional to the two initial air masses. The mixture point will be closer to the initial point with the larger mass. In the figure, if the volume at point *A* is 7000 ft³/min and the volume at point *B* is 3000 ft³/min, then line *AC* will be 3 units long and line *BC* will be 7 units long. The state point values for *C* can then be read from the chart. They can also be calculated from the tables, but the graphical solution is much faster unless a high degree of accuracy is required.

19.5.2 Sensible heating and cooling

The word *sensible* implies that the heating or cooling takes place at a constant humidity ratio. These processes are shown as horizontal lines — constant value of *w* — with the dry-bulb temperature increasing for heating (line *AB* in Fig. 19.4) and decreasing for cooling (line *CD* in Fig. 19.4). Note that although the humidity ratio remains constant, there is a change in the relative humidity. As the dry-bulb temperature increases, the air will hold more moisture at saturation.

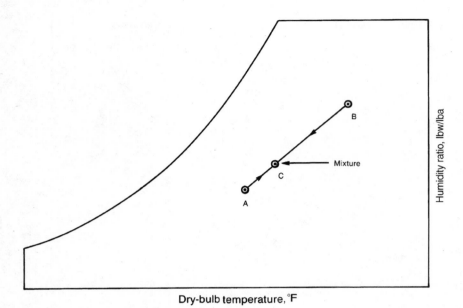

Figure 19.3 Mixing of two airstreams.

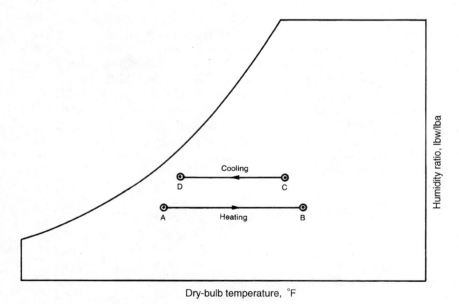

Figure 19.4 Sensible heating and cooling.

19.5.3 Cooling and dehumidifying

Most refrigerated cooling processes also include dehumidification (Fig. 19.5). The process is shown as a straight line sloping down and to the left from the initial state point. As discussed in Sec. 9.7.2, the real process involves sensible cooling to saturation, then further cooling down the saturation curve to an apparatus dew point (ADP). Some air is "bypassed" through the cooling coil without being cooled. The final state point is therefore a mixture of the initial state and the ADP state, usually very close to the ADP.

19.5.4 Adiabatic saturation

If an airstream is passed through a water spray (Fig. 19.6) in such a way that the leaving air is saturated adiabatically, then the process can be shown on the chart as a constant-wet-bulb process (Fig. 19.7), and the final wet- and dry-bulb temperatures are equal. In practice, this process is called *evaporative cooling,* and saturation is not achieved (Fig. 19.8). The *efficiency,* denoted eff, of an air washer or evaporative cooler is the ratio of the dry-bulb temperature difference from point 1 to point 2 to the initial difference between the dry- and wet-bulb temperatures:

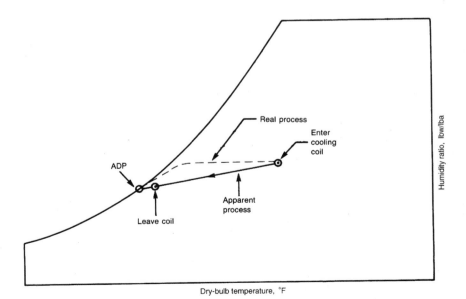

Figure 19.5 Cooling and dehumidifying.

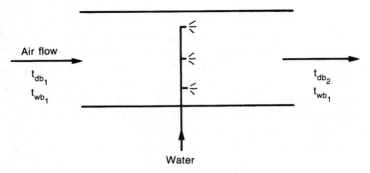

Air flow

t_{db_1}
t_{wb_1}

t_{db_2}
t_{wb_1}

Water

Figure 19.6   Adiabatic saturation process.

$$\text{Eff} = \frac{t_{db_1} - t_{db_2}}{t_{db_1} - t_{wb_1}} \tag{19.4}$$

The evaporative cooling or air washer process creates a sensible cooling effect by lowering the dry-bulb temperature, but increases the relative humidity in so doing.

19.5.5   Humidification

As noted above, moisture may be added and humidity increased by the evaporative cooling process. This usually requires reheat or mixing

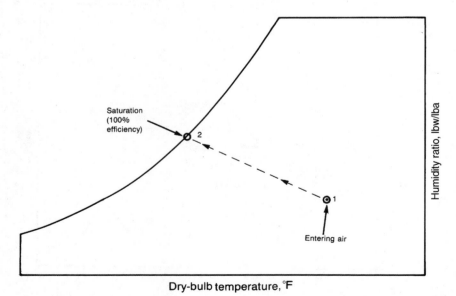

Saturation
(100%
efficiency)

2

Entering air

Humidity ratio, lbw/lba

Dry-bulb temperature, °F

Figure 19.7   Adiabatic saturation.

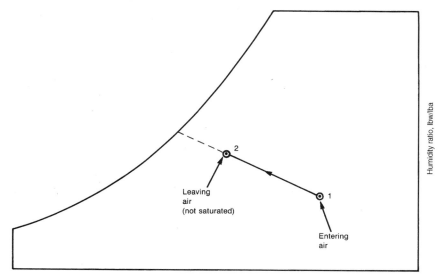

Figure 19.8 Evaporative cooling.

for accurate temperature control. The more common humidification process involves the use of steam or sometimes a heated evaporative pan (see the discussion in Sec. 10.19). Humidification by means of steam humidifier is shown in Fig. 19.9 as a straight line sloping upward (increasing humidity ratio) and to the right (heat added by steam). The slope of the line can be calculated from the masses of the airstream and the added water vapor together with their heat contents, as shown in the examples in Secs. 10.19.2 and 10.19.3.

19.5.6 Chemical dehumidification

This process is described in Sec. 11.7.2.

19.6 The Protractor on the ASHRAE Psychrometric Chart

Figure 19.1 includes a protractor above and to the left of the main chart. For a full discussion of this tool, see Ref. 1.

One of the most important uses of the protractor is in determining the slope of the *condition line* for the air being supplied to a space to offset sensible and latent cooling loads. First, the sensible heat/total heat ratio S/I, based on design load calculations, is calculated. For example, if the total cooling load is 125,000 Btu/h and the sensible

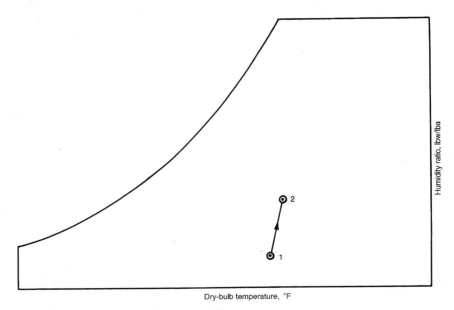

Figure 19.9 Steam or heated-pan humidification.

load is 100,000 Btu/h, the ratio is 0.80. Second, a line is plotted on the protractor from the origin to the value of the ratio, as shown in Fig. 19.10. The state point corresponding to the design room condition of, say, 76°F db and 50 percent RH is located on the chart. Then a line is drawn from this state point toward the saturation curve, parallel to the line on the protractor. The state point of the air supplied to the room must be somewhere on the line on the chart. In this example, there is an ADP at about 52°F, so this process can be accomplished without reheat. If the sensible/total heat ratio were 0.60, as shown by the dashed line on the protractor, then the process on the chart, also shown dashed, would have no ADP and would be impossible to accomplish directly. An arbitrary ADP could be established, and reheat would be needed, as shown.

The other scale on the protractor, based on the enthalpy divided by the humidity ratio, can be used to determine the slope of a humidification process.

19.7 Effects of Altitude

The tables and the chart of Fig. 19.1 are based on a standard atmospheric pressure of 29.92 inHg. The partial pressure of water vapor is

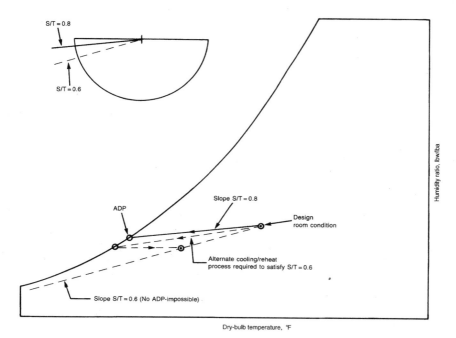

Figure 19.10 Using the protractor.

a function of temperature only, while the total atmospheric pressure decreases with altitude. The rule of thumb is that the standard chart and tables are sufficiently accurate up to about 2000 ft above sea level. At higher elevations, new tables and charts are needed.[2]

High-altitude charts are available from several sources. ASHRAE publishes charts for 5000 and 7500 ft. The U.S. Bureau of Mines publishes a *composite* chart for various elevations below sea level, down to 10,000 ft.

The general effect of increasing altitude is to *expand* the chart (Fig. 19.11). That is, for a uniform grid of enthalpy and humidity ratio, as the altitude increases (and atmospheric pressure decreases), the lines defining the other properties change as follows:

1. Dry-bulb temperature lines are unchanged.
2. Wet-bulb temperature lines expand up and to the right.
3. Relative-humidity lines, including saturation, expand up and to the left.
4. Volume lines expand up and to the right.

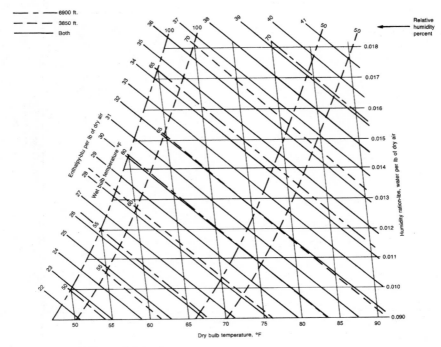

Figure 19.11 Effects of altitude.

5. For a given combination of dry-bulb and wet-bulb temperatures, the change in relative humidity is very small and for most air conditioning processes can be neglected.

19.8 Summary

This discussion of psychrometrics has been very brief. The subject is very important to the HVAC designer, and further study of Ref. 1 and other sources is recommended. Every set of HVAC design calculations should include one or more psychrometric charts, reflecting the anticipated performance of the system being designed.

References

1. ASHRAE Handbook, *1997 Fundamentals,* Chap. 6, "Psychrometrics."
2. R. W. Haines, "How to Construct High-Altitude Psychrometric Charts," *Heating/ Piping/Air Conditioning,* October 1961, p. 144.

20

Engineering Fundamentals: Part 5

Sound and Vibration

20.1 Introduction

The HVAC designer cannot neglect consideration of the sound and vibration generated by HVAC equipment. This chapter briefly discusses the fundamentals of sound and vibration control. References for further study are cited at the end of the chapter.

20.2 Definitions

Sound is a form of energy, detected as a variation in pressure and stress in an elastic or viscous medium. The traditional concept is that sound is generated by a source and is transmitted through a path to a receiver (Fig. 20.1). Diminishment of the sound during transmission is called *attenuation*. The receiver is usually a human ear or a microphone.

Vibration is a form of energy, detected as cyclic movement in a machine or structure. Sound and vibration are mutually convertible, and many transmission problems and solutions depend on this fact.

Noise is unwanted sound. One person's sound may be another person's noise, e.g., heavy-metal rock music. In general, however, noise is random sound. White noise, used for masking unwanted sound, is random sound in the speech interference range.

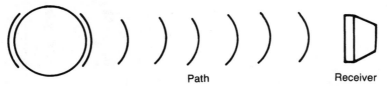

Path   Receiver

Figure 20.1 Sound transmission.

20.3 Methods of Specifying and Measuring Sound

There are several ways of describing the characteristics of sound: sound power, sound pressure, intensity, loudness, frequency, speed, and directivity.

20.3.1 Sound power

An acoustical source radiates energy in the form of sound. This acoustical power is expressed in watts. A *watts exponential* scale of sound power has been developed. A sound power level of 10^{-12} W represents the threshold of hearing for excellent young ears. This is given the value of 0 dB (decibels) and is the reference level for Table 20.1. Table 20.1 lists the decibel value corresponding to a given watts exponential, together with an example of this sound power, over a scale from 0 to 200 dB.

20.3.2 Intensity and sound pressure level

Sound power cannot be measured directly but must be calculated from pressure measurements. If an imaginary sphere is placed around a sound source (with the source at the center of the sphere), all the energy from the source must pass through the sphere. Power flow through a unit area of the sphere is the *intensity,* expressed in watts per unit area. Intensity varies inversely as the square of the distance from the source. The intensity and the sound pressure level are nearly identical numerically if proper units are used. The ASHRAE Handbook uses watts per square meter and micropascals (μPa). The measuring system converts the pressure to decibels, corresponding to the sound power levels of Table 20.1.

20.3.3 Loudness and frequency

Sound may be visualized as traveling in a wave pattern similar to that of alternating electric current (Fig. 20.2). Variation above and below the reference is called the *amplitude* and determines the loudness. The

TABLE 20.1   Sound Power Outputs

Source	Approximate power output		
	W linear	W exponential	Decibel re: 10^{-12} W
Saturn rocket	100,000,000	10^8	200
Turbojet engine*	100,000	10^5	170
Jet aircraft at takeoff†	10,000	10^4	160
Turboprop at takeoff	1,000	10^3	150
Prop aircraft at takeoff‡	100	10^2	140
Large pipe organ	10	10^1	130
Small aircraft engine	1	10^0	120
Blaring radio	0.1	10^{-1}	110
Automobile at highway speed	0.01	10^{-2}	100
Voice, shouting	0.001	10^{-3}	90
Garbage disposal unit	0.0001	10^{-4}	80
Voice, conversation level	0.00001	10^{-5}	70
Electronic equipment ventilation fan	0.000001	10^{-6}	60
Office air diffuser	0.0000001	10^{-7}	50
Small electric clock	0.00000001	10^{-8}	40
Voice, soft whisper	0.000000001	10^{-9}	30
Rustling leaves	0.0000000001	10^{-10}	20
Human breath	0.00000000001	10^{-11}	10
Threshold of hearing	0.000000000001	10^{-12}	0

*With afterburner.
†Four jet engines.
‡Four propeller engines.
SOURCE: Reprinted by permission from ASHRAE Handbook *1987 HVAC Systems and Applications; 1997 Fundamentals,* Chap. 7, Table 2, is similar.

distance from one wave peak to the next is the *wavelength,* the reciprocal of which is the *frequency* or *pitch* of the sound. Frequency is measured in cycles per second (cps). The term *hertz* (abbreviated Hz) is used instead of cycles per second. Figure 20.2 represents a pure tone—one single frequency. Most sound is made up of several frequencies (tones), with each frequency having a different loudness. For

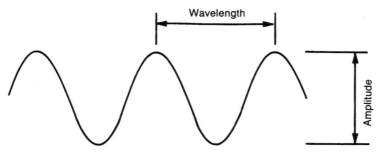

Figure 20.2   Wavelength and amplitude of sound.

example, air noise in a duct is made up of several high-frequency tones generated by turbulence of the air due to fittings and obstructions as well as that due to straight-line flow and friction against duct walls. This air noise will usually be accompanied by sound transmitted from the fan, with a predominant frequency which is a function of the fan speed and number of blades.

20.3.4 Frequency spectrum

A good human ear can hear, and distinguish, a wide range of frequencies—from a low of about 20 Hz to a high of about 20,000 Hz. For the purposes of analysis, this range is divided into several octaves. Two tones are said to be an *octave* apart when the frequency of one is twice that of the other. A common example is the musical octave on the piano or other instrument. On the musical scale, the A below middle C has a frequency of 440 Hz. A sound-level measuring instrument is usually equipped with filters so that the sound level of each octave band, or in some cases ⅓ octave band, can be measured. The commonly used scale for octave and ½ octave bands is shown in Table 20.2.

The quality of a sound is determined by the harmonics (other frequencies) and their relationship to the dominant frequency. The tonal quality of each of the various musical instruments is determined by the harmonics typical of that instrument.

20.3.5 Speed and directivity

The speed of sound varies with the medium used as the path and is a function of that medium's density and modulus of elasticity. The speed of sound is highest in high-density materials such as steel (16,000 ft/s) and water (5000 ft/s). In air at room temperature and sea level pressure, it is about 1100 ft/s.

Directivity means that a real sound source does not generate a uniform pattern; sound intensity varies with direction from the source. This can be particularly effective when the sound is generated as a result of vibration of the source. In the direction of vibrational movement, the sound level will be much higher than in other directions.

20.4 Sound and Vibration Transmission

In HVAC practice, the sound and vibration sources are the elements of the HVAC system: fans, pumps, compressors, air flowing in ducts, and water and steam flowing in pipes. The paths for transmission are

TABLE 20.2 Center Approximate Cutoff Frequencies for Octave and One-Third Octave Band Series Frequency, Hz

1/1 Octave bands			1/3 Octave bands		
Lower	Center	Upper	Lower	Center	Upper
			22.4	25	28
22.4	31.5	45	28	31.5	35.5
			35.5	40	45
			45	50	56
45	63	90	56	63	71
			71	80	90
			90	100	112
90	125	180	112	125	140
			140	160	180
			180	200	224
180	250	355	224	250	280
			280	315	355
			355	400	450
355	500	710	450	500	560
			560	630	710
			710	800	900
710	1,000	1,400	900	1,000	1,120
			1,120	1,250	1,400
			1,400	1,600	1,800
1,400	2,000	2,800	1,800	2,000	2,240
			2,240	2,500	2,800
			2,800	3,150	3,550
2,800	4,000	5,600	3,550	4,000	4,500
			4,500	5,000	5,600
			5,600	6,300	7,100
5,600	8,000	11,200	7,100	8,000	9,000
			9,000	10,000	11,200
			11,200	12,500	14,000
11,200	16,000	22,400	14,000	16,000	18,000
			18,000	20,000	22,400

SOURCE: Reprinted by permission from ASHRAE Handbook, *1997 Fundamentals,* Chap. 7, Table 1. (Original source: ANSI Standard SI.6.)

the ducts and the building structure. The receivers are the people in the building.

20.4.1 Transmission through ducts

The sound transmitted through ducts comes from at least three sources: fan room equipment, air noise generated in the ducts, and sound generated outside the ducts and transferred through the duct walls.

Some fan room equipment noise is in the third category. Noise from the fan is transmitted directly through the ducts and is dominated by a frequency determined by the fan speed and number of fan blades. Air noise is generated as part of the energy loss due to friction; more noise is generated at points of turbulence such as elbows, transitions or branches, and particularly dampers. Poorly constructed lightweight dampers have a tendency to vibrate. Dampers at terminal units, e.g., VAV boxes, may be noisy if the entering air pressure is too high. Sound is always generated when a fluid passes from a high-pressure region to a lower-pressure one. Improperly braced duct sidewalls may flex in resonant motion, creating a rumbling sound.

All these things apply also to the return air system, and because the return air path is usually shorter than the supply air path, the return air duct/plenum system is often an acoustic problem.

20.4.2 Transmission through the building structure

Most equipment-generated sound—from fans, pumps, and com-pressors—is transmitted through the building structure as vibration and is heard or felt by the receiver. The path may not be obvious and may be complex, as in Fig. 20.3. In this real example, a refrigeration compressor in the basement of a church was properly isolated from

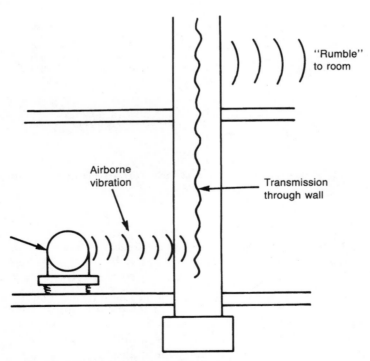

Figure 20.3 Sound transmission through structure.

the floor. But vibration energy was transmitted several feet through the air to a 12-in concrete wall. The wall had a natural frequency which was excited by the compressor vibration, amplified, and transmitted upward through the wall. In the nave of the church, the vibration of the wall generated an audible rumble.

This is an example of the most common method of transmission through a building structure: The natural frequency of the structural member is often "in tune" with one or more frequencies of the sound source. When this happens, the sound is readily transmitted, and even amplified, over a considerable distance. In a multistory building, often the sound will be heard (or felt) several floors away from the source, while the sound is barely noticeable near the source. In the situation described, the problem was solved by building an enclosure of 2-in-thick acoustical board around the sides and top of the compressor. This eliminated the airborne vibration. A similar classic example identified a flagpole 50 ft away from a building waving in harmony with a hard mounted reciprocating compressor in the basement of the building. Installation of vibration isolators on the compressor stabilized the flagpole.

20.5   Ambient Sound-Level Design Goals

The ambient sound level which is acceptable in an acoustical environment varies with the function being served in that environment. To design, specify, and construct facilities to these acoustical requirements, it is necessary to have some standard criteria. The standard criteria used in acoustical design are *noise criteria* (NC) and *room criteria* (RC) curves.

A minimum level of background sound is often desirable; e.g., in open-plan offices, a fairly high background noise level in the speech interference range (250 to 4000 Hz) will mask crosstalk and afford privacy.

20.5.1   Noise criteria curves

Noise criteria curves were the standard for many years, and they define acceptable limits for sound pressure level in each octave band. Figure 20.4 shows the standard NC curves. The actual environment must not exceed the specified curve at any point, but can be at any level below the curve. The resulting sound may be too quiet in some frequencies. A higher sound power level is acceptable at lower frequencies. These curves emphasize the fact that high frequencies sound louder than low frequencies when sound power levels are equal.

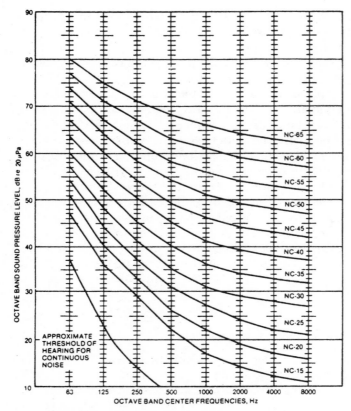

Figure 20.4 NC curve. (*Reprinted by permission from ASHRAE Handbook,* 1987 HVAC Systems and Applications.)

20.5.2 Room criteria curves

Room criteria curves (Fig. 20.5) were introduced in the ASHRAE Handbook in 1980. They provide guidance when a minimum level of background sound is needed for masking or other purposes. According to the Handbook, "the shape of the RC curve is a close approximation to a well-balanced blandsounding spectrum." Such a background, with no hisses or rumbles, is usually unobtrusive and acceptable, even when at a fairly high level, as long as it is essentially constant.

20.5.3 Design goals

Design goals for background noise levels for various environments are given in Table 20.3. These are stated in terms of NC curves. The table omits criteria for concert halls, theaters, and recording studios; these

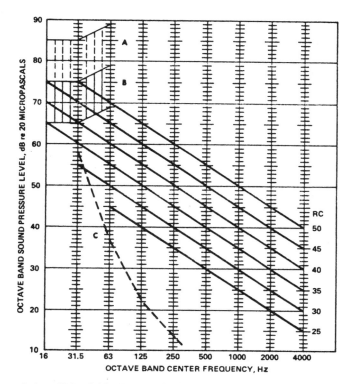

Region A: *High probability that noise-induced vibration levels in lightweight wall and ceiling constructions will be felt; anticipate audible rattles in light fixtures, doors, windows, etc..*
Region B: *Noise-induced vibration levels in lightweight wall and ceiling constructions may be felt; slight possibility of rattles in light fixtures, doors, windows, etc.*
Region C: *Below threshold of hearing for continuous noise.*

Figure 20.5 RC curve. (*Reprinted by permission from ASHRAE Handbook,* 1995 HVAC Applications, *Chap. 43, Fig. 5.*)

are generally 25 NC or less. Industrial environments are not listed but tend to be higher, due to industrial processes. Government agencies, such as the Occupational Safety and Health Administration (OSHA), specify maximum noise levels and duration of exposure for industrial environments. Because most of these exceed the sound level of the HVAC systems, the HVAC designer is not generally very concerned with sound attenuation in these environments.

20.5.4   A-weighted sound-level criteria

The A-weighted sound level (abbreviated dbA) is a simple, single-number method of stating a design goal. Its usefulness is limited because it conveys no information on the sound spectrum—the sound

TABLE 20.3 Recommended Indoor Design Goals for
Air Conditioning Sound Control

Type of area	Recommended RC or NC criteria range
1. Private residences	25–30
2. Apartments	25–30
3. Hotels/motels	
a. Individual rooms or suites	30–35
b. Meeting/banquet rooms	25–30
c. Halls, corridors, lobbies	35–40
d. Service/support areas	40–45
4. Offices	
a. Executive	25–30
b. Conference rooms	25–30
c. Private	30–35
d. Open-plan areas	35–40
e. Computer equipment rooms	40–45
f. Public circulation	40–45
5. Hospitals and clinics	
a. Private rooms	25–30
b. Wards	30–35
c. Operating rooms	35–40
d. Corridors	35–40
e. Public areas	35–40
6. Churches	25–30
7. Schools	
a. Lecture and classrooms	25–30
b. Open-plan classrooms	30–35
8. Libraries	35–40
9. Concert halls	
10. Legitimate theaters	
11. Recording studios	
12. Movie theaters	30–35
13. Laboratories with fume hoods	

NOTE: These are for *unoccupied* spaces, with all systems operating.
SOURCE: Reprinted by permission from ASHRAE Handbook, *1987 HVAC Systems and Applications;* subsequent editions are similar.

levels at the various frequencies. The measuring device contains a weighting network which deemphasizes the lower frequencies. It tells little or nothing about the quality of the sound, which may hiss or rumble or have a dominant tone (frequency).

20.6 Reducing Sound and Vibration Transmission

Sound transmission may be reduced by containment or absorption. Containment implies an enclosed space with sound barriers all around. This is not as simple as it sounds. Massive barriers will contain most frequencies, but some low frequencies may be transmitted. More lightweight barriers transmit more frequencies and may have a fairly high natural frequency. The best barriers combine mass with sound-absorbing material. For example, a standard panel for use in constructing a sound-absorbing plenum (Fig. 20.6) is made of a high-density sound-absorbing material, 4 in thick with a perforated sheet-steel face on one side and solid sheet-steel face on the other.

For any type of enclosure, more serious difficulties are posed by openings or penetrations through the barrier. Duct or pipe penetrations are typical problems. These act as sound leaks, often conveying sound as though there were an actual opening.

All penetrations should be carefully sealed, by using methods similar to those shown in Fig. 20.7. When the sleeve is properly installed and the space between sleeve and duct (or pipe) is packed and caulked as shown, then the only sound transmission will be directly through the duct or pipe. This must be attenuated in other ways.

For very noisy machinery, it is often desirable to provide a separate room, with sound-absorbing walls, similar to a sound-absorbing plenum.

20.6.1 Sound attenuation in ducts

An unlined sheet-metal duct system has considerable natural attenuation. Reflection occurs at turns and transitions. At every branch and

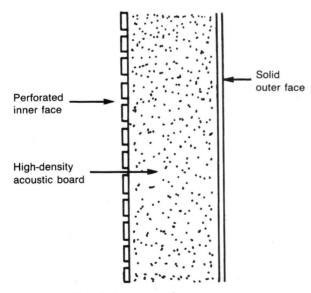

Figure 20.6 Sound plenum panel.

Perforated inner face

High-density acoustic board

Solid outer face

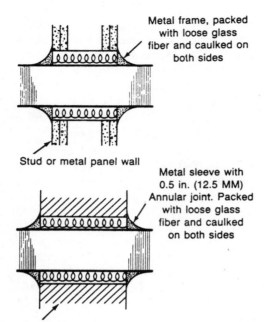

Metal frame, packed with loose glass fiber and caulked on both sides

Stud or metal panel wall

Metal sleeve with 0.5 in. (12.5 MM) Annular joint. Packed with loose glass fiber and caulked on both sides

Masonry or concrete wall

Figure 20.7 Sound control at wall penetration. (*Reprinted by permission from ASHRAE Handbook, 1995* HVAC Applications, *Chap. 43, Fig. 28.*)

outlet, some sound energy is lost by division. Sound attenuation can be increased by an interior absorptive lining. Any soft, flexible material will absorb sound. The principal criterion for absorptive duct lining, besides softness, is that it must resist erosion by the airstream. In some environments, such as hospitals, erosion can create problems, and internal insulation is not allowed. This material will also act as thermal insulation, reducing or eliminating the need for exterior insulation. The metal duct must be oversized to allow for the lining and the higher friction loss of the material. Where the duct passes through an equipment room or a space where sound is being generated, exterior insulation should be provided to minimize flanking noise transmission. This noise may be carried through the duct wall for some distance. References 1 and 2 contain details and data for calculating sound transmission in ducts.

Sound traps may also be used. A sound trap is an attenuation device inserted in the duct (Fig. 20.8). It is made with convoluted passages to minimize direct sound transmission, and it has an absorptive lining. A sound trap is rated by the manufacturer for absorption (sound pressure) loss in each of the various octave bands. Insertion loss ratings should be used; this is the loss as installed in the duct and with air flowing. The other criterion for sound trap selection is the static pressure loss at a design flow rate, usually 0.10 to 0.20 inH_2O.

Figure 20.8 Sound trap. (*Courtesy of Titus Products, Division of Phillips Industries, Inc.*)

Ductwork should always be isolated from fans and other vibrating equipment by flexible connections. A typical flexible connection (Fig. 20.9) is made of heavy canvas or synthetic fabric with metal flanges at each edge for connection to equipment and duct.

20.6.2 Sound attenuation in piping

Sound attenuation in water piping is somewhat more difficult than in air ducts because the solid column of water and metal transmits sound farther and faster than through air. In addition, sound is generated by fluid flowing in a pipe much as air flows in a duct. In most cases, piping noise is attenuated by distance, branching, external insulation, and installation of piping in concealed spaces. It is essential to avoid transmission of vibration from equipment to which the piping connects. Most vibrating equipment is provided with flexible mountings (see below). The piping must also flex. Flexible connectors are made of rubber, metal, fabric reinforced with metal braid, and in other ways. It is preferable to use two flexible connectors at a 90° angle to each other, as shown in Fig. 20.10. A single isolator is not always effective. Spring hangers are also needed, as shown, for a short distance beyond

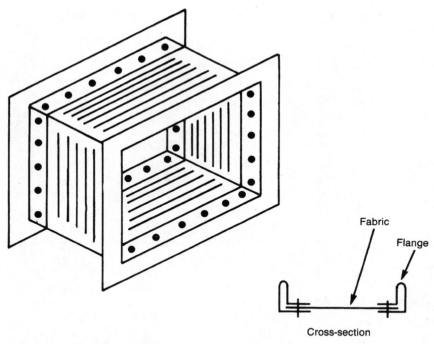

Fabric

Flange

Cross-section

Figure 20.9 Flexible connector for duct.

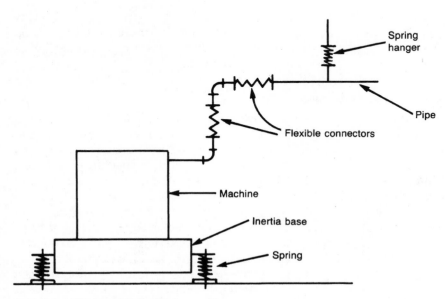

Spring hanger

Pipe

Flexible connectors

Machine

Inertia base

Spring

Figure 20.10 Piping with flexible connections.

the flexible connectors; two or three such hangers are usually specified.

20.6.3 Vibration transmission and isolation

The simplest way to minimize vibration transmission is to isolate the machinery that causes the vibration. This is done by means of flexible isolators, often combined with inertia bases. A flexible isolator is most often a spring. For less demanding applications, it may be a block of cork or rubber, or a rubber-in-shear device, in which the rubber is used as a spring. In general, the best isolation is provided by the "softest" support, allowing the vibrational energy to be dissipated in movement.

An inertia base (Fig. 20.11) is usually made of concrete, from 4 to 8 in or more thick. The machinery is mounted on the base, and the whole assembly is mounted on spring isolators. The purpose of the inertia base is to increase the mass of the system so that the imbalance of moving machinery will be partially neutralized. The base is usually sized to have a weight 3 to 4 times that of the machinery.

Spring isolators are made in various styles. They are cataloged by style, open (uncompressed) height, spring rate (in pounds per inch of deflection), and efficiency. *Efficiency* refers to the amount of vibrational energy attenuated at a specified load and deflection. The manufacturers' catalogs generally provide a great deal of engineering data. A typical spring for a piece of equipment on an inertia base might require a 5- to 6-in open height and a 2-in static deflection with the machinery not operating. The value of the inertia base and springs can be completely negated by allowing alternate paths for travel of the vibration. Piping isolation is described above.

Electric conduit can conduct vibration. The final conduit connection between the mounted equipment and the power source should be made with a flexible conduit with a 360° turn, as in Fig. 20.12. Drain lines either should not touch the floor below the inertia base or should be provided with flexible connectors.

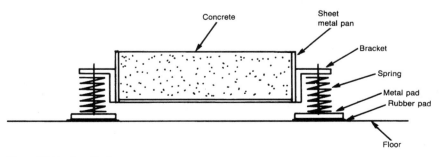

Figure 20.11 Inertia base.

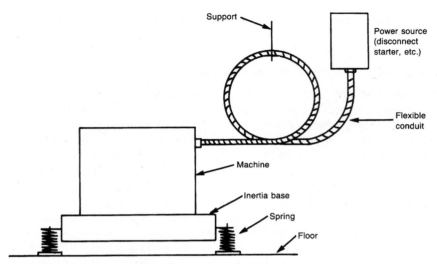

Figure 20.12 Providing electrical service to a vibrating machine.

The equipment room floor, on which the inertia bases are mounted, should be sufficiently stiff to avoid acting as a diaphragm and amplifying the vibration. The structural engineer must be consulted.

To illustrate some of the problems that occur, consider the following real-life incident. A judge in a newly constructed federal building complained that her courtroom was excessively noisy. Because the room was directly below a fan and equipment penthouse, it was felt that direct sound transmission was the probable cause. An engineer with a sound meter was sent to check on this. The unoccupied courtroom seemed quiet enough, but then the engineer sat at the judge's bench to make sound measurements. At this point the entire bench—desk, platform, and chair—began to vibrate. An additional noise could be heard, not too noticeable, but the vibration was excessive. Then it stopped. After a short time the cycle was repeated. Investigation revealed a large duplex air compressor in the penthouse. The unit had been properly mounted on spring isolators. Then a drain line was added, from the storage tank to the floor, making a solid contact and completely negating the isolators. Natural frequencies caused this vibration to be transmitted through the floor to the structure, down to the floor below, and out the judge's bench. The drain line was changed so as to avoid its touching the floor, and the problem was solved.

20.7 Summary

This has been a brief overview of a complex subject. The HVAC designer must be aware of the possibility of unacceptable noise and vibration being produced by HVAC equipment. Proper acoustical design should eliminate these problems.

References

1. ASHRAE Handbook, *1997 Fundamentals,* Chap. 7, "Sound and Vibration Fundamentals."
2. ASHRAE Handbook, *1995 HVAC Applications,* Chap. 43, "Sound and Vibration Control."

Index

ABOUT THE AUTHORS

ROGER W. HAINES, P.E., has been in the field of HVAC systems design since 1953. Widely recognized as one of the foremost authorities in the field of HVAC engineering, he is the author of *Roger Haines on HVAC Controls*, also published by McGraw-Hill, which is now in its fifth edition. He is a frequent contributor to *Heating/Piping/Air Conditioning* magazine. A California-based consulting engineer since 1981, Mr. Haines is a Fellow and Life Member of ASHRAE.

C. LEWIS WILSON, P.E., is chairman and chief executive officer of the Heath Engineering Company, based in Salt Lake City, Utah, which specializes in mechanical/electrical engineering analysis and technical design. A former adjunct professor of HVAC at the University of Utah, Brigham Young University, and the University of New Mexico, he has written numerous articles on HVAC and energy-related topics.